R. Margesin · F. Schinner (Eds.)

Biotechnological Applications of Cold-Adapted Organisms

Professor Rosa Margesin
Professor Franz Schinner

Institute of Microbiology
University of Innsbruck
Technikerstraße 25
A-6020 Innsbruck
Austria

rosa.margesin@uibk.ac.at
franz.schinner@uibk.ac.at

ISBN 978-3-642-63663-9

Library of Congress Cataloging-in-Publication Data

Biotechnological applications of cold-adapted organisms / Rosa
  Margesin, Franz Schinner, editors.
       p. cm
   Includes bibliographical references and index.
   ISBN 978-3-642-63663-9   ISBN 978-3-642-58607-1 (eBook)
   DOI 10.1007/978-3-642-58607-1
   1. Biotechnology.   2. Cold adaptation.   I. Margesin, Rosa, 1962.
II. Schinner, Franz, 1947-
TP248.2.B3796 1999
660.6--dc21   98-55124

Cover design: Design & Production, Heidelberg
Typesetting: Camera ready by editors
SPIN 10690962   31/3137 - 5 4 3 2 1 0 - Printed on acid-free paper

Preface

Cold adapted microorganisms, plants and animals are widely distributed in nature since more than 80% of the biosphere show temperatures below 5°C. These organisms play a major role in the processes of nutrient turnover and primary biomass production in cold ecosystems and have adapted to their environment in such a way that metabolic processes, reproduction and survival strategies are optimal.

Cold adapted organisms have received little attention so far. Only within the past few years it has been recognized that these organisms and their products provide a large potential for biotechnological applications. Biochemical reactions can be performed at low temperatures and terminated by mild heat treatment, without the risk of contaminations by mesophiles or thermophiles. Thus, the practical utilization of cold-adapted organisms would constitute a considerable progress towards the saving of energy.

In this book, prominent authors from industries and universities present new concepts and developments. Their articles summarize the actual and potential applications of cold-adapted organisms and viruses in several fields of biotechnology, such as enzymes, health, food, agriculture and environment. Further chapters cover the application of cold-adapted microorganisms for bio-mining and as genetic tools.

We are most grateful to the authors for their excellent contributions, to Brigitte Marschall and Alexander Koren (vanillapaint) for skilful preparation of the layout, and to the publishers, especially to Dr. Dieter Czeschlik, for their cooperation.

Innsbruck, October 1998   R. Margesin and F. Schinner

Contents

Development of regulatable expression systems for cloned genes in cold-adapted bacteria

E. Remaut[1]*, C. Bliki[2], M. Iturriza-Gomara[1] and K. Keymeulen[1]

[1] Molecular Biology Department, Flemish Inter-university Institute for Biotechnology, Universiteit Gent, K.L. Ledeganckstraat 35, B-9000 Gent, Belgium
[2] Belgian Co-ordinated Collections of Microorganisms/Laboratory of Molecular Biology Plasmid Collection, K.L. Ledeganckstraat 35, B-9000 Gent, Belgium

1
Introduction

In considering potential biotechnological applications of cold-adapted bacteria, the aspects that are most frequently highlighted relate to the exploitation of these organisms for waste water treatment and biodegradation of pollutants at ambient temperatures and to the energy saving benefits that could be derived from using their cold-active enzymes in many industrial processes, ranging from the food industry to the laundry business.[1,2]

Our interest in bacteria that still have considerable metabolic activity at very low temperatures originated from a different consideration. For many years our laboratory had been engaged in the development of efficient bacterial expression systems for cloned eukaryotic genes. Although others and we succeeded in generating powerful systems for massive production of essentially any protein in *Escherichia coli*, there remained one serious drawback. Upon over-expression many proteins accumulated inside the cell in an insoluble, non-native form, the so-called inclusion bodies.[3] Shein and Noteborn were first to observe that lowering the usual growth temperature from 37 to below 30°C in some cases resulted in an increased recovery of soluble, correctly folded protein.[4] In view of the rapidly decreasing protein synthetic capacity, *E. coli* would not be the organisms of choice for evaluating still lower temperatures. Clearly, here was a niche in which organisms that thrive well at very low temperatures, could be exploited as cell factories for the manufacturing of adventitious proteins. Unlike the situation in *E. coli* where a wealth of efficient expression systems are known in great molecular detail, no such examples have yet been identified in cold-adapted bacteria.

We took the approach of introducing well established *E. coli*-derived expression-controlling elements into psychrotrophic hosts. As vector system we opted for

---

* Corresponding author

broad-host-range plasmids. Transfer of genetic material between mesophilic donor bacteria and psychrotrophic strains had already been established and it had been shown that plasmid-borne genes were successfully expressed at temperatures as low as 4°C.[4,5] In this chapter we will focus on the use of a regulatable promoter whose activity can be induced by addition of a synthetic inducer.

2
The strains

The strains used in this study are listed in Table 1. All are antarctic isolates, obtained from the collection of C. Gerday, Liège, Belgium. When we originally set out on this project, the psychrophilic and psychrothrophic strains were routinely maintained at 4°C on agar plates made up in the medium recommended by the supplier. The colonies were re-suspended and re-streaked every 14 d. We argued that for the purpose of determining which strains were most useful for genetic engineering experiments it would be advantageous to try and establish conditions allowing more rapid propagation of the strains as well as a more convenient procedure for permanent storage. More or less arbitrarily we decided to adopt 15°C as the standard temperature for growth in liquid culture. The following basal media compositions were evaluated: marine broth, nutrient broth and LB medium (1% tryptone, 0.5% yeast extract and NaCl at final concentrations of 0.5, 1, 1.5 and 3.5%). For all strains tested the LB medium gave the best results in that it supported the shortest doubling times as well as allowed the cultures to reach the highest cell densities at saturation. The optimal concentration of NaCl was 0.5% for strains TA1, TA20 and TAD1. For strains TA40 and TA114, slightly higher cell densities were obtained when the concentration of NaCl was increased to 1.5 and 3.5%, respectively.

From our previous experience with *E. coli* strains harboring recombinant plasmids, we had learned that cryopreservation at -80°C in the presence of 50% glycerol was a satisfactory method to maintain good viability and, more importantly, excellent stability of the plasmid content. In our hands lyophilization of certain host-plasmid combinations resulted in loss of the plasmid. Saturated cultures were thoroughly mixed with an equal volume of highly pure glycerol and stored in a freezer at -80°C. The viable count was determined prior to the freezing step and at regular intervals during several months of maintenance at this temperature. We considered the method adequate for a given strain when the survival rate, determined as colony forming ability at 4°C, did not drop below 20% of the input value. Strains TA40 and TA114 failed to pass this criterion. In fact their viability was reduced by several orders of magnitude. It is perhaps surprising that precisely two truly psychrophilic strains, displaying very short doubling times at 15°C, did not survive the freezing step. On the other hand, strain TA20, likewise a true psychrophile, survived quite well after prolonged storage at -80°C. The psychrotrophic strains, TA1 and TAD1, showed nearly 100% colony forming ability.

All strains are aerobic, gram-negative rods. No attempts to come to a taxonomic classification were undertaken. Our major concern was to identify a number of parameters that could be easily scored and that would allow to unambiguously dif-

Table 1. Overview of the properties of the strains

	TA1	TAD1	TA20	TA40	TA114
Doubling time (min) in LB medium					
at 15°C	140	135	240	100	100
at 20°C	120	80	400	not determined	
at 28°C	60	50	NG	NG	NG
Cryopreservation at -80°C	good	good	good	bad	bad
Natural resistance to antibiotics					
ampicillin (100 µg µl⁻¹)	R	R	S	S	S
chloramphenicol (25 µg ml⁻¹)	S	R	S	S	S
erythromycin (5 µg ml⁻¹)	R	R	R	S	S
kanamycin (50 µg ml⁻¹)	S	S	S	S	S
streptomycin (20 µg ml⁻¹)	R	R	S	S	R
tetracyclin (10 µg ml⁻¹)	S	R	S	S	S
Electroporation					
pJB3tet	yes	yes	no	not determined	
pKT240	yes	yes	no	not determined	
pBBR122	yes	no	no	not determined	
Conjugation					
pJB3tet	yes	yes	no	not determined	
pKT240	yes	yes	no	not determined	
pBBR122	yes	no	no	not determined	

NG no growth, *S* sensitive, *R* resistant

ferentiate the strains from one another. A very useful criterion proved to be the natural resistance of the strains to various antibiotics commonly used as selection markers in cloning vectors (Table 1). Remarkably, resistance to a number of common antibiotics appears to be rather widely spread in these natural, antarctic isolates. We have not established the exact nature of the resistance mechanism. This could be due to a permeability barrier in the cell wall or alternatively indicate the presence of an enzymatic activity that inactivates the antibiotic. The genes involved could be located on the chromosome or be carried by plasmids. We have used several procedures for isolation of extrachromosomal elements (see below). In no instance could we demonstrate the presence of a plasmid. We favor the hypothesis that the observed resistances are due to permeability barriers.

3
The plasmids

Three different replicons, all having broad-host-range properties, were evaluated for their proficiency to replicate in the psychrotrophic and psychrophilic hosts at low temperature. Plasmid pJB3 is a minimal replicon derivative of the natural broad-host range plasmid, RK2, belonging to the IncP-1 incompatibility group. The minimal replicon consists of the origin for vegetative replication *(oriV)* and the gene *trfA* whose protein product acts as an initiator at the iterons in *oriV*. In addition, pJB3 has retained a functional *oriT* so that it can be mobilized by RK2 mobilization and transfer functions. The plasmid specifies resistance to ampicillin and carries the *lac* promoter and operator from *E. coli*.[7]

Plasmid pKT240 was derived from the broad-host-range plasmid, RSF1010, belonging to the IncP-4 incompatibility group. The plasmid contains the *rep* gene and *oriV* for vegetative replication as well as an *oriT*, allowing it to be mobilized by the *mob* function present on the plasmid and RK2 transfer functions. The plasmid carries resistance genes for the antibiotics ampicillin and kanamycin.[8] Since pJB3 and pKT240 belong to two different incompatibility groups, they can co-exist in the same cell.

Plasmid pBBR122 was derived from a natural plasmid first isolated from the bacterium *Bordetella bronchisepta*. It has broad-host-range properties and does not belong to any of the known incompatibility groups. pBBR122 was engineered to contain resistance genes for the antibiotics chloramphenicol and kanamycin.[9]

3.1
Introduction of the plasmids into the psychrotrophs by electroporation

Electroporation has proven to be a very valuable technique for introducing plasmids into a wide variety of bacterial species. A typical procedure worked out for the strains under study is outlined below. A freshly saturated culture was diluted 100-fold in 200 ml LB medium and grown at 15°C with vigorous shaking in a baffled Erlenmeyer flask to a cell density of 0.2 OD_{650} units. Samples of 20 ml were then allowed to cool on ice for 30 min, and the cells were collected by centrifugation in a refrigerated centrifuge at 5,000 rpm for 15 min. It is very critical that the cells remain cooled during all further manipulations; all solutions and recipients should be kept on ice. The cell pellet was re-suspended in 20 ml of an osmoprotective solution (OPS), consisting of 137 mM sucrose, 1mM HEPES buffer pH 7.8 and 10% glycerol, re-pelleted and washed with 10 ml OPS. The pelleted cells were carefully but thoroughly resuspended in 400 µl OPS, re-pelleted and finally thoroughly resuspended in 60 µl OPS. At this stage the cells can either be used immediately for electroporation or they may be flash-frozen on dry ice-methanol and stored at –80°C. The cells maintain their competent state for at least 6 months. For the actual electroporation a sample is thawed slowly on ice. Highly purified plasmid DNA, dissolved in 1–3 µl water at a concentration of 1 µg µl⁻¹, is thoroughly mixed with the cell slurry by vortexing. The cells are left on ice for 1 min before transfer to a cooled electroporation cuvette. Trapped air bubbles should be carefully removed. We performed the electroporation in 0.2-mm cuvettes in a GenePulser II apparatus (Bio-Rad Laboratories, Hercules, CA, USA) applying the following settings: capacitance, 25 µ farad; resistance, 200 ohm; field strength, 2500 volts. Following the electrical pulse the cells were gently mixed by pipetting up and down in 1 ml LB medium supplemented with the required NaCl concentration and transferred to a cultivation tube. The cells were grown with good aeration for 2 h before plating on selective 1.2% agar plates. The number of colonies was scored after 3 d of incubation at 15°C.

The cold-adapted strains TA1, TA20 and TAD1 were used as hosts for electroporation with the plasmids pKT240, pBBR122 and pJB3tet2, a derivative of pJB3 in which the *lac* promoter was deleted and replaced by a fragment carrying the tetracycline resistance gene from pBR322. Strain TA1 could be transformed with all three plasmids. The average efficiency obtained was 10^2 transformants μg^{-1} plasmid DNA. Strain TAD1 could be transformed with pKT240 and pJB3tet2 at about the same efficiency as TA1. In contrast, we were unable to introduce pBRR122 into strain TAD1. Using sequential introduction by electroporation, i.e. electroporation with the first plasmid, preparation of electrocompetent cells from this new strain followed by eletroporation with the second plasmid, plasmids pJB3tet2 and pBRR122 were shown to co-exist in TA1 in a stable manner at 15°C. Finally, despite many attempts in which several experimental conditions such as cell density at collection, composition of the washing solutions, and field strength during electroporation were varied, strain TA20 remained completely recalcitrant to electroporation with any of the plasmids tested.

The maximal efficiency of electroporation obtained with strain TA1 and plasmid pKT240 is about seven orders of magnitude lower than the efficiency scored in most *E. coli* strains. One demonstrated course for the low efficiency is the presence of an efficient restriction barrier in TA1. Indeed, pKT240 plasmid DNA prepared from a transformed TA1 culture now showed an increase in electroporation efficiency of at least three orders of magnitude. Other factors affecting the efficiency of electroporation are probably inherent to the structure of the cell wall of these organisms. The reasons for the recalcitrance of TA20 to electroporation remain unclear. The strain does not secrete detectable amounts of DNase, a frequent cause for failure of electroporation, nor could we detect the presence of resident plasmids that might present problems of incompatibility with the incoming plasmid. A very important factor in obtaining good electroporation is the density of the culture when the cells are collected and prepared for electrocompetency. For the strains used in this study the density is critical at around 0.2 OD_{650} units. Once the cell density exceeds 0.3 OD_{650} units the efficiency drops steeply to around 10 transformants μg^{-1} plasmid DNA. At still higher cell densities the cells can no longer be made electrocompetent.

In order to determine the stability of the several host-plasmid combinations, transformed cells were grown at 15 and 4°C in LB medium without antibiotics for at least 50 generations. This was achieved by repeated 100-fold dilutions of a freshly saturated culture into fresh LB medium. Suitable dilutions were then plated on selective and non-selective agar plates and the number of colonies compared. Plasmid pKT240 displayed a stability of 100% in strains TA1 and TAD1 both at 15 and 4°C. The pJB3 derivatives were less stable with values of 20% for TAD1 and 10% for TA1, irrespective of the growth temperature. Plasmid pBRR122 was virtually 100% stable in strain TA1 growing at 15°C, but was rapidly lost at 4°C with only 2% of the cells still expressing the resistance marker after 50 generations of non-selective growth.

3.2
Introduction of the plasmids into the psychrotrophs by mobilization

Since all plasmids used in this study have broad-host-range properties and have retained their *oriT* as well as *mob* functions, it was obvious to attempt to introduce them into the psychrotrophic strains by mobilization. An important concern here was that the conjugation phase preferentially had to be carried out at temperatures not higher than 15°C. The usual temperature for conjugation with mesophilic organisms is 37°C which is obviously too high for cold-adapted bacteria. Initial experiments using appropriately marked *E. coli* strains indicated that the conjugation phase could be carried out at 15°C while still resulting in virtually 100% transfer of the mobilizable plasmid from the donor to the acceptor strain.

The general procedure adopted in this study was as outlined below. Approximately 10^8 cells of the *E. coli* donor strain, harboring anyone of the plasmids pJB3tet2, pKT240 or pBRR122 together with the helper plasmid pRK2073 (a ColE1 derivative which supplies the transfer functions of plasmid RK2,[10] were mixed with the acceptor strain in a 1:1 ratio and spotted on the surface of an LB-agar plate. Following incubation at 15°C for 20 h, a sample of the full-grown spot was re-suspended in LB medium and suitable dilutions were plated out on selective plates at 4°C. At this temperature, the *E. coli* cells do not form colonies. The colony count of ex-conjugants was assessed after incubation at 4°C for one week. Plasmids pJB3tet2, pKT240 and pBBR122 were all successfully mobilized into strain TA1. Strain TAD1 accepted pJB3tet2 and pKT240, but not pBRR122. None of the replicons could be introduced into strain TA20. In addition to the *E. coli* [pRK2073] donor system, we also used the *E. coli* strain S17-1 as donor in conjugation experiments. This strain has the broad-host-range plasmid RP4 integrated in the chromosome.[11] The results obtained were in all aspects identical to the ones reported above.

The results obtained with conjugation are remarkably parallel to those obtained with electroporation of the strains. Since during conjugation the plasmid DNA is transferred into the acceptor cell as a single-stranded molecule, it is not likely (or at least much less than double-stranded DNA) to be subjected to host-specific restriction.[12] It is therefore highly likely that strain TA20 either does not support replication of any of the replicons tested or that it does not express the resistance genes for the antibiotics, kanamycin and tetracyclin, at a sufficiently high level. Plasmid pBBR122 cannot be propagated in strain TAD1. In this case the failure must in all likelihood be attributed to the origin of vegetative replication not being functional in this host. The possibility that the selection marker is expressed at a too low level is very unlikely in view of the consideration that the kanamycin resistance gene on pBRR122 is the same gene as present on pKT240 whose presence in TAD1 can be selected for using this antibiotic. The kanamycin resistance gene used here derives from transposon 903 and is known to be expressed at high levels in *E. coli*. In a comparison of the protein profiles of strains TAD1 and TAD1, transformed with pKT240, the neomycinefosforyl transferase, product of the Kan[R] gene, could be easily identified as a prominent new protein band on sodium dodecyl sulfate (SDS) containing polyacrylamide gels. It follows that both the promoter and the ribosome binding site of the gene are efficiently recognized in this psychrotrophic strain at low temperatures (15°C).

3.3
Analysis of plasmid DNA introduced into the psychrotrophic strains

The results obtained so far establish the functionality of a given replicon as well as the functionality of the selection markers, kanamycin and tetracyclin, in the psychrotrophic host cultivated at temperatures as low as 4°C. In a next step we wanted to verify the physical intactness and identity of the plasmids after several generations in the antarctic hosts. Various methods described for isolation of plasmid DNA were applied to the transformed strains TA1 and TAD1 including the cleared lysate technique using lysozyme and the neutral detergent Triton X-100[13], and SDS-mediated lysis at neutral[14] or alkaline pH.[15,16] In our hands none of these methods proved adequate in yielding sufficiently pure plasmid DNA in sufficient quantities to allow detailed restriction enzyme analysis. We therefore adopted the following procedure. Plasmid DNA was prepared from a small volume of culture (1–3 ml) essentially using the SDS-alkaline lysis procedure.[15] Following purification of the supercoiled plasmid DNA over a resin as commercially available from e.g., Qiagen GmbH (Hilden, Germany) or Promega (Madison, WI, USA), the DNA was transformed into the restriction-negative *E. coli* strain MC1061.[17] Plasmid DNA was then recovered from ten individual transformants using the standard Birnboim method[15] and analyzed on agarose gels following restriction enzyme digestion. In this way we could show that all plasmids tested remained structurally unaltered after several generations in TA1 or TAD1 cultivated at 15 or 4°C. The restriction pattern was identical to that of the original *E. coli*-derived plasmid. These results indicate that once the restriction-modification barrier is surmounted, the plasmid DNA does not suffer any rearrangements in the cold-adapted hosts.

4
The expression system

4.1
Firefly luciferase as the reporter gene

In order to assess the functionality of *E. coli*-derived transcription and translation regulatory elements in the psychrotrophs, we made use of a reporter gene. Because of ease of detection and quantification we opted for the eukaryotic luciferase from the beetle, *Photinus pyralis*. In this system, generation of light depends solely on the interaction of a single polypeptide with the substrate luciferin in an ATP consuming reaction. The different expression modules used to drive the synthesis of luciferase will be detailed in later sections of this text. Luciferase-producing transformants growing on luciferin-containing plates were easily identified in the darkroom as light-emitting colonies. Although the light intensities could also be measured on intact, living cells using a Luminoscan apparatus (Labsystems, Helsinki, Finland), we soon became aware of problems of linearity in the intensity of the emitted light signal. In part, this was shown to be caused by permeability barriers for luciferin but also intracellular concentrations of ATP may be rate limiting. Therefore we used cell lysates to quantify the amount of luciferase synthesized.

Optimal results were obtained when the cells were subjected to sonication for 10 s in a buffer consisting of 7 mM $MgCl_2$, 33 mM Tris-HCl pH 8, and 2 mM ATP. The reaction was started by the addition of luciferin. Emitted light intensities were recorded in a Luminoscan microtiter plate reader.

4.2
The *E. coli* trc promoter is functional in psychrotrophs

Plasmid pJB3tet2 was used as the acceptor vector for the luciferase gene under the transcriptional control of the *trc* promoter. Trough several intermediate constructs the expression plasmid pJBtrcLUC was generated (Fig. 1, Fig. 4). In this plasmid the coding region of the luciferase gene was fused in the correct reading frame to the initiator ATG codon of the *lacZ* ribosome binding site, preceded by the *trc* promoter. The *trc* promoter is a hybrid construct consisting of the −10 region and operator sequence of the *E. coli lac* operon and the −35 region of the *E. coli trp* operon.[18] The activity of the promoter is repressed by the product of the *lacI*q gene, which is also cloned onto the expression plasmid. The *lacI*q gene carries a promoter-up mutation which allows increased levels of repressor molecules to be synthesized.[19] Repression on the *trc* promoter can be lifted by the addition of the synthetic inducer isopropyl-β-thiogalactoside (IPTG).

Plasmid pJBtrcLUC was introduced by electroporation into strain TA1. A representative transformed colony was used for the study of repression and induction kinetics of luciferase in the psychrotrophic strain growing at 15°C (Fig. 2). The level of luciferase detected in a non-induced culture remained very low throughout the logarithmic growth phase and well into the stationary phase, which was reached at

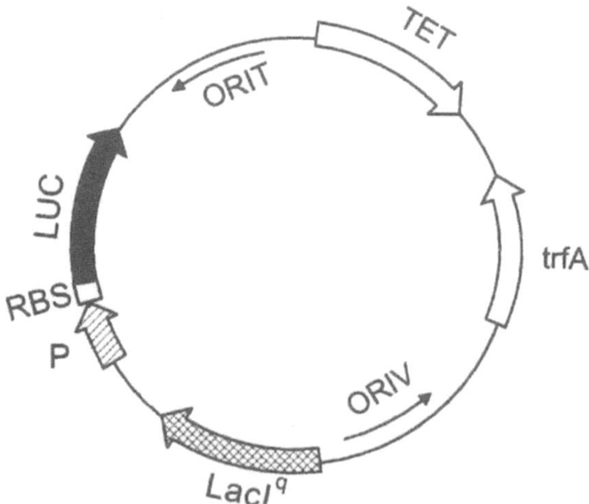

Fig. 1. Schematic representation of the functional elements of a model expression plasmid
P promoter; *RBS* ribosome binding site, *LUC* luciferase coding region, *ORIT* origin of transfer during mobilization, *TET* tetracyclin resistance gene, *trfA* replication protein, *ORIV* origin of vegetative replication, *LacI*q, repressor gene

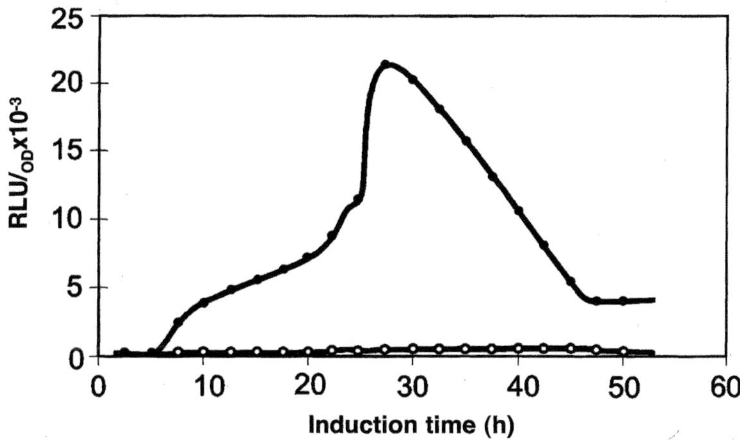

Fig. 2. Induction kinetics of luciferase in strain TA1
Strain TA1 transformed with pJBtrcLUC was grown in LB at 15°C to a cell density of 0.2 OD_{650} units. One half of the culture was induced with IPTG at a final concentration of 1 mM. Samples were taken at several time points after induction. The luciferase content of the cells was determined as described in the text. Relative light units (RLU) are expressed per OD_{650} unit

about 15 h after the start of the experiment. Induction kinetics of luciferase after addition of 1mM IPTG were very slow. Detectable amounts were first scored at about 7 h after induction, corresponding to three doublings of the bacteria. The maximal level of activity was reached after 27 h, which is well after the culture had entered stationary phase. Continued incubation resulted in virtually complete loss of the culture's light-producing capacity. No visual lysis of the cells was observed.

In a next experiment we compared the efficiency of the pJBtrcLUC expression module in strains TA1, TAD1 and *E. coli* strain MC1061 at different temperatures (Fig. 3). In TA1 growing at 4 and 15°C, the *trc* promoter is repressed at least as stringently as it is in *E. coli* at 15 or 37°C. Strain TAD1 at 15°C shows a slightly elevated non-induced expression level. This strain also reaches the highest level after induction with IPTG. In strain TA1, relative light units measured at 4 and 15°C were comparable to those measured in *E. coli* at 15°C. The remarkably low value obtained in *E. coli* at 37°C most likely reflects an artefact in recovering active luciferase, which may be due to thermal denaturation of the protein or to the formation of inclusion bodies.

Several conclusions can be drawn from these experiments. Clearly, the *lacI*q gene is functional in the psychrotrophic strains both at the level of transcription and of translation. At least in strain TA1 the system allows the synthesis of a sufficiently high level of repressor molecules to virtually completely silence the activity of the *trc* promoter. Repression on the promoter can be lifted by the addition of the synthetic inducer IPTG. The *trc* promoter is recognized by the host polymerases at temperatures as low as 4°C. The results further imply that the ribosome binding sites of the *lacI*q gene and the *lacZ* gene are recognized by the translational machin-

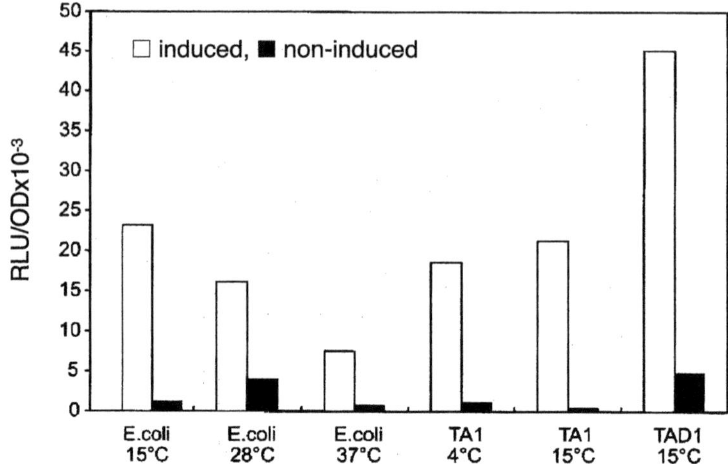

Fig. 3. Repression levels in strain TA1, TAD1 and *E. coli* at 15 and 4°C
The cultures were grown at the temperatures indicated. Induction of luciferase was started at a cell density of 0.2 OD_{650} units by adding IPTG to 1 mM final concentration. Relative light units (RLU) are expressed per OD_{650} unit

ery of the psychrotrophic hosts. As noted earlier, however, the induction kinetics of luciferase are extremely slow.

At this stage several reasons can be put forward to explain this phenomenon. Obviously one could argue that the *E. coli*-derived transcriptional and translational control signals may be operative in the new hosts at only a fraction of their strength in *E. coli*. We will discuss this possibility in greater detail further in the text. Another possibility would be that the luciferase polypeptide has a short half-life as a consequence of proteolytic attack. However, using a polyclonal antibody against firefly luciferase (Cortex Biochem, San Leandro, CA, USA) we detected the polypeptide in TA1 as a sharply defined band after SDS-polyacrylamide gel electrophoresis. No breakdown products could be visualized. In contrast, minor amounts of breakdown products were seen in *E. coli* cells at 15°C. An alternative explanation for the slow onset of induction could be that the binding of repressor molecules to the operator sequence is much stronger at the lower temperatures such that higher concentration of inducer are needed to lift repression. Also, the cell walls of the psychrotrophic strains may be less permeable to IPTG. Below we will present evidence that several of the discussed factors are at play in determining the final result.

4.3
Comparison of different ribosome binding sites

Numerous studies employing various expression systems in *E. coli* have pointed towards the rate of initiation of a particular polypeptide as perhaps the most deter-

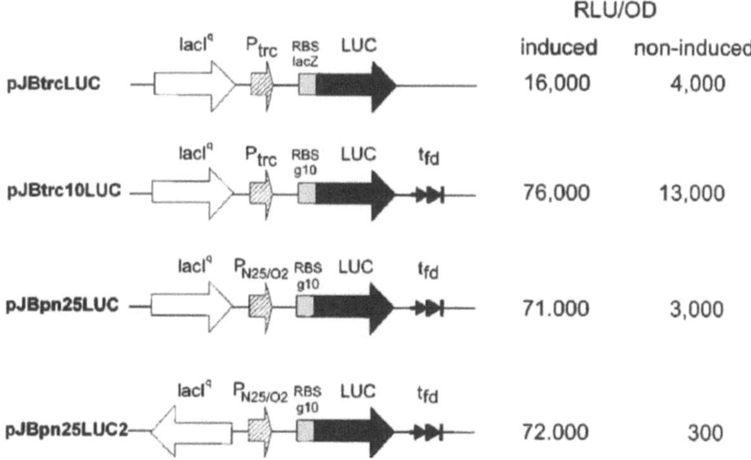

Fig. 4. Schematic representation of structural features on expression plasmids and their effects on regulated expression of luciferase
Luciferase synthesis is expressed as relative light units (RLU) per OD_{650} unit. *RBS* ribosome binding site, *g10* gene-10 from phage T7, *lacZ* β-galactosidase gene, t_{fd} direct repeat of the central transcription terminator from phage fd

mining factor in high-level heterologous gene expression.[20] For instance, in a study on inducible synthesis of β-galactosidase driven by the strong P_L promoter from phage λ, it was shown that subtle changes in the nucleotide sequence of the ribosome binding site could lead to dramatic differences in the level of synthesis of the polypeptide.[21]

In order to study the problem in the psychrotrophic strains we constructed a plasmid, designated pJBtrc10LUC (Fig. 4), in which the *lacZ* ribosome binding site in front of the luciferase coding region was replaced by the ribosome binding site from gene–10 of phage T7. Gene–10 codes for the major capsid protein of the phage. When cloned and expressed in *E. coli* from e.g. the phage λ P_L promoter, the polypeptide may account for over 50% of total protein.[22] The ribosome binding site of gene–10 is peculiar in that, apart from the Shine-Dalgarno sequence itself, the nucleotide sequence in front of the initiator ATG is unusually rich in A- and T-residues. It has been speculated that the untranslated mRNA stretch is therefore less likely to become involved in secondary structure formation. Examples in *E. coli* indicate that failure to efficiently synthesize a cloned gene product may in certain cases be caused by the formation of a too stable secondary structure of the mRNA such that the ribosome binding site is rendered inaccessible to initiating ribosomes.[21] We argued that sequestering of the ribosome binding site in unfavorable secondary structure might be particularly disadvantageous for expression at very low temperatures. A comparison of the maximal levels of luciferase production using either the *lacZ*-derived or the gene–10-derived ribosome binding site, shows that the latter is nearly 5-fold as efficient.

During the course of these experiments we also obtained evidence that 1 mM IPTG is a sub-optimal concentration for rapid onset of induction. Both for strain TA1 and TAD1 a concentration of 5 mM was found to be optimal. At this concentration maximal induction was obtained at 5 h after induction at 15°C. Still higher concentrations resulted in an inhibition of luciferase synthesis.

4.4
Comparison of the trc and the phage T5 derived N25/O2 promoter

The *N25/O2* promoter is a chimerical construct consisting of the strong, constitutive *N25* promoter from phage T5 and two copies of the *lac* operator sequence. The latter feature renders the promoter regulatable by the LacI repressor much in the same way as the natural *lac* promoter and the hybrid *trc* promoter. The *N25/O2* promoter is about twice as strong as the *trc* promoter.[23]

In plasmid pJBpn25LUC, the *N25/O2* promoter drives expression of the luciferase coupled to the gene-10 ribosome binding site (Fig. 4). Strain TA1 transformed with this plasmid and induced with 5 mM IPTG at 15°C, synthesized the same amount of luciferase as did pJBtrc10LUC. From this result it would appear that, in contrast to the situation in *E. coli*, the *N25* promoter is not superior over the *trc* promoter in the psychrotroph at low temperature. However, in order to substantiate this conclusion one would have to compare actual transcription rates. It remains entirely possible that the efficiency of translation initiation of the gene–10-luciferase mRNA is a rate limiting factor. Excess mRNA which is not efficiently entering the translation pathway may be subjected to rapid nucleolytic attack.

The presence of two operator copies in the *N25/O2* module had a pronounced beneficial effect on the level of repression. The basal level of non-induced luciferase synthesis was reduced by a factor of 4.

4.5
Uncontrolled read-trough by adventitious promoters

In the hitherto described expression modules the *lacI*q gene is adjacent to the main promoter, *trc* or *N25/O2*, and is transcribed in the same direction. In order to evaluate possible interference from this promoter we constructed plasmid pJBpn25LUC2 in which the orientation of the *lacI*q gene is inverted (Fig. 4). The plasmid directed the same high level of luciferase synthesis in TA1 at 15°C as did pJBpn25LUC. Remarkably, the repressed level was further decreased by a factor of 10. Essentially the same results were obtained in strain TAD1. These results are somewhat puzzling and imply that the *lacI*q promoter is able to transcribe over two operator sequences, as in *N25/O2*, even when these sequences are complexed with repressor molecules.

Such a situation is not unprecedented in *E. coli*, however. We have shown earlier that the strong P_L promoter is able to displace the LacI repressor bound to its operator sequence (unpubl. data). It should be remembered that the *lacI*q gene carries a promoter-up mutation and should be considered a reasonably strong promoter. Alternatively the read-trough transcription may originate from an unidentified

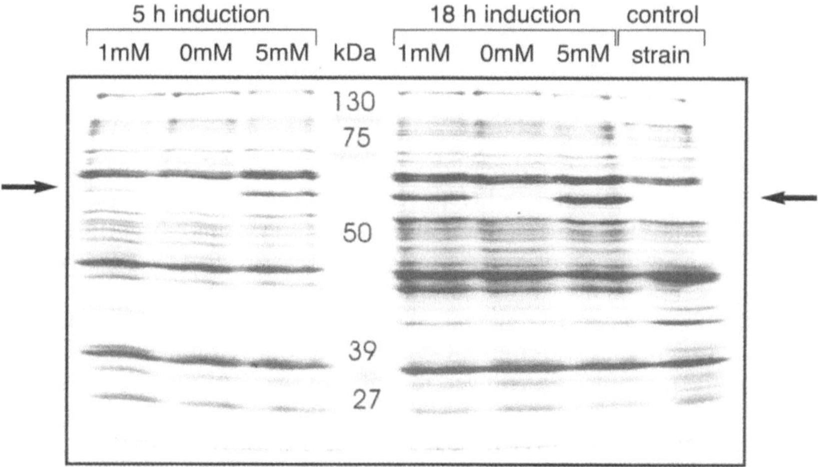

Fig. 5. Accumulation of luciferase polypeptide in strain TA1 transformed with plasmid pJBpn25LUC
Strain TA1[pJBpn25LUC] was induced with the indicated concentrations of IPTG. At 5 and 18 h after induction, proteins were separated on SDS – 15% polyacrylamide gels and revealed by coomassie staining. The arrows indicate the position of the luciferase band

promoter present on the pJB3 vector. The beneficial effect of inverting the orientation of the *lacI*q gene would then be the indirect result of colliding promoters. An example of the extinguishing effect of colliding transcription has been described in *E. coli*.[24]

4.6
Accumulation of luciferase in the psychrotrophs

As a complementary approach to the measurement of relative light units we also determined the accumulation of luciferase protein in the cells by SDS-polyacrylamide gel electrophoresis. Not unexpectedly the results essentially corroborate our findings obtained by measuring enzymatic activity. The expression modules that make use of the *trc* promoter and the *lacZ* ribosome binding site produced an amount of luciferase that was barely detectable after coomassie staining of the gels but could be clearly identified by Western blotting. Introduction of the gene–10 ribosome binding site in the expression modules allowed the luciferase to accumulate to much higher levels. Using optimal inducer concentration, the polypeptide could be induced to a level of about 7% of total cell protein (Fig. 5).

5
Conclusions

In addition to the obvious requirement for strong expression signals, a major concern in designing an efficient expression system relates to the degree of experimental control that can be exerted over the activity of the several elements. Studies in *E. coli* have shown that the ability to control the activity of the promoter is an absolute prerequisite for maintaining the expression plasmid in a stable fashion in the host.[25] The problem is of course most severe in cases were the expressed protein is toxic to the host. But even the continued expression of a seemingly innocuous polypeptide may also be detrimental, most likely because the cell is experiencing a metabolic burden. This will result in spontaneous non-producing mutants rapidly overtaking the growing culture. Even though this may often not be experienced as a serious problem on a laboratory scale, the situation becomes disastrous when production in large-scale fermentation is envisaged.

The *E. coli lacI*^{q}-*Ptrc* repressor-promoter system was shown to be operative at 30°C in other gram-negative genera, such as *Rhizobium, Agrobacterium* and *Pseudomonas*.[26] We show here that this regulon is also functional at temperatures as low as 4°C in two different psychrotrophic species. A major improvement in lowering the basal, repressed level of expression was obtained trough a combination of two factors: (i) choice of a promoter-operator sequence that is intrinsically better repressed; (ii) designing a plasmid in which read-trough transcription from adventitious promoters is abolished. With this system the difference in expression between the repressed and the induced state was 250-fold.

In the present study we focused on vectors that incorporate an accessible ribosome binding site, designed for in frame fusion of an appropriately engineered coding region. The strong gene–10 ribosome binding site from coliphage T7 was also efficiently recognized in the psychrotrophs at low temperatures. It is tempting to correlate this observation with the predicted low potential of this sequence to form secondary structure, a feature which may be particularly advantageous at low temperatures. It would be interesting to compare native ribosome binding sites derived from well-translated psychrophilic genes. The preferred way to do this is to introduce the coding regions preceded by their native 5'-untranslated sequences immediately downstream of the promoter. Several vectors, displaying the basic expression-related features of the exemplified plasmids but specifically adapted for the cloning of native gene sequences, have been constructed. The vectors have been deposited in the plasmid collection of BCCM™/LMBP, K.L. Ledeganckstraat 35, B-9000 Gent, Belgium (fax no. ++32 9 264-5348). Additional information on available plasmids, their general properties and full restriction maps can be obtained at this address.

Acknowledgements. We are most grateful to C. Gerday for providing us with the antarctic isolates used in this study. We thank M. Casadaban and A. Pühler for their kind gifts of the *E. coli* strains MC1061 and S17-1, respectively. Thanks are due to M. Bagdasarian, C. Locht and S. Valla for kindly providing the plasmids pKT240, pBBR122 and pJB3 respectively. K. Keymeulen was the recipient of a pre-doctoral grant from the Fonds voor Wetenschappelijk Onderzoek (FWO). M. Iturriza held a Human Capital & Mobility (HCM) pre-doctoral research training fellowship. This work was supported by the Flemish Inter-university Institute for Biotechnology (VIB) and by the FWO (contract no. G.3097.93 and G.0050.97), and carried out within the framework of the Concerted Action EUROCOLD.

6
References

1. Gounot AM. Bacterial life at low temperature: physiological aspects and biotechnological applications. J Appl Bacteriol 1991; 71:386-397.
2. Margesin R, Schinner F. Properties of cold-adapted micro-organisms and their potential role in biotechnology. J Biotechnol 1994; 33:1-14.
3. Makrides SC. Strategies for achieving high-level expression of genes in *Escherichia coli*. Microbiol Rev 1996; 60:512-538.
4. Schein CH, Noteborn MHM. Formation of soluble recombinant proteins in *E. coli* is favored by lower growth temperature. Bio/Technology 1988; 6:291-294.
5. Feller G, Thiry M, Arpigny LL, Mergeay M, Gerday C. Lipases from psychrotrophic antarctic bacteria. FEMS Micriobiol Lett 1990; 66:239-294.
6. Kolenc RJ, Innis WE, Glick BR, Robinson CW, Mayfield CJ. Transfer and expression of mesophylic plasmid-mediated degradative capacity in a psychrotrophic bacterium. Appl Environ Microbiol 1988; 54:638-641.
7. Blatny JM, Brautauset T, Winther-Larsen HC, Haughan K, Valla S. Construction and use of a versatile set of broad-host-range cloning and expression vectors based on the RK2 replicon. Appl Environ Microbiol 1997; 63:370-379.
8. Bagdasarian MM, Amman E, Lucz R, Ruckert B, Bagdasarian M. Activity of the hybrid *trp-lac (tac)* promoter of *Escherichia coli* in *Pseudomonas putida*. Construction of broad-host-range, controlled-expression vectors. Gene 1983; 26:273-282.
9. Antoine R, Locht C. Isolation and molecular characterization of a novel broad-host-range plasmid from *Bordetella bronchiseptica* with sequence similarities to plasmids from gram-positive organisms. Mol Microbiol 1992; 6:1785-1799.
10. Figurski DR, Helinski DR. Replication of an origin-containing derivative of plasmid RK2 dependent on a plasmid function provoded in *trans*. Proc Natl Acad Sci USA 1979; 76:1648-1652.
11. Simon R, Priefer U, Pühler A. A broad-host-range mobilization system for *in vivo* genetic engineering: transposon mutagenesis in gram-negative bacteria. Bio/Technology 1983; 1:784-790.
12. Willets N, Wilkins B. Processing of plasmid DNA during bacterial conjugation. Microbiol Rev 1984; 48:24-41.
13. Clewell DB, Helinsli DR. Super-coiled circular DNA-protein complex in *Escherichia coli*: purification and induced conversion to an open circular form. Proc Natl Acad Sci USA 1969; 62:1159-1166.
14. Guerry P, LeBlanc DJ, Falkow S. General method for the isolation of plasmid deoxyribonuleic acid. J Bacteriol 1973; 116:1064-1066.
15. Birnboim HC, Doly J. A rapid alkaline extraction procedure for screening recombinant plasmid DNA. Nucleic Acids Res 1979; 7:1543-1517.
16. Kado CI, Liu ST. Rapid procedure for detection and isolation of large and small plasmids. J Bacteriol 1981; 145:1365-1373.
17. Casadaban MJ, Cohen SN. Analysis of gene control signals by DNA fusion and cloning in *Escherichia coli*. J Mol Biol 1980; 138:179-207.
18. Brosius J, Erfle M, Storella J. Spacing of the −10 and −35 regions in the *tac* promoter. Effect on its *in vivo* activity. J Biol Chem 1985; 26:3639-3541.
19. Müller-Hill B, Crapo L, Gilbert W. Mutants that make more *lac* repressor. Proc Natl Acad Sci USA 1968; 59:1259-1264.

20. Gold L, Stormo GD. High-level translation initiation. Methods Enzymol 1990; 185:89-93.
21. Stanssens P, Remaut E, Fiers W. Alterations upstream from the Shine-Dalgarno region and their effect on bacterial gene expression. Gene 1985; 36:211-223.
22. Mertens N, Remaut E, Fiers W. Increased stability of phage T7*g10* mRNA is mediated by either a 5'- or 3'-terminal stem-loop structure. Biol Chem 1996; 377:811-817.
23. Deuschle U, Kammerer W, Gentz R, Bujard H. Promoters of *Escherichia coli:* a hierarchy of *in vivo* strength indicates alternate structures. EMBO J 1986; 5:2987-2994.
24. Hopkins AS, Murray NE, Brammar, DJ. Characterization of λ*trp* transducing bacteriophage made *in vitro*. J Mol Biol 1976; 107:549-569.
25. Mertens N, Remaut E, Fiers W. Tight transcriptional control mechanism ensures stable high-level expression from T7 promoter-based expression plasmids. Bio/Technology 1995; 13:175-179.
26. Cebolla A, Vasquez ME, Palomares AJ. Expression vectors for the use of eukaryotic luciferases as bacterial markers with different colors of luminescence. Appl Environ Microbiol 1995; 61:660-668.

Biotechnology of enzymes from cold-adapted microorganisms

S. Ohgiya[1]*, T. Hoshino,[1] H. Okuyama,[2] S. Tanaka[1] and K. Ishizaki[1]

[1] Hokkaido National Industrial Research Institute, 2-17-2-1 Tsukisamu-higashi, Toyohira-ku, Sapporo 062-8517, Japan
[2] Graduate School of Environmental Earth Science, Hokkaido University, Kita-10 Nishi-5, Kita-ku, Sapporo 060-0010, Japan

1
Introduction

One of the main goals in enzyme research is industrial application. Nowadays, we are surrounded by enzymes as well as chemicals produced by enzymes in our daily life. Since papain (EC 3.4.22.2) was used, probably as the first exogenous enzyme, to prevent the formation of chill hazes in beer,[1] many enzymes isolated from various species have been developed for industrial use. These enzymes are used as biological catalysts in various industries such as detergent, food, chemical, textile, pharmaceutical, and paper industries. For instance, proteolytic enzymes are used for detergents; pharmaceutical agents; leather bating; enzymatic conversion of peptidyl substances; and food processing such as cheese production, meat tenderizing, dough conditioning in baking, and protein recovery from waste food materials.[2] Most of the industrial enzymes have been isolated from mesophiles and thermophiles, since innumerable kinds of mesophiles are easily obtained from environmental sources, and enzymes obtained from thermophiles are suitable for industrial processes due to their thermostability. Heat-stable enzymes isolated from thermophiles also have an advantage in terms of storage stability, because they can be transported and stored at the ambient temperature. On the other hand, although many enzymes have also been isolated from cold-adapted microorganisms,[3] psychrophiles and psychrotrophs, there have been few reports on the industrial use of such enzymes. It is reasonable to expect that cold-adapted microorganisms produce enzymes that are active even at a low temperature, i.e., "cold-active enzymes." They would not only be more active at a low temperature than enzymes isolated from mesophiles and thermophiles but would also presumably have distinct characteristics.

Enzymes expressed in cold-adapted microorganisms living in polar regions,[4] high alpine regions,[5] glaciers,[6] and in the deep sea[7] have been screened so far. However, in past studies, psychrophiles in collected samples may have died due to rela-

---

* Corresponding author

tively high temperatures during transportation, storage, or cultivation.[8,9] Industrial companies are now interested in useful extremophiles including cold-adapted microorganisms.[10] Although cold-active enzymes are not widely used in industry at present, elucidation of the characteristics of cold-active enzymes would enable the currently used mesophilic enzymes to be replaced by cold-active counterparts in low-temperature industrial processes.

In this chapter, we describe enzymatic processes at low temperatures in industry and possible industrial applications of cold-active enzymes.

2
The enzyme business market and the possibility of enzymatic processes at a low temperature in industry

According to the enzyme market report by Novo Nordisk,[11] worldwide sales of industrial enzymes in 1996 were approximately US$ 1.5 billion, an approximately 20% increase from that in 1995. Major fields of total sales were in the detergent industry (34%), starch industry (12%), and textile industry (11%). In Japan, detergent and food companies are the major users of industrial enzymes. Sales totaling US$ 150 million, two-thirds to three-quarters of the total domestic enzyme sales in 1997, were to such companies.[12] The market for recombinant enzymes was estimated to be about half of the total domestic enzyme market in Japan in 1997. Recently, the proportion of recombinant enzymes in the total enzyme market has been increasing throughout the world.

Most industrial enzymes are produced by microorganisms. In fact, almost all bacterial enzymes for industrial use are prepared from mesophiles and thermophiles. However, in many cases, the enzymatic reaction has to be carried out at a low temperature due to instability of sources or products, thermodynamic characteristics of the enzymatic reactions, or difficulty in heating the reaction system. In such cases, cold-active enzymes would have great advantages over mesophilic and thermophilic enzymes.[13] The potential advantages of cold-active enzymes over the currently used enzymes in industrial processes are listed in Table 1. Due to their superiority over other enzymes, it is expected that the cold-active enzymes will be utilized for various industrial processes in the near future. In Japan, more than half of the currently used industrial enzymes in the detergent and food industries could potentially be replaced by cold-active counterparts.

Enzymes isolated from cold-adapted microorganisms can be classified into three groups (Fig. 1):

- Group I: Heat-sensitive. Other enzymatic characteristics are similar to mesophilic enzymes.
- Group II: Heat-sensitive and relatively more active than mesophilic enzymes at a low temperature.
- Group III: Same thermostability as mesophilic enzymes but more active than mesophilic enzymes at a low temperature.

Since most of the enzymes so far isolated from cold-adapted microorganisms have

Table 1. Industrial application of enzymes at low temperatures

Applications	Enzymes	Advantages
Detergent	Protease, lipase, cellulase, etc.	Use in tap water
Food industries		
Modification of constituents	Galactosidase, lipase	Keeping freshness
Improvement of taste and flavor	Protease, lipase, etc.	Keeping freshness
Removal of fish skin	Protease	Maintaining product quality
Clarification of fruit juce	Pectinase, cellulase	Keeping fragrance
Preservation	Lysozyme, glucose oxidase	Improved preservative effect
Enzymatic synthesis	Lipase, nitrile hydratase, etc.	For volatile and heat sensitive materials
Treatment of wastewater	Catalase	Less energy consumption
Biotechnology		
Protoplast formation	Cell wall digesting enzymes	High viability
Molecular biology	Phosphatase, uracil DNA glycosylase	Complete heat inactivation

shown characteristics similar to those of mesophilic enzymes, they are classified as group I or II. They rapidly lose their enzymatic activities even with moderate heat treatment.[9] This enzymatic characteristic is sometimes desirable in food processing.

Enzymes in groups II and III could be regarded as "true" cold-active enzymes. The relatively high activity of enzymes in groups II and III at a low temperature is thought to be due to low activation energy and high affinity with substrates.[14]

3
Application of enzymes in the detergent industry

The largest application of industrial enzymes is as additives in detergents.[2,11,12,15] Many detergent brands contain combinations of proteases, lipases (EC 3.1.1.3), cellulases (EC 3.2.1.4), and other degradation enzymes. Generally, proteases and lipases are added to detergent to digest proteins and lipids from human sweat, foods, and soils in cloth fibers. Cellulases remove microfibrils and amorphous celluloses, which trap dirt particles and cause a loss of surface smoothness. Amylases, laccase (EC 1.10.3.2), and cellulase are now used for desizing, bleaching, and "bio-stoning" of denim in jeans, respectively. A combination of amylases, protease, and lipase is an important component of detergents for automatic dishwashing machines.

In Asian countries such as Japan, Korea, and China, people have traditionally used cold tap water for washing. From the viewpoint of energy-saving, it is expected that this trend will spread. Since the use of phosphate in detergents has been decreased and the washing performance of detergents is more reduced at a low temperature, cold-active enzymes will be needed to compensate for the reduced efficacy of detergents.[16]

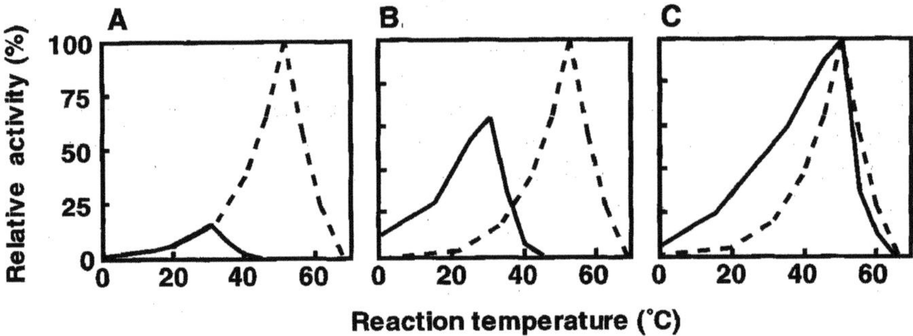

Fig. 1. Three types of enzymes isolated from cold-adapted microorganisms
Solid lines indicate typical thermoprofiles of enzymes isolated from cold-adapted microorganisms.
Dotted lines indicate typical thermoprofiles of enzymes isolated from mesophiles or thermophiles
(**A**) Group I, (**B**) Group II, (**C**) Group III

Several proteases that are active even at low temperatures (10–20°C) have been used in detergent manufacturing (Table 2). Some cold-active proteases have been proved to be more effective than mesophilic enzymes for washing in low-temperature water.[17–19] A protease from antarctic krill *(Euphausia superba)* was added to a detergent.[20] Interestingly, the optimal temperature of the antarctic krill protease was 37°C, but the optimal temperature was decreased to 20°C in the presence of 2-mercaptoethanol.[21] Moreover, protease activity at 5°C was increased 50-fold by the addition of 1-propanol. These results indicate that, in some cases, the high activity of cold-active enzymes at low temperatures might be regulated by coexisting substances.

Alkaline cellulases that can hydrolyze carboxymethyl cellulose even at 10°C were isolated from *Bacillus* sp.[22,23] One of them, alkaline cellulase K, has been added to laundry detergents as an effective component.[24] As the alkaline cellulase is active toward amorphous celluloses but less active toward crystalline celluloses, the enzyme improves the washing performance of the detergent with little damage to cloth.

A cold-active lipase originating from *Humicola lanuginosa* has been used for laundry detergents to hydrolyze triglycerides from skin.[15,19] The optimal tempera-

Table 2. Cold-active proteases used in detergents

Source (strain)	Optimal temperature	Reference
Bacillus sp.	50°C	Fragrance J 1985; 73:79-81
Bacillus sp.	45°C	Curr. Microbiol 1986; 14:7-12
Antarctic krill	20 or 40°C	Activity was stimulated by mercaptoethanol, Agric Biol Chem 1987; 51:3363-3368
Bacillus subtilis	50°C	Jpn Res Assn Text End-Uses 1987; 28:167-172
Paecilomyces maquandii	45°C	Novo industy, patent (WO 88/03948), 1988
Nocardiopsis dassonvillei	60°C	Novo industy, patent (WO 88/03947), 1988

ture of the lipase is 35°C, and 60% of its maximal activity remains at 5°C. In addition, the lipase was found to be resistant to a protease that is added to the same detergent.

Enzymes that are used as detergent additives have been improved for practical uses. Some commercial proteases and amylase have been endowed with improved stability and resistance to bleaching agents by protein engineering.

4
Application of enzymes in the food industry

Numerous enzymes are used in various processes in the food industry[11,12,15]; for example, the conversion of various kinds of sugar compounds; extraction of lipids; processing of fruit juice and cheese; modification of food texture; and improvement in the quality of milk, bread, and alcoholic drinks. Enzymatic starch conversion is the second biggest market for industrial enzymes, with world-wide annual sales totaling US\$ 200 million. Low temperature is a necessary condition in many cases in the food industry, since the quality of food tends to decrease even at a moderate temperature. Typically, cold-active enzymes are active at low temperatures and easily lose their catalytic activities with moderate heat treatment; thus, foods can retain their freshness and own flavor during the enzymatic reaction and following enzyme inactivation. Therefore, activity at low temperatures and heat-lability would be desirable for enzymes in the food industry.

Alpha-amylases (EC 3.2.1.1), glucoamylases (EC 3.2.1.3), and glucose isomerases (EC 5.3.1.18) are used for the production of glucose from starch. Since these processes are carried out at high temperatures (55–115°C), thermostable enzymes have been used for the enzymatic conversion.[15,25–27] Most sugar compounds, such as maltose, functional oligosaccharides, and cyclodextrin, have also been produced by enzymatic processes at high temperatures (40–60°C), in which thermostable enzymes have been predominantly used. Alpha-amylase also facilitates fermentation of bread. In this case, mesophilic α-amylase from fungus is commonly used instead of bacterial thermostable enzymes because its activity ceases as the temperature increases, resulting in appropriate breakdown of starch. On the other hand, enzymatic processes at low temperatures (5–25°C) have been developed for the production of functional oligosaccharides, such as palatinose and cyclodextrin.[28,29] Demand for the functional mono-, di-, and oligosaccharides has been gradually increasing, because they have attracted considerable attention as sweeteners that cause less tooth decay and selectively stimulate bifidobacteria growth in the human gut.

Beta-galactosidase (EC 3.2.1.23) is known to catalyze not only the hydrolysis of lactose but also the transfer of galactoside residues to lactose to form galactooligosaccharides.[30,31] Beta-galactosidase from psychrotrophic *Bacillus subtilis* KL88 has been reported to produce oligosaccharides from lactose at 10°C.[32] Similarly, β-galactosidase from *Cryptococcus laurentii*, of which a strain is a psychrotrophic yeast, was used to produce galactooligosaccharides due to its transgalactosidation activity.[33] Beta-galactosidase has been widely used for the breakdown of lactose in milk to produce low-lactose milk for people who suffer intestinal problems from drinking normal milk due to their poor production of β-galactosidase.[34] This enzymatic activity is also useful for improving the quality of ice cream and whey for various kinds of foods.[34] A cold-active β-galactosidase gene has been isolated from psychrophilic *Arthrobacter* sp. B7.[35] The enzyme expressed in *Escherichia coli* retained 25% of its maximal activity at 10°C.[36] Cold-active β-galactosidase was isolated from psychrotrophic *Buttiauxella agrestis* NC4.[37] This enzyme had maximal activity at 50°C and its activation energy was estimated to be 39.1 kJ mol^{-1}. Another cold-active β-galactosidase was recently purified from an unidentified bacterium isolated from seawater under drift ice and had a larger V_{max} value at 0°C than an *E. coli* counterpart (K. Kawasaki, unpubl. data). These cold-active β-galactosidases are possible candidates for milk processing.

Pectin, polymers of galacturonic acid and its methyl ester, is converted from an insoluble form to a soluble form by fruit ripening. Thus, a part of pectin in fruit contaminates the juice during the pressing process, resulting in an increase in viscosity and poor color. The addition of an enzyme mixture containing pectolytic enzymes, such as pectin methyl esterase (EC 3.1.1.11) and polygalacturonases (EC 3.2.1.15), prior to the pressing process not only reduces contamination of the juice by pectin but also increases the juice yield due to improvement of filtration. Enzymatic clarification of apple juice by the pectolytic enzymes may preferably be carried out at 15°C because gelatin fining is effectively carried out at a low temperature.[38] Cold-active polygalacturonase was isolated from a psychrophilic fungus, *Sclerotinia borealis*, which causes snow mold diseases.[39] The secreted enzyme showed maximal activity at 40–50°C and retained 30% of its maximal activity at 5°C.

Chymosin (EC 3.4.23.4), an aspartyl protease originally found in the fourth stomach of the unweaned calf, has been used for cheese production. However, the supply of chymosin has been depleted due to the increase in cheese production, and the chymosin source has been shifted to recombinant bacteria as well as chymosin-producing bacteria such as *Mucor miehei*.[40] Recombinant calf chymosin is produced by *E. coli* as a prochymosin form followed by conversion to chymosin.[41,42]

Characteristics of the proteolytic coagulants strongly affect the aging process and the final nature of produced cheese.[43] A chymosin-like protease produced by *Mucor pusillus* has been improved by genetic engineering to have less thermal stability.[44] This heat-labile protease will be useful for the production of long-life cheese and protease-less whey, because its activity is easily diminished. A lactic acid bacterium, *Streptococcus lactis* IAM 1198, which can grow at a low temperature, produced two intracellular proteases that had different optimal temperatures.[45] One of them had maximal activity at 6°C and was thought to be useful for aging cheese.[46] Immobilization of enzymes is suitable for large-scale and continuous production in industry. In many cases, it overcomes problems such as enzyme stability and enzyme recovery. For cheese production, although practical problems exist, immobilized chymosin substitutes can be used to digest casein at a low temperature, because digested casein does not gel at a low temperature.[34] Gel formation is initiated by an increase in temperature after processing with the immobilized chymosin substitutes. Soymilk-curd, a cheese analog, is produced by limited proteolysis of

soymilk. Commercial plant proteases, papain and bromelain (EC 3.4.22.5), catalyzed the limited proteolysis of soy-protein at 5°C and produced curd after heat
treatment.[47]

Lysozyme (EC 3.2.1.17) is an antimicrobial enzyme that is isolated mainly from
egg white. Lysozyme can legally be added to cheese as a preservative in some European countries.[34] An enzyme mixture containing this enzyme has been developed
for preservation of rice products in Japan.[33] Cold-active lysozyme has been isolated from the clam shell, *Chlamys islandica.*[48] Since this enzyme shows high activity
at 0°C, it would be useful as a enzymatic preservative for various foods that must
be stored at a low temperature.

Some of the steps in the processing of fish and other seafoods, such as removal
of skin and preparation of roe from fish, are laborious and, in some cases, cannot
be performed by automated machines. Therefore, enzymatic processing methods
have been developed for fish and other seafoods to reduce damage to the products
during processing.[48] For instance, cold-active pepsins from Atlantic cod *(Gadus
morhua)* and orange roughy *(Hopolostethus atlanticus)* have been used in the preparation of roe from salmon *(Salmo salar)* and orange roughy. Proteases and carbo-
hydrases have been used in the de-skinning process of starry ray *(Raja radiata)* at
0–10°C.[48] Similarly, alkaline proteases from *Aspergillus* sp. have been used in the
removal of skin from squid.[49]

Beta-glucosidase (EC 3.2.1.21) is used to enhance the aroma of wine by releasing terpenes from terpene glucosides.[15] Similarly, acetolactate decarboxylase (EC
4.1.1.5) isolated from mesophiles, which has maximal activity at a relatively low
temperature, has been used as a starter enzyme for brewing at low temperatures.[15]
This enzyme directly catalyzes the conversion of α-acetolactic acid to acetoin, without passing through the intermediate compound diacetyl, which emits an imma-
ture smell. Since such aromatic products are highly volatile, enzymatic reactions for
bouquet modification should be carried out at a low temperature. In the case of
cheese production, some lipases are used to generate a strong flavor such as in Stil-
ton, Parmesan, Gorgonzola, and Roquefort cheese by releasing free fatty acids from
lipids.[50]

Transglutaminase (EC 2.3.2.13) catalyzes the cross-linking reaction between two
amino acids in a protein, resulting in a change in food texture.[51] This enzyme can
also be used for the modification of proteins with other proteins or chemical substances. Transglutaminase produced from *Streptoverticillium* sp. was shown to be
useful in strengthening gels of surimi, minced fish flesh, at 10 and 25°C.[52] On the
other hand, the genes encoding transglutaminases have been isolated from various
fishes for the production of enzymes by genetic engineering.[53,54]

Glucose oxidase (EC 1.1.3.4) has been used as a preservative to remove oxygen
from foods.[55] This enzyme has also been used as a hydrogen peroxide-generating
enzyme for milk, since, in the presence of hydrogen peroxide, hydrogen peroxide
itself and lactoperoxidase (EC 1.11.1.7) inhibit the growth of various bacteria in
raw milk.[34,56] The preservative effect of glucose oxidase from *Aspergillus niger* has
been demonstrated. In the presence of glucose and catalase (EC 1.11.1.6), this
enzyme improved preservation of shrimp for 10 days at 0–2°C in terms of inhibition of psychrotroph growth and maintenance of sensory factors.[56]

Polyunsaturated fatty acids (PUFAs), including eicosapentaenoic acid (EPA) and docosahexaenoic acid (DHA), now attract considerable attention as sources of pharmaceutical agents, functional foods, and supplemental nutrition. The market for PUFAs has been steadily growing. PUFA-enriched substances have been produced by lipases from various sources. Lipase from psychrotrophic *Pseudomonas fluorescens*, for which the optimal temperature was 40°C, produced lipids enriched with PUFAs by catalyzing the selective incorporation of PUFAs in lipids at a low temperature.[57] In addition, lipase from *Pseudomonas* sp. produced a high concentration of monoglycerides containing 40% PUFA at the sn-2 position at 10°C.[58] Selective digestion of fish oils by a lipase from *Candida cylindracea* yielded PUFA-rich oils composed of up to 50% PUFA.[12] Lipase and phospholipase A2 (EC 3.1.1.4) from the porcine pancreas transferred PUFAs from fish oils to phospholipids.[50]

5
Application of enzymes in chemical synthesis and treatment of wastewater

One of the most successful examples of enzymatic synthesis of chemical compounds is the synthesis of acrylamide by nitrile hydratase (EC 4.2.1.84). Nitto Chemical has developed an enzymatic method by which 20,000 tons acrylamide per year can be produced.[12,59,60] Acrylamide is used as a paper strengthener. It is also used in the production of R-mandelic acid and nicotinamide, which are the intermediates of pharmaceutical agents. *Rhodococcus* sp. N-774 produced nitrile hydratase, which is active at low temperatures.[61,62] This strain was immobilized and used for the continuous production of acrylamide at a low temperature. An immobilized cell system of *Pseudomonas chlororaphis* strain B-23, which is resistant to amide, has been developed and utilized for the industrial production of acrylamide at 0–5°C.[63]

The optical resolution of compounds by enzymes to produce chiral intermediates has attracted the attention of pharmaceutical companies. Lipases are used in the synthesis of various chemical compounds such as optically active esters.[64] Many lipases from mesophiles showed hydrolytic activity at 4°C.[65] Cold-active lipases could be suitable for the synthesis of volatile compounds such as perfumery. The optimal temperature of lipase from psychrophilic *Ancinobactor* sp. isolated from Siberian soil was 20°C for hydrolytic activity, and it showed 90% of its maximal activity at 4°C.[66] In the ester synthesis reaction in n-hexane, this lipase showed higher activity at -25°C than at 4 and 30°C.[67] Moreover, the ratio of activity of this lipase at -25°C to 30°C was higher than those for other lipases such as *C. cylindracea*, indicating that this lipase is highly adaptable to a cold environment. Scallop hepatopancreas lipase, which showed high activity at a low temperature (optimal temperature 10°C) did not distinguish between positions sn-1 and sn-3 at 30°C. Interestingly, this enzyme showed selective hydrolysis activity at the sn-1 position at temperatures lower than 30°C and at the sn-3 position at temperatures higher than 30°C.[68,69] *Candida antarctica* produces two lipases, one of which catalyzes esterification of fatty acids to alkylglucopyranoside.[15] The product, glycolipid, is expected to be a powerful detergent.

Hydrogen peroxide is currently used for washing semiconductor circuits. Catalase produced by recombinant bacteria is used for degradation of excess hydrogen peroxide in wastewater tanks in semiconductor factories.[12] Hydrogen peroxide may also be used for sterilizing contact lenses.[15] The World Food and Agriculture Organization and the United Nations Agriculture Organization permit the addition of hydrogen peroxide to milk at a concentration of 0.05–0.25% as a preservative if all of hydrogen peroxide is destroyed by catalase after processing.[70] This "cold pasteurization" system is thought to be useful for developing countries. Therefore, cold-active catalases are potential enzymes for degrading the remaining hydrogen peroxide at the ambient temperature. Immobilized catalase would be effective for the decomposition of residual hydrogen peroxide in milk.[71] Heat-stable catalases isolated from *Pseudomonas* sp. and *Micrococcus* sp. might be useful for degrading excess hydrogen peroxide in wastewater tanks.[72] Catalase is also used to decompose residual hydrogen peroxide from herring roe after bleaching by hydrogen peroxide. We isolated a catalase-producing bacterium, *Vibrio* sp. S-1, from a drain pool of a fish product-processing plant.[73] The S-1 catalase showed approximately three-times higher activity than that of bovine liver catalase at 30–40°C and was inactivated at 60°C (I. Yumoto, unpubl. data).

6
Application of enzymes in biotechnology research

In plant biotechnology, protoplasts must be prepared under mild conditions. Since cold-active cell-wall-digestive enzymes, such as cellulase and pectinase, are suitable for protoplast preparation, they would be very useful in the field of plant biotechnology. A polysaccharide-digestive enzyme (optimal temperature 40°C) from *Acremonium alcalophilum* JCM7366 is expected to be used in protoplast preparation at a low temperature because it retains 40% of its maximal cellulase activity and 50% of its maximal xylanase activity at 10°C.[74]

Heat-labile alkaline phosphatases (EC 3.1.3.1) were isolated from bacteria in antarctic seawater.[75] The optimal temperature for the enzymatic reaction was 25°C, and the enzyme was irreversibly inactivated by heat treatment at 55°C for 10 min. Psychrophilic *Arthrobacter* sp. was reported to produce two alkaline phosphatases that had different heat stabilities.[76] Another alkaline phosphatase was isolated from Atlantic cod *(Gadus morhua)*.[77] The catalytic efficiency (k_{cat}/K_m) of the cod enzyme was 2.5-fold greater than that of the calf intestinal enzyme but was unstable at 40°C. In molecular biology, alkaline phosphatase, which catalyzes dephosphorylation of nucleic acids, must be inactivated after the reaction, because the residual active enzyme would abolish the subsequent kinase or ligase reaction. Therefore, heat-labile alkaline phosphatase is a preferable enzyme for molecular biology experiments.

Uracil DNA glycosylase (EC 3.2.2.3) has also been used in molecular biology to prevent carry-over contamination in the polymerase chain reaction (PCR).[78] However, this enzyme must be inactivated after treatment since the residual enzyme will degrade DNA produced by the subsequent PCR. Thus, heat-labile uracil-DNA glycosylase purified from a psychrophilic marine bacterium, BMTU3346, would be useful.[79] This enzyme was rapidly inactivated at 40°C, with a half-life of 2 min.

7
Potential enzyme candidates from cold-adapted microorganisms for industrial application

Enzymes used as detergent additives must function not only at a low temperature but also at high pH. To obtain such double-extremophilic enzymes, double-extremophiles have been screened with media at a high pH and a low temperature.[80] Pullulanase (EC 3.2.1.41) and α-amylase have been purified from alkalopsychrotrophic *Micrococcus* sp. 207, and they showed maximal activity at 50 and 55–60°C, respectively.[81,82]

The deep-sea is one of the prospective treasure houses of cold-active enzymes. A cold-active amylase ws isolated from a psychrotrophic *Vibrio* sp. in the deep-sea mud at a depth of 1,100–1,200 m.[83] Although the optimal temperature of this cold-active amylase is as high as that of mesophilic amylases (35°C), the activity of the enzyme at 10°C was 40% of its maximum, which is much higher than that of mesophiles. Genes encoding malat dehydrogenases (EC 1.1.1.37) from deep-sea psychrophilic bacteria, *Vibrio* sp. 5710 and *Photobacterium* sp. SS9, have also been reported.[84,85]

Although enzymes classified into group III (Fig. 1) are very useful, there have been few reports of the isolation of such enzymes. We have isolated a lipase from the snow mold fungus *Typhula ishikariensis*.[86] This lipase showed maximal activity at 15°C and more than 40% of its maximal activity at 0–40°C. Compared with other previously reported cold-active enzymes, this lipase is relatively stable at a

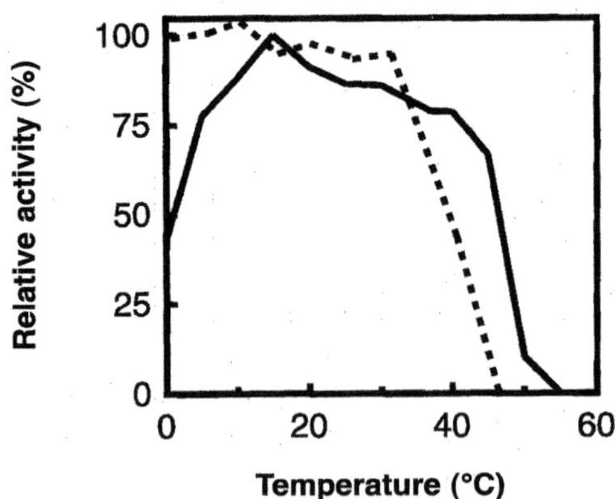

Fig. 2. Enzymatic activity and stability of a lipase isolated from *Typhula ishikariensis*
Solid line: enzymatic activity at different temperatures, *dotted line:* remaining activity of lipase after preincubation at indicated temperatures. Maximal activity in each experiment is defined as 100%

high temperature and shows high activity over a wide temperature range (Fig. 2). Recently, Morita et al.[87] isolated a lipase from a crab intestine that showed high activity over a wide temperature range and an amylase from a salmon intestine that showed high activity at a low temperature. Moreover, chymotrypsin from Atlantic cod was reported to be more active at a low temperature and more stable at a high temperature than bovine α-chymotrypsin.[88] It is expected that these enzymes will be used in detergent manufacturing and chemical industries due to their high activity over a wide temperature range as well as their stability.

Enzymes from cold-adapted microorganisms have another potential application. L-Glutamate dehydrogenase was purified from a psychrophile L101.[89] The enzyme was fixed with ferrocene on the surface of an electrode by albumin and glutaraldehyde. The constructed biosensor sensitively responded to L-glutamate even at 4°C. Such a novel biosensor that can function at a low temperature could be used in on-line monitoring of low-temperature bioprocesses and quality control in the preservation of foods at a low temperature.

8

Improvement in enzymatic characteristics by protein engineering and the expression system of cold-active enzymes

The successful application of an enzyme in industry sometimes requires an improvement in the enzyme's disadvantages and the endowment of useful characteristics to the enzyme by protein engineering. In some cases, enzymatic activity at a low temperature may need to be enhanced, while in other cases, heat-labile or heat-stable characteristics are required. Stability at a high pH and tolerance to bleach agents are the most important characteristics of enzymes to be used in detergents. Novo Nordisk succeeded in producing stable and bleach-tolerant protease and amylase by genetic engineering.[12] Moreover, highly active proteases have been created by genetic engineering at Novo Nordisk and Gist-Brocades, and highly active lipase has been created by Genenvor.[12]

Mucor pusillus protease was modified by genetic engineering.[44] This enzyme, in which Tyr75 was substituted for Asn, showed high activity for milk aggregation at a low temperature and rapidly lost proteolytic activity during the heating process. This heat-labile protease is thought to be suitable for the production of long-life cheese and protease-less whey.

Feller et al. isolated α-amylases from psychrophilic *Alteromonas haloplanctis* and compared its characteristics with that of α-amylase from mesophilic *Bacillus amyloliquefaciens*.[90] The cold-active amylase had higher activity than the mesophilic amylase at temperatures lower than 40°C. A comparison of their thermodynamic activation parameters indicated that a more flexible enzyme structure would reach an activated conformation with less heat content and that an increase in free energy corresponding to the activated complex would be attained through a minimum change of entropy. In a similar comparative study in which subtilisin from antarctic psychrophilic *Bacillus* TA41 was used, amino acids responsible for the flexibility of the cold-active enzyme were identified.[91] Rentier-Delrue et al. cloned triosephosphate isomerase (EC 5.3.1.1) genes from psychrophilic *Moraxella* sp. TA137 and thermophilic *Bacillus stearothermophilus*, and expressed them in *E. coli*.[92] A comparison of the structures of these triosephosphate isomerases and a chicken counterpart revealed that the adaptability of the enzyme to low temperatures seems to be due to the nature of helix-capping residues. A comparison of free energy, enthalpy, and entropy among four kinds of lactate dehydrogenase (EC 1.1.1.27) was also reported.[93] These comparative studies have provided important information on natural strategies to obtain cold-adaptation at the molecular level. For a review, see refs. 94–98.

Recently, evolutional engineering has been used to make subtilisin BPN' more active at a low temperature.[99] Subtilisin BPN' was first mutated to create a primary protease lacking activity at a low temperature, since the native enzyme had proteolytic activity even at 10°C, which resulted in an unclear distinction between the activities of the native enzyme and artificially improved enzymes. Random mutagenesis was applied to the gene of the primary subtilisin BPN' mutant, and then *E. coli* transformants containing additionally mutated subtilisin BPN' genes were screened on skim-milk plates. Enzymes that showed higher activity at a low temperature were identified by larger clear zones formed on the skim-milk plates. The generated novel enzymes, with a single-point mutation G131D at the second mutagenesis, acquired 1.5-fold higher activity (k_{cat}/K_m). Another subtilisin BPN' mutant M-15, V84I, was obtained by similar mutation-screening steps.[100] The M-15 mutant acquired higher activity than that of the native enzyme due to a smaller K_m at 5°C. However, K_m at 25°C and k_{cat} at 5 and 25°C were similar in the two enzymes. In addition, the secondary structure of the M-15 mutant seemed to be similar to the native structure at 10°C. These results suggest that K_m at a low temperature is also an important factor in high enzymatic activity at a low temperature.

Structural analysis, protein engineering, and evolutionary molecular engineering would be useful for production of much more desirable enzymes. Furthermore, an expression system suitable for the production of cold-active enzymes should be established. Since cold-active enzymes are usually thermolabile, ordinary expression systems such as the *E. coli* expression system may not be the most suitable for the expression of cold-active enzymes. Over the past decade, cold-shock responses have been reported in various species.[101] CspA is the major cold-shock protein in *E. coli*.[102] Recently, an expression system with a *cspA* promoter that is controlled by cold-shock treatment has been developed.[103] Other cold-inducible promoters in *E. coli* have also been reported.[104] The cold-shock inducible expression systems would be suitable for the expression of thermolabile enzymes. In addition, it is well known that many proteins aggregate in insoluble forms in *E. coli*; however, the cultivation of transformants at a low temperature increases the proportion of soluble proteins in some cases.[105] Since the *cspA* promoter can direct the synthesis of proteins even at 10°C, the cold-shock inducible expression system could be useful for the expression of aggregation-prone recombinant proteins at a low temperature.[106] Although host-vector systems in cold-adapted microorganisms have not been established yet, it would be a powerful tool for evolutional molecular engineering. One example has been reported. The TOL plasmid pWWO, which harbored a toluene utilization gene in mesophilic *Pseudomonas putida* PaW1, was transferred to psychrotrophic *P. puti-*

da Q5 by conjugation.[107] The resultant *P. putida* Q5 transformant could utilize toluate as a sole carbon source at low temperatures, indicating that the mesophilic plasmid was replicated and that the genes of the mesophilic enzymes were transcribed, even at low temperatures at which the original mesophilic host cannot grow.

9
Conclusions

At present, the most desirable characteristic of industrial enzymes is stability. Thermophilic enzymes from various microorganisms, including cold-adapted microorganisms, are therefore used for most industrial processes. However, it is expected that more cold-active enzymes will be discovered and used in the near future, since the need for characteristic enzymes in our daily lives and in industries are becoming increasingly diverse. In addition, from the viewpoint of energy-saving and environmental protection, cold-active enzymes are much more preferable than energy-consuming mesophilic and thermophilic enzymes. Cold-active enzymes from cold-adapted microorganisms will probably be the most useful in the detergent industry. Although cold-active enzymes have various advantages over other enzymes in the food industry, introduction of cold-active enzymes in the food industry will probably be slow due to the safety problem of cold-adapted microorganisms and their cold-active enzymes. For instance, although *E. coli* β-galactosidase has useful characteristics, this enzyme has not been used in the food industry because it is not yet regarded as absolutely safe by food law.[40] Indeed, some cold-adapted microorganisms are known pathogens. Therefore, research on the safety of cold-adapted microorganisms is needed to expand the application of cold-active enzymes not only in the food industry but also in other industries. In addition, genetic engineering, including evolutionary molecular engineering, will be increasingly used as a powerful tool for obtaining attractive cold-active enzymes, such as cold-active, stable, and highly active enzymes.

10
References

1. Lea AGH. Enzymes in the production of beverages and fruit juices. In: Tucker GA, Woods LFJ, eds. Enzymes in Food Processing, 2nd ed. Bishopbriggs: Blackie Academic & Professional, 1995:223-249.
2. Poldermans B. Proteolytic enzymes. In: Gerhartz W, ed. Enzymes in Industry. Weinheim: VCH Verlagsgesellschaft mbH, 1990:108-118.
3. Fairbairn DJ, Law BA. Proteinases of psychrophilic bacteria: their production, properties, effects and control. J Dairy Res 1986; 53:139-177.
4. Vazquez SC, Rios Merino LN, MacCormack WP. Protease-producing psychrotrophic bacteria isolated from Antarctica. Polar Biol 1995; 15:131-135.
5. Schinner F, Margesin R, Pümpel T. Extracellular protease-producing psychrotrophic bacteria from high alpine habitats. Arctic Alpine Res 1992; 24:88-92.
6. Margesin R, Palma N, Knauseder F, Schinner F. Proteases of psychrotrophic bacteria isolated from glaciers. J Basic Microbiol 1991; 31:377-383.

7. Hamamoto T, Horikoshi K. Psychrophilic microorganisms and their enzymes. Novo Nordisk Enzyme Symp 1993:2-7.

8. Morita RY. Psychrophilic bacteria. Bacteriol Rev 1975; 39:144-167.

9. Araki T. Ecology, isolation and cultivation of cold-adapted microorganisms. In: Ohshima Y, ed. Handbook of Extremophiles. Tokyo: Science forum, 1991:149-162 (in Japanese).

10. Pennisi E. In industry, extremophiles begin to make their mark. Science 1997; 276:705-706.

11. Novo Nordisk A/S World Wide Web page 1998: URL http://www.novo.dk.

12. Kawada T, ed. Industrial Enzymes Nikkei Biotechnology Annual Report 1998. Tokyo:Nikkei BP, 1997:611-626 (in Japanese).

13. Margesin R, Schinner F. Properties of cold-adapted microorganisms and their potential role in biotechnology. J Biotechnol 1994; 33:1-14.

14. Feller G, Narinx E, Arpigny JL, Aittaleb M, Baise E, Genicot S, Gerday C. Enzymes from psychrophilic organisms. FEMS Microbiol Rev 1996; 18:189-202.

15. Uwajima T. Development and use of industrial enzymes. In: Sangyo-you kouso (Industrial Enzymes). Tokyo: Maruzen, 1995:15-58 (in Japanese).

16. Maase FWJL, Van Tilburg R. The benefit of detergent enzymes under changing washing conditions. J Am Oil Chem Soc 1983; 60:1672-1675.

17. Onouchi T. A new protease for low temperature washing. Fragrance J 1985; 73:79-81 (in Japanese).

18. Okamoto I, Minagawa M. Removal of proteins from fabrics by protease: the effect of *Bacillus* alkaline protease in low temperature washing. Jpn Res Assn Text End-Uses 1987; 28:167-172.

19. Sakaguchi H. Detergents containing enzymes have been popularized with more than 80% of consumers: noteworthy alkaline cold-active lipases. Yushi (Oils and Fats) 1988; 41:56-63 (in Japanese).

20. Murakami K, Miyake Y. Protease added detergent. Kokai Tokkyo Koho 1988:S63-221199 (in Japanese).

21. Yamada K, Murakami K, Tachibana H, Watanabe Y, Miyake Y, Omura H. Stimulation by 2-mercaptoethanol and alcohols of protease in Antarctic krill, *Euphausia superba*, at a low temperature. Agric Biol Chem 1987; 51:3363-3368.

22. Kawai S, Okoshi H, Ozaki K, Shikata S, Ara K, Ito S. Neutrophilic *Bacillus* strain, KSM-522, that produces an alkaline carboxymethyl cellulose. Agric Biol Chem 1988; 52:1425-1431.

23. Ito S, Shikata S, Ozaki K, Kawai S, Okamoto K, Inoue S, Takei A, Ohta Y, Satoh T. Alkaline cellulase for laundry detergents: production by *Bacillus* sp. KSM-635 and enzymatic properties. Agric Biol Chem 1989; 53:1275-1281.

24. Ito S, Ozaki K. Development of alkaline cellulases for laundry detergents. Gendai-kagaku Zoukan 1995; 28:176-186 (in Japanese).

25. Woods LFJ, Swinton SJ. Enzymes in the starch and sugar industries. In: Tucker GA, Woods LFJ, eds. Enzymes in Food Processing, 2nd ed. Bishopbriggs: Blackie Academic & Professional, 1995:250-267.

26. Goldstein WE. Enzymes in starch processing and baking. In: Gerhartz W, ed. Enzymes in Industry. Weinheim: VCH Verlagsgesellschaft mbH, 1990:92-102.

27. Hagen HA, Pedersen S. Glucose isomerization. In: Gerhartz W, ed. Enzymes in Industry. Weinheim: VCH Verlagsgesellschaft mbH, 1990:102-108.

28. Suzuki K, Nakajima Y. Characterization and utilization of palatinose. Shokuhin-Kogyo 1983; 4:34-39 (in Japanese).

29. Rendleman JA Jr. Enzymic conversion of malto-oligosaccharides and maltodextrin into cyclodextrin at low temperature. Biotech Appl Biochem 1996; 24:129-137.

30. Mozaffar Z, Nakanishi K, Matsuno R. Formation of oligosaccharides during hydrolysis of lactose in milk using β-galactosidase. J Food Sci 1985; 50:1602-1606.

31. Mozaffar Z, Nakanishi K, Matsuno R. Continuous production of galactooligosaccharides from lactose using immobilized β-galactosidase from *Bacillus circulans*. Appl Microbiol Biotechnol 1986; 25:224-228.

32. Rahim KAA, Lee BH. Specificity, inhibitory studies, and oligosaccharide formation by

β-galactosidase from psychrotropic *Bacillus subtilis* KL88. J Dairy Sci 1991; 74:1773-1778.
33. Kobayashi T, Fukumoto H. Market of each enzyme. In: Market Prospect of Industrial Enzymes in Japan. Tokyo: CMC, 1995:55-137 (in Japanese).
34. Law BA, Goodenough PW. Enzymes in milk and cheese production. In: Tucker GA, Woods LFJ, eds. Enzymes in Food processing, 2nd ed. Bishopbriggs: Blackie Academic & Professional, 1995:114-143.
35. Loveland J, Gutshall K, Kasmir J, Prema P, Brenchley JE. Characterization of psychrotrophic microorganisms producing β-galactosidase activities. Appl Environ Microbiol 1994; 60:12-18.
36. Trimbur DE, Gutshall KR, Prema P, Brenchley JE. Characterization of a psychrophilic *Arthrobacter* gene and its cold-active β-galactosidase. Appl Environ Microbiol 1994; 60:4544-4552.
37. Amarita F, Alkorta F, Lescan du Plessix M, Cantabrana T, Rodriguez-Fernandez C. Isolation and properties of free and immobilized β-galactosidase from the psychrotropic enterobacterium *Buttiauxella agrestis* (strain NC4). J Appl Bacteriol 1995; 78:630-635.
38. Lea AGH. Enzymes in the production of beverages and fruit juices. In: Tucker GA, Woods LFJ, eds. Enzymes in Food Processing, 2nd ed. Bishopbriggs: Blackie Academic & Professional, 1995:223-249.
39. Takasawa T, Sagisaka K, Yagi K, Uchiyama K, Aoki A, Takaoka K, Yamamoto K. Polygalacturonase isolated from the culture of the psychrophilic fungus *Sclerotinia borealis*. Can J Microbiol 1997; 43:417-424.
40. Reimerdes EH. Dairy products. In: Tucker GA, Woods LFJ, eds. Enzymes in Food Processing. 2nd ed. Bishopbriggs: Blackie Academic & Professional, 1995:119-126.
41. Uwajima T. Production of enzymes by genetic engineering. In: Sangyo-you-kouso (Industrial Enzymes). Tokyo: Maruzen, 1995:101-120 (in Japanese).
42. Kawaguchi Y, Shimizu N, Nishimori K, Uozumi T, Beppu T. Renaturation and activation of calf prochymosin produced in an insoluble form in *Escherichia coli*. J Biotechnol 1984; 1:307-316.
43. Kindstedt PS, Guo MR. Recent developments in the science and technology of pizza cheese. Aust J Dairy Tech 1997; 52:41-43.
44. Yamashita T, Higashi S, Higashi T, Machida H, Iwasaki S, Nishiyama M, Beppu T. Mutation of a fungal aspartic proteinase, *Mucor pusillus* rennin, to decrease thermostability for use as a milk coagulant. J Biotechnol 1994; 32:17-28.
45. Akuzawa R, Yokoyama K. Purification, crystallization and some properties of low-temperature active intracellular proteinase from *Streptococcus lactis*. Jpn J Zootech Sci 1982; 53:814-821.
46. Akuzawa R, Ito O, Yokoyama K. Degradation of casein by crystalline low-temperature active proteinase from *Streptococcus lactis*. Jpn J Zootech Sci 1985; 56:56-61.
47. Kamata Y, Chiba K, Yamauchi F, Yamada M. Selection of commercial enzymes for soymilk-curd production by limited proteolysis with immobilized enzyme reactor. Nippon Shokuhin Kogyo Gakkaishi 1992; 39:102-105 (in Japanese).
48. Vilhelmsson O. The state of enzyme biotechnology in the fish processing industry. Trends Food Sci Tech 1997; 8:266-270.
49. Kobayashi T, Fukumoto H. Trends of enzyme manufacturers. In: Market Prospect of Industrial Enzymes in Japan. Tokyo: CMC, 1995:139-181 (in Japanese).
50. Tombs MP. Enzymes in the processing of fats and oils. In: Tucker GA, Woods LFJ, eds. Enzymes in Food Processing, 2nd ed. Bishopbriggs: Blackie Academic & Professional, 1995:268-291.
51. An H, Peters MY, Seymour TA. Roles of endogenous enzymes in surimi gelation. Trends Food Sci Tech 1996; 7:321-327.
52. Lee HG, Lanier TC, Hamann DD, Knopp JA. Transglutaminase effects on low temperature gelation of fish protein sols. J Food Sci 1997; 62:20-24.
53. Sano K, Nakanishi K, Nakamura N, Motoki M, Yasueda H. Cloning and sequence analysis of a cDNA encoding salmon *(Onchorhynchus keta)* liver transglutaminase. Biosci Biotechnol Biochem 1996; 60:1790-1794.

54. Yasueda H, Nakanishi K, Kumazawa Y, Nagase K, Motoki M, Matsui H. Tissue-type transglutaminase from red sea bream *(Pagrus major)*. Sequence analysis of the cDNA and functional expression in *Escherichia coli*. Eur J Biochem 1995; 232:411-419.

55. Uhlig H. Survey of industrial enzymes. In: Gerhartz W, ed. Enzymes in Industry. Weinheim: VCH Verlagsgesellschaft mbH, 1990:77-92.

56. Dondero M, Egaña W, Tarky W, Cifuentes A, Torres JA. Glucose oxidase/catalase improves preservation of shrimp *(Heterocarpus reedi)*. J Food Sci 1993; 58:774-779.

57. Kojima Y, Shimizu A. A new lipase, production method and application. Kokai Tokkyo Koho 1996:H8-154674 (in Japanese).

58. Zuyi L, Ward OP. Lipase-catalyzed alcoholysis to concentrate the *n*-3 polyunsaturated fatty acid of cod liver oil. Enzyme Microbiol Technol 1993; 15: 601-606.

59. Yamada H, Kobayashi M. Nitrile hydratase and its application to industrial production of acrylamide. Biosci Biotechnol Biochem 1996; 60:1391-1400.

60. Kobayashi M, Nagasawa T, Yamada H. Enzymatic synthesis of acrylamide: a success story not yet over. Trends Biotechnol 1992; 10:402-408.

61. Nakai K, Watanabe I, Sato Y, Enomoto K. Development of acrylamide manufacturing process using microorganisms. Nippon Nougeikagaku Kaishi 1988; 10:1443-1450 (in Japanese).

62. Watanabe I, Satoh Y, Enomoto K, Seki S, Sakashita K. Optimal conditions for cultivation of *Rhodococcus* sp.N-774 and for conversion of acrylonitrile to acrylamide by resting cells. Agric Biol Chem 1987; 51: 3201-3206.

63. Ashina Y, Watanabe I. Dramatic development of enzymatic process for acrylamide production. Kagaku to Kogyo 1990; 43:1098-1101 (in Japanese).

64. Sonomoto K, Tanaka A. Application of lipase in organic solvents for the preparation of optically active terpene alcohol esters. Ann N Y Acad Sci 1988; 542:235-239.

65. Hoshino T, Yamane T, Shimizu S. Selective hydrolysis of fish oil by lipase to concentrate *n*-3 polyunsaturated fatty acids. Agric Biol Chem 1990; 54:1459-1467.

66. Esaki, N. Psychrophilic enzymes from psychrophiles. In: Abstract book, Biological Molecular Response under Extreme Environments. 1996 (in Japanese).

67. Suzuki T, Akitoh M, Kurihara T, Esaki N, Sayuda K. Characterization of lipase produced by psychrophilic *Acinetobacter* sp. Nippon Nogeikagaku Kaishi 1996; 70:308 (in Japanese).

68. Itabashi Y, Ota T. Lipase activity in scallop hepatopancreas. Fish Sci 1994; 60:347.

69. Itabashi Y, Nishihara H, Ohta T, Suzuki T. Effects of temperature and pH on stereoselectivity of the lipase from scallop hepatopancrease. Proc Jpn Conference Biochem Lipids 1994; 36:354-357 (in Japanese).

70. FAO report on the meeting of experts on the use of H_2O_2 and other preservatives in milk, 1957; FAO/57/11/8655.

71. Tarhan L. Use of immobilised catalase to remove H_2O_2 used in the sterilisation of milk. Process Biochem 1995; 30:623-628.

72. Boismenu D, Lépine F, Gagnon M, Dugas H. Heat inactivation of catalase from cod muscle and from some psychrophilic bacteria. J Food Sci 1990; 55:581-582.

73. Yumoto I, Yamazaki K, Kawasaki K, Ichise N, Morita N, Hoshino T, Okuyama H. Isolation of *Vibrio* sp. S-1 exhibiting extraordinarily high catalase activity. J Ferment Bioeng 1998; 85:113-116.

74. Hayashi K., Nimura Y, Miyaji T, Ohara N, Uchimura T, Suzuki H, Komagata K, Kozaki M. Purification and properties of a low-temperature-active enzyme degrading both cellulose and xylan from *Acremonium alcalophilum* JCM 7366. Seibutsukogaku Kaishi 1997; 75:9-14 (in Japanese).

75. Kobori H, Sullivan CW, Shizuya H. Heat-labile alkaline phosphatase from Antarctic bacteria: rapid 5'-end-labeling of nucleic acids. Proc Natl Acad Sci USA 1984; 81:6691-6695.

76. De Prada P, Loveland-Curtze J, Brenchley JE. Production of two extracellular alkaline phosphatases by a psychrophilic *Arthrobacter* strain. Appl Environ Microbiol 1996; 62:3732-3738.

77. Asgeirsson B, Hartemink R, Chlebowski JF. Alkaline phosphatase from Atlantic cod *(Gadus*

morhua). Kinetic and structural properties which indicate adaptation to low temperatures. Comp Biochem Physiol 1995; 110B:315-329.

78. Longo MC, Berninger MS, Hartley JL. Use of uracil DNA glycosylase to control carry-over contamination in polymerase chain reactions. Gene 1990; 93:125-128.

79. Sobek H, Schmidt M, Frey B, Kaluza K. Heat-labile uracil-DNA glycosylase: purification and characterization. FEBS Lett 1996; 388:1-4.

80. Kimura T, Horikoshi K. Isolation of bacteria which can grow at both high pH and low temperature. Appl Environ Microbiol 1988; 54:1066-1067.

81. Kimura T, Horikoshi K. Characterization of pullulan-hydrolysing enzyme from an alkalo psychrotrophic *Micrococcus* sp. Appl Microbiol Biotechnol 1990; 34:52-56.

82. Kimura T, Horikoshi K. Purification and characterization of a-amylases of an alkalopsychrotrophic *Micrococcus* sp. Starch 1990; 42:403-407.

83. Hamamoto T, Horikoshi K. Characterization of an amylase from a psychrotrophic *Vibrio* isolated from a deep-sea mud sample. FEMS Microbiol Lett 1991; 84:79-84.

84. Ohkuma M, Ohtoko K, Takada N, Hamamoto T, Usami R, Kudo T, Horikoshi K. Characterization of malate dehydrogenase from deep-sea psychrophilic *Vibrio* sp. strain 5710 and cloning of its gene. FEMS Microbiol Lett 1996; 137:247-252.

85. Welch TJ, Bartlett DH. Cloning, sequencing and overexpression of the gene encoding malate dehydrogenase from the deep-sea bacterium *Photobacterium* species strain SS9. Biochim Biophys Acta 1997; 1350:41-46.

86. Hoshino T, Sakamoto T, Ohgiya S, Shimanuki T, Ishizaki K. Low-temperature-active lipase of *Typhula ishikariensis.* In: Abstract book, Beijerinck Centennial. Microbial Physiology and Gene Regulation: Emerging Principles and Application 1995:183-184.

87. Morita Y, Nakamura T, Hasan Q, Murakami Y, Yokoyama K, Tamiya E. Cold-active enzymes from cold-adapted bacteria. J Am Oil Chem Soc 1997; 74:441-444.

88. Raae AJ. Effect of low and high temperatures on chymotrypsin from Atlantic cod (*Gadus morhua* L.); comparison with bovine α-chymotrypsin. Comp Biochem Physiol 1990; 97B:145-149.

89. Yamamura A, Murakami Y, Sakaguchi T, Yokoyama K, Tamiya E. Biosensor using new cold-active L-glutamate dehydrogenese. In: Abstract book, Annual Meeting of Japan Biotechnology Society 1996; 176 (in Japanese).

90. Feller G, Lonhienne T, Deroanne C, Libioulle C, Beeumen JV, Gerday C. Purification, characterization, and nucleotide sequence of the thermolabile α-amylase from the antarctic psychrotroph *Alteromonas haloplanctis* A23. J Biol Chem 1992; 267:5217-5221.

91. Davail S, Feller G, Narinx, E, Gerday C. Cold adaption of proteins. J Biol Chem 1994; 269:17448-17453.

92. Rentier-Delrue F, Mande SC, Moyens S, Terpstra P, Mainfroid V, Goraj K, Lion M, Hol WGJ, Martial JA. Cloning and overexpression of the triosephosphate isomerase genes from psychrophilic and thermophilic bacteria. J Mol Biol 1993; 229:85-93.

93. Low PS, Bada JL, Somero GN. Temperature adaptation of enzymes: role of the free energy, the enthalpy, and the entropy of activation. Proc Natl Acad Sci USA 1973; 70:430-432.

94. Gerday C, Aittaleb M, Arpigny JL, Baise E, Chessa J-P, Garsoux G, Petrescu I, Feller G. Psychrophilic enzymes: a thermodynamic challenge. Biochim Biophys Acta 1997; 1342:119-131.

95. Feller G, Gerday C. Psychrophilic enzymes: molecular basis of cold adaptation. Cell Mol Life Sci 1997; 53:830-841.

96. Jaenicke R, Závodszky P. Proteins under extreme physical conditions. FEBS Lett 1990; 268:344-349.

97. Jaenicke R. Protein structure and function at low temperatures. Philos Trans R Soc Lond B Biol Sci 1990; 326:535-551.

98. Zuber H. Temperature adaptation of lactate dehydrogenase. Structural, functional and genetic aspects. Biophys Chem 1988; 29:171-179.

99. Tange T, Taguchi S, Kojima S, Miura K, Momose H. Improvement of a useful enzyme (subtilisin BPN') by an experimental evolution system. Appl Microbiol Biotechnol 1994; 41:239-244.

100. Kano H, Taguchi S, Momose H. Cold adaptation of a mesophilic serine protease, subtilisin, by in vitro random mutagenesis. Appl Microbiol Biotechnol 1997; 47:46-51.
101. Jones PG, Inouye M. The cold-shock response – a hot topic. Mol Microbiol 1994; 11:811-818.
102. Jones PG, VanBogelen RA, Neidhardt, FC. Induction of proteins in response to low temperature in *Escherichia coli*. J Bacteriol 1987; 169:2092-2095.
103. Vasina, JA, Baneyx F. Recombinant protein expression at a low temperatures under the transcriptional control of the major *Escherichia coli* cold shock promoter *cspA*. Appl Environ Microbiol 1996; 62:1444-1447.
104. Qoronfleh MW, Debouck C, Keller J. Identification and characterization of novel low-temperature-inducible promoters of *Escherichia coli*. J Bacteriol 1992; 174:7902-7909.
105. Schein CH, Noteborn MHM. Formation of soluble recombinant proteins in *Escherichia coli* is favored by lower growth temperature. Bio/Technol 1988; 6:291-294.
106. Vasina JA, Baneyx F. Expression of aggregation-prone recombinant proteins at low temperatures: a comparative study of the *Escherichia coli cspA* and *tac* promoter systems. Protein Expr Purif 1997; 9:211-218.
107. Kolenc RJ, Inniss WE, Glick BR, Robinson CW, Mayfield CI. Transfer and expression of mesophilic plasmid-mediated degradative capacity in a psychrophilic bacterium. Appl Environ Microbiol 1988; 54:638-641.

Low temperature organic phase biocatalysis using cold-adapted enzymes

R. K. Owusu Apenten

Department of Food Science, School of Physical Sciences, University of Leeds, Leeds, Ls 2 9JT, United Kingdom

1
Introduction

Enzymes from ectothermic organisms are adapted to function within specific niches in the environment. There seems to be parity in respect of conformational stability, flexibility and activity for enzymes at the growth temperature for the source organisms.[1] This makes cold-adapted enzymes particularly suited for catalysis at low temperatures. Organic phase biocatalysis (OPB) refers to the performance of enzymatic reactions in media comprising wholly or partly of organic solvents. This approach extends the well-known advantages of enzymatic catalysis (e.g., high specificity and catalytic efficiency, mild reaction conditions) to reactions involving water insoluble substrates. Replacing an aqueous reaction medium with an organic solvent allows novel reactions, e.g. reverse hydrolysis and transesterification, to be carried out. OPB grew rapidly in the mid-1980's with applications of enzymes, whole cells and tissues, as biocatalysts in water-organic solvent phase systems.

The purpose of this review is to advocate low or sub-zero temperature organic phase biocatalysis (LTOPB) using enzymes from cold-adapted organisms. At this time, there are two examples of the use of cold-adapted enzymes for LTOPB.[2,3] There is an urgent need to stimulate research in what is, scientifically and technologically, a neglected aspect of biocatalysis. In principle, all major classes of enzymes (hydrolases, oxidases, lyases, synthethatases, and dehydrogenases) can be employed for LTOPB. Biocatalysis in aqueous-organic solvent systems using normal and thermophile derived enzymes has been extensively reviewed.[4-15]

A significant number of important issues surrounding OPB at 25–100°C have as their point of convergence, enzyme stability and activity (Table 1). Many of these issues could be overcome at low temperatures. The possible advantages of LTOPB are listed in Table 2. Some advantages accrue from the use of a low or sub-zero reaction temperatures, e.g. increased solvent choice, increased product yield, greater compatibility with heat sensitive reagents. Other improvements (increased enzyme flexibility and activity, increased resistance to cold-denaturation) may arise from the use of cold-adapted enzymes for LTOPB.

Table 1. Enzyme stability and rigity in relation to organic phase biocatalysis

1. Enzyme destabilization by organic solvents
2. Decreased activity of dry enzymes owing to their high rigidity
3. Decreased activity of freeze-dried enzymes due to cold denaturation
4. Restricted organic solvent choice due enzyme instability in polar (water miscible) organic solvents

Table 2. Advantages of low temperature organic phase biocatalysis (LTOPB)

1. Improved biocatalysis stability at low temperatures
2. Wider biocatalyst choice to include mesophile and cold-adapted enzymes (the two groups comprise the majority of enzymatic proteins in the biosphere)
3. Increased solvent choice to include (so-called denaturing) polar, water-miscible, organic solvents
4. Increased product yield
5. Improved compatibility with heat sensitive reactants and products
6. Improved oxygen solubility for oxidase catalyzed reactions

2
Organic phase biocatalysis at low temperatures is cryoenzymology

Theoretical and experimental support for LTOPB is derived from cryoenzymology. This is defined as the study of enzyme structure and catalysis at sub-zero temperatures.[16] By the early 1980's over 25 mesophile-derived enzymes and proteins had been studied at sub-zero temperatures. At such temperatures, enzyme reaction intermediates are more long-lived and therefore easier to characterize.[17–21] The most common solvents for cryoenzymology are:

1. water-miscible organic co-solvent systems (1-phase systems),
2. water-in-oil microemulsions or reverse micelles,
3. water-in-oil macroemulsions.

Co-solvents are usually weakly protic, polar organic solvents with relatively low viscosity.[22,23] Water co-solvent mixtures containing about 50% (w/w) organic component can be cooled to between -44°C and -66°C without freezing. At a co-solvent concentration of about 90% (w/w) temperatures of -100°C can be attained.

Some important solvent characteristics include density (d), dielectric constant (ε), freezing point (fp), apparent viscosity (η) and protonic activity (pH*).[24] The temperature dependence of some of these parameters is also important. Solvent dielectric constant varies with temperature according to Akerlof's equation:[25]

$$\log \varepsilon = a - b\mathrm{T} \tag{1}$$

where a and b are empirical constants. Eq. (1) is valid over a temperature range of -100°C to +100°C. Solvent dynamic viscosity increases with decreasing temperature according to:

$$\log \eta = c + e/T \tag{2}$$

where c and e are constants and T is the absolute temperature.[26] Regardless of the arrangements for avoiding ice formation, organic solvents may affect the following enzymatic properties:[16-20]

1. structure,
2. substrate binding (cf. Km or Ks) and,
3. catalytic properties (k_{CAT})

3
Enzyme stability in organic solvents at low temperature

For convenience, organic solvents can be classed as water miscible or water immiscible. In terms of their effects on enzyme structure the two classes of organic solvent can be considered together. Enzyme stability varies with co-solvent concentration (C_{ORG}) according to Eq. (3):

$$Tm - Tm_o = (R\ Tm\ Tm_o\ /\Delta h\ )\ n\ K_b\ C_{ORG} \tag{3}$$

where Tm is the enzyme denaturation temperature in a wholly aqueous system, Tm_o is the denaturation temperature in the presence of an organic solvent, Δh is the enthalpy change for the rapture of an "interesidue" hydrophobic interaction , n is the average number of co-solvent binding sites per amino acid residue and K_b is a binding constant.[27-30]

Derivation of Eq. (3) assumes that there is a stoicheometric binding of co-solvent molecules to independent sites on the enzyme. Only dissolved organic solvent molecules may bind to the enzyme. This description for co-solvent mediated denaturation of enzymes has generally fallen out of favor because of:

1. uncertainties surrounding values for n, Δh and K_b for different proteins and organic solvents,
2. an inability to account for the enzyme *stabilization* by moderate concentrations of co-solvent (see below).

However, Eq. (3) is useful as the basis for a discussion of the joint effects of temperature and organic solvents on enzyme stability. It can be predicted from Eq. (3) that a plot of $Tm - Tm_o$ versus C_{ORG} will give a straight-line graph. The gradient of the plot will increase with the magnitude of K_b. Increasing the co-solvent concentration will result in Tm_o falling below the ambient temperature *and* the appearance of solvent-induced unfolding of the structure of an enzyme.[27-30] The co-solvent concentration necessary to unfold 50% of enzyme molecules (C_m) is expressed by a variation of Eq. (3):

$$C_m = (Tm - Tm_o)\ \Delta h\ /(Tm\ Tm_o\ 2.303\ R\ n\ K_b) \tag{4}$$

A value for C_m can be found by setting Tm_o equal to the prevailing temperature, say 298°K; using typical values for parameters in Eq. (4) (Δh = 1250 cal mole^{-1} , n = 1 and $K_b = 1.3. 10^{-2} M^{-1}$ methanol) then C_m depends only on the Tm, which is the heat unfolding temperature for a specified enzyme.

Assuming, for example, that Tm = 333 K then (from Eq. (4) C_m will be equal to 13 M (33% v/v methanol).[29,30] Providing that K_b and Δh are weakly temperature dependent, a reduction of the reaction temperature from $Tm_o = 25°C$ to $Tm_o = 0°C$ or -20°C increases C_m from 13 M (33% v/v) to 24 M (62% v/v) or 34.5 M (~90% v/v) methanol, respectively. In conclusion, Eq. (3 & 4) lead to two potentially useful strategies for increasing the tolerance of enzymes towards organic solvents:

1. increase Tm by using a more heat stable enzymes – specifically Eq. (4) accounts for observed correlation between enzyme heat stability and the resistance to organic solvent denaturation,[31,32]
2. decrease Tm_o by using a lower reaction temperature.

The former strategy has been advocated by Don Cowan at the University College London (UK) over the past decade.[8,31,32] The second strategy, what we have termed low temperature organic phase biocatalysis (LTOPB), has only begun to receive attention, as described below.

Enzyme stability can be improved by suspending the dry enzyme powder in 100% immiscible organic solvent.[33] The marked stability enhancements observed in such cases is due to increases in the (protein) glass-transition temperatures (Tg) at low water activity.[34,35] As an example, the Tm for dry lysozyme was 117–140°C compared to 68.8–72°C for lysozyme in dilute aqueous solution.[36] Dry enzymes contain adsorbed moisture.[35] Therefore, C_{ORG} and C_m refer to the organic solvent concentration in the adsorbed water phase. A concentration equal to C_m might not be attainable for a co-solvent with limited solubility in an aqueous phase. In general the conditions $C_{ORG} << C_m$ is consistant with enzyme stability.

A detailed study of co-solvent effects on enzyme stability will consider parameters like the enthalpy (ΔH), entropy (ΔS), Gibbs free energy (ΔG), and heat capacity (ΔCp) change for protein unfolding. Schrier et al. showed using UV difference spectrophotometry (UVDS) that the Tm for ribonuclease decreased but that ΔH did not vary with C_{ORG}.[27,28] In contrast, Brandts and Hunts revealed that ΔG, ΔH, ΔS and ΔCp shows a complex dependence on C_{ORG} and that the trends change with temperature.[37] Parody et al. examined the denaturation of lysozyme by UVDS and found that Tm decreased linearly with C_{ORG} whereas ΔH increased, reached a maximum, and then decreased with increasing C_{ORG}.[30]

Studies using differential scanning calorimetry (DSC) have verified that Tm decreases linearly with C_{ORG} whilst ΔH at first increases to a maximum and then decreased as function of C_{ORG}.[38-41] Using high precision DSC measurements, it was further shown that ΔCp decreases linearly with co-solvent concentration.[42-44]

Hydrophobic interactions contribute greatly to the stabilization of globular proteins.[45] The driving force, for the sequestration of nonpolar amimo acid sidechains within the protein interior, may be the (thermodynamically unfavorable) negative hydration entropy change associated with the insertion of a nonpolar group into water. In the liquid state, water forms a tetrahedral, 4-coodinate, 3-dimensional molecular network with an average bond length and angle that are greater than those found in ice. The unfavorable entropy change for hydrating nonpolar solutes is partly due to the redistribution of water molecules around nonpolar solutes in an attempt to retain an optimum 3-dimensional H-bonding network in liquid water.[46,47]

The hydration of polar as well as nonpolar solutes leads to a decrease in entropy.[48,49] In contrast, ΔCp for solvating nonpolar solutes is positive compared to the negative ΔCp for the hydration of polar solutes.[50] Using computer simulations, it was shown that hydration of nonpolar solutes leads to perturbation of water structure within the first hydration shell. The resulting H-bonding pattern is more ice-like (decreased average bond length and angle) thereby accounting for the positive ΔCp. For polar or charged solutes, hydration produces less ice-like water structure (increased average H-bond length and average root mean square bond angle) and a negative ΔCp.[48,49]

The consensus is that water-structure is central to discussions of protein stabilization. The disruption of hydrophobic interactions occur via the perturbation of water structure by added co-solvent molecules. This is seen as the basis for enzyme destabilization by organic solvents.

Though probably of secondary importance, direct binding of organic solvent molecules to accessible sites on a denatured enzyme, would shift the native-unfolded state equilibrium, leading to destabilization. A decrease in the dielectric constant (ε) for a water-co-solvent system, as compared to purely aqueous solvent, would also destabilize enzymes by altering their ionization potential, the balance of surface electrostatic interactions, and their degree of hydration stabilization. Such changes could be sufficient to induce enzyme unfolding and/or aggregation.

Enzyme denaturation by dielectric effects may be annuled at low temperatures. For an aqueous solvent, the empirical constants for Askerlof's equation are $a = 1.945$ and $b = -2\text{x}10^{-3}$ with ε ranging from 88.1 (at 0°C) to 55.6 (at +100°C). By comparison, $a = 1.863$ and $b = -1.92\text{x}10^{-3}$ for a water DMSO (50% w/w) system and consequently, $\varepsilon = 88.1$ at a temperature of -40°C. Therefore, at sub-zero temperatures, the *thermal stability* of enzymes could reach levels observed in a simple aqueous buffer.[16,17,20,24] However, some enzymes are subject to cold instability.[47]

With an immiscible solvent phase, destabilisation can arise from enzyme adsorption at the water-organic solvent interface. The process involves diffusion of the enzyme molecule to the interface, adsorption, and unfolding or relaxation of the native structure. This is followed by irreversible processes such as sulfhydry/disulfide exchange leading to interfacial aggregation. Therefore, the degree of enzyme denaturation at an oil-water interface is related to the interfacial tension at water-immiscible solvent interface, the area of the interface and shear or agitation rate.[32]

The structure of enzymes within reverse micelles developed from sodium bis (2-ethylhexyl) sulphosuccinate (AOT) have been widely studied by Luisi and co-workers using conventional spectrophotometric techniques.[51–53] Effects on enzyme stability depend on the ratio of AOT to water and on the isoelectric point of the enzyme (or effective charge). Enzyme destabilization occurs when there is strong interaction with the anionic headgroups of the AOT. Enzymes that are negatively charged experience less structural perturbation as a result of being repulsed from the water-surfactant interface.

Cytochrome C (having a net charge of +8) was modified by acylation or succinylation to produce variants with net charges of -3 and -8. The cationic form was perturbed within AOT reverse micelles resulting in a loss of stability. By contrast the structure of the negatively charged protein was apparently unchanged in AOT micelles.[54] Entrapment of bovine liver catalase within AOT reverse micelles had little effect on its stability and catalytic properties.[55] A recent study of the heat unfolding of ribonuclease T-1 in AOT reverse micelles provides a further informative example of the effect of these systems on enzyme structure.[56] There is a need for more exhaustive studies on enzymes in reverse micelles especially at low and sub-zero temperatures.

4
Enzyme activity in organic solvents at low temperature

Observations from cryoenzymology show that organic solvents generally increase K_m values but that V_{max} values remain unchanged. The effect on substrate binding is due to the lower dielectric constant in water co-solvent systems. However, the ε-value increases with decreasing temperatures and the negative effect of co-solvent on enzyme-substrate interactions can be ameliorated at sub-zero temperatures.[16,17,20,24]

From basic Michaelian kinetics, $V_{max} \propto k_{CAT}$ (s^{-1}) whilst K_m, to a first approximation, is the dissociation constant for the enzyme-substrate complex. Analysis of the temperature dependence of k_{CAT} (s^{-1}) leads to activation parameters associated with the formation of an activated transion state complex (ES* or EP*):[57]

$$E + S \leftrightarrows ES^* \rightarrow ES \leftrightarrows EP^* \rightarrow E + P \tag{5}$$

For a multistage reaction, the activation free energy ($\Delta G^{\#}$) is associated with the formation of the least stable (high-energy) transitions state along the reaction progress axis. $\Delta G^{\#}$ is given by the Eyring relation:

$$\Delta G^{\#} = R\,T \ln (k_{CAT}\, h\, /\, K_B\, T) = \Delta H^{\#} - T\Delta S^{\#} \tag{6}$$

where h is the Planks constant, K_B is the Boltzman constant, T is the absolute temperature, $\Delta H^{\#}$ is the activation enthalpy and $\Delta S^{\#}$ is the activation entropy. After a slight re-arrangement Eq. (6) becomes,

$$\ln (k_{CAT}\,/T) = -\Delta H^{\#}\,/(RT) + \Delta S^{\#}\,/R - \ln (h\,/\,K_B) \tag{7}$$

and a plot of $\ln (k_{CAT}/T)$ versus $1/T$ will produce a straight-line graph with a slope of $-\Delta H^{\#}/\,R$ and a $y = 0$ intercept of $\Delta S^{\#}/R - \ln(h/K_B)$. Hence $\Delta S^{\#}/R$ can be readily determined. The reaction activation energy ($\Delta E^{\#}$) is simply $\Delta H^{\#} + RT$.

Analyses of the effect of temperature on K_m (actually K_s) lead to thermodynamic parameters for the reversible binding of enzyme to substrate,

$$E + S \leftrightarrows ES \tag{8}$$

The binding enthalpy (ΔH^{o}), entropy (ΔS^{o}) and free energy change (ΔG^{o}) can be determined from standard thermodynamic relations. Assuming that these parameters are independent of temperature;

$$\Delta G^\circ = - RT \ln (1/K_m) = \Delta H^\circ - T \Delta S^\circ \qquad (9)$$

and

$$\ln (1/K_m) = \Delta S^\circ/R - \Delta H^\circ/RT \qquad (10)$$

To obtain a full appreciation of the effect of decreasing temperatures on enzyme catalysis the preceeding analyses ought to be carried for all substrates in a given reaction and for the forward as well as the reverse reaction.[57]

5
Low and subzero temperature organic phase biocatalysis using normal enzymes

In a low water environment, reactions such as reverse hydrolysis (esterifications, peptide bond formation etc.) can be performed. Peptide bonds synthesis via protease catalyzed aminolysis of amino acid esters has been widely discussed.[58,59] The amino acid ester (starting substrate) undergoes nucleophilic substitution to form an acylated enzyme intermediate with the release of an alcohol. This is then hydrolyzed in the presence of water, to give a free enzyme and amino acid. A range of nucleophiles (Nu) can compete with water for the acyl-enzyme leading, for example, to aminolysis (Fig. 1).

The effect of subzero temperatures on dipeptide synthesis by OPB with chymotrypsin immobilized on celite was examined by Jonsson et al.[60] The esterase and

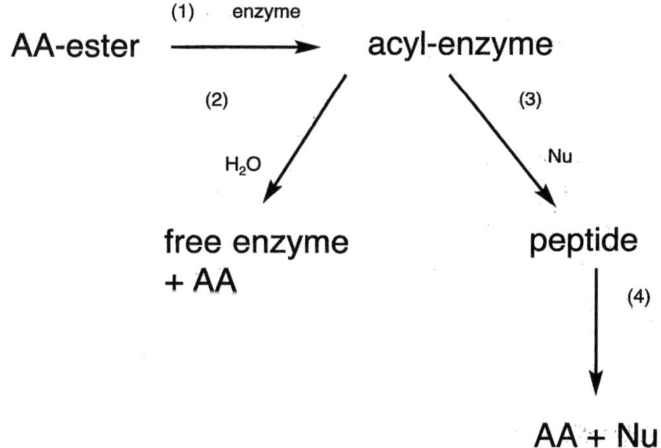

Fig. 1. Protease catalyzed peptide synthesis
(1) Esterase activity leads to formation of an acyl enzyme intermediate (acyl-enzyme) from the breakdown of the amino acid ester (AA-ester) with release of an alcohol. Acyl-enzyme undergoes (2) hydrolysis or (3) reaction with nucleophile (Nu) leading to peptide synthesis. (4) Re-hydrolysis of product peptide due to enzyme amidase activity.

amidase activities of chymotrypsin were studied using N-acetyl-L-phenylalanine ethyl ester (Ac-PheOEt) and N-acetyl-L-phenylalanyl-L-alaninamide (Ac-PheAla-NH$_2$) as substrates. The organic solvent phases were acetonitrile, acetone, tetrahydrofuran (THF) or diglyme containing 2.5–10% (v/v) water. Michaelian parameters for the hydrolysis of the donor ester and dipeptide were determined. The yield of product was also examined at temperatures between -30°C and 25°C.

The yield of dipeptide increased with decreasing temperature. With a solvent phase comprising of acetonitrile with 10% water, the maximal yield was 99% at -20°C compared with 84% at 25°C. The favorable effects of decreasing temperature on product yield were ascribed to three factors:

1. a decrease in p-value defined as the apparent concentration of Nu at which the rate of aminolysis is equal to the rate of hydrolysis,
2. an increase in the α-value, i.e., the ratio-enzyme specificity [(amino acid ester): enzyme specificity (peptide)],
3. ice formation in some of the reaction systems.

In a further report from the same group the influence of reaction temperatures on synthetic activity, product yield and nucleophile specificity for celite immobilized subtilisin Carlsberg and α-chymotrypsin were studied.[61] Acyl-transfer reactions between Ac-PheOEt and 14 different nucleophiles (11 amino acid amides and 3 dipeptides) were examined. Once again a decrease of temperatures (from 25°C to -1°C) had a positive effect on the peptide yield. The order of aminolytic efficiency with different nucleophiles was not changed. However, all nucleophiles became more effective in competition with water.

The above illustrates that product yield can be improved by performing OPB reactions at low temperature. The reasons for such improvements are manifold. First, the Nu reaction is more favored (cf. hydrolysis) at low temperature. Second, a decreasing reaction temperature enhances protease esterolytic activity at the expense of the amidase activity. Third, the rate of re-hydrolysis of the peptide produced by aminolysis is lower at low temperature.[60,61]

The above study shows that there can be quite subtle phenomenon arising from the combined effects of low temperature and co-solvent on the enzyme. An understanding of such phenomena requires an investigation of changes in K_m and k_{CAT} values for the four substrates – amino acid ester, water, Nu and peptide – shown in Figure 1.[60,61] Equally important, there may be changes in the ionization state of other reagents (e.g., buffer salts) and changes in water activity.

What are the possible effects of decreasing temperature on the ionization of buffer salts and Nu and how might this affect product yield? As illustrated in Figure 1 the nucleophile for aminolysis is unionized. First, consider the simple relations,

$$Nu + H^+ \leftrightarrows Nu\text{-}H^+ \tag{11a}$$

$$Tris + H^+ \leftrightarrows Tris\text{-}H^+ \tag{11b}$$

$$pH = pKa + \log\left([Tris]/[Tris\text{-}H^+]\right) \tag{11c}$$

Any shift in the ionization equilibrium towards Nu Eq. (11a) would promote peptide synthesis. For the reactions under discussion, Nu was alanine (Ala-NH$_2$).[60,61] Commercial alanine hydrochloride was neutralized at pH 10, extracted into an organic phase and dried. Aminolysis was performed with a medium consisting of 90% co-solvent and 10% water (actually 50 mM Tris-HCl buffer, pH 7.8 at 25°C).

As the reaction temperature was reduced from 25°C to -20°C, there would be an increase in the solvent pH* by about 1.3-1.75 pH* units. This is due to the large heat of neutralization (ΔH_{NEU}) associated with nitrogen-based buffers. With a Tris buffer (70% v/v) methanol co-solvent system, I estimate (from the literature data) that $\Delta H_{NEU} \approx$ -53.1 kJ mol^{-1} for the temperature interval between +20°C and -20°C.[24] Owing to this exothermicity, formation of Tris-H$^+$ (Eq. (11b)) will be encouraged at low temperatures. The pKa for Tris will increase by between 0.028 and 0.039 units °C^{-1} (cf. Eq. (11c)). A similar treatment for Nu leads to a modest increase in the pKa (and the nucleophilic character) at low temperatures because for Ala-NH$_2$, ΔH_{NEU} is only -3.1 kJ mol^{-1}. But there is still the not so small pH change. A solvent pH* rise of 1.3-1.75 units would increase the apparent concentration of Nu by 20–50 fold for aminolysis at -20°C as compared to the same reaction at room temperature.

Temperature induced shifts in medium pH* are an important "issue" in extreme temperature biology. To avoid ambiguity, the pH* of a reaction medium should be pre-adjusted to the desired value at the intended reaction temperature. Choosing a less temperature-sensitive buffer (e.g., phosphate buffer) is also advisable when this is compatible with the enzyme.

The other factor sometimes responsible for the improvement of product yields during LTOPB is ice formation. This leads to the sequestration of water (which solidifies) and the concentration of reagents in the nonaqueous liquid part of the system. The reduction in water activity increases the rate of synthesis compared to hydrolysis. Studies by Jakubke and co-workers offer an intriguing perspective on the effect of freezing on enzymic peptide synthesis since co-solvents did not feature greatly in their early work.[62-66]

High yields of peptide synthesis were achieved by the simple procedure of freezing an *aqueous* solution of protease, acyl donor and nucleophile. Chymotrypsin catalyzed peptide synthesis at -18°C with a yield of 80% after a 4 h reaction time. Reverse hydrolysis was also catalyzed by proteinase zymogens and range of other hydrolases such as elastase, glucosidases and ribonuclease. Synthesis was hampered by the presence of liquid water.[62,63] There was evidence that the use of organic solvents and hydrophobic immobilization supports at subzero temperatures would further enhance product synthesis.

6

Low and subzero temperature organic phase biocatalysis using cold-adapted enzymes

Recent reviews on the current research in the field of cold-adapted enzymes include those by Herbert,[67] Teichgraber et al.,[68] Margesin and Schinner,[69] Feller et al.,[70] Brenchley,[71] Feller and Gerday,[72] and also Marshall.[73] From recent structural studies, it seems that the adaptation mechanisms leading to cold adaptation are subtle. Consequently, the 3-D structures for cold-adapted enzymes are similar to those for the corresponding mesophile derived enzymes. Compared to mesophile or thermophile derived enzymes, cold-adapted enzymes, are thought to exhibit:

1. reductions in the extent of electrostatic and dipole-dipole interactions,
2. reductions in the hydrophobicity of the protein core,
3. increased protein surface interactions with water,
4. decreased heat stability,
5. increased molecular flexibility.

The k_{CAT}/K_m ratios for cold-adapted enzymes are also greater than the corresponding values for mesophile or thermophile derived enzymes, when such parameters are compared at low temperature. It may not be essential to ascribe the increased catalytic efficiency of cold-adapted enzymes to the increased flexibility, though a degree of flexibility is required for catalysis.

The above characteristics make cold-adapted enzymes potentially well suited to LTOPB. Klibanov recently produced a list of seven factors thought to be responsible for the lowering of enzymatic activity in organic solvents.[74] Four of the most important factors were each capable of up to 100-fold activity reduction. They include a low conformational mobility or flexibility for enzymes suspended in dry organic solvents, cold denaturation to produce inactive conformations during lyophilization to produce dry enzyme powders for OPB, sub-optimal reaction pHs, unfavorable substrate desolvation, i.e., unfavorable transfer of nonpolar substrates from a nonpolar solvents to an enzyme active site. A lack of thermostability was not mentioned as constraint for OPB.

The limitations of OPB at 25–100°C could be largely circumvented by performing such reactions at low temperature in conjuction with cold-adapted enzymes. Some possible advantages of LTOPB as opposed to OPB at ambient or elevated temperatures, are listed in Table 2. In addition to these, further advantages can be expected by virtue of using cold-adapted enzymes for LTOPB. These include:

1. higher enzyme flexibility,
2. improved resistance towards cold-induced unfolding,
3. higher values for k_{CAT}/K_m,
4. increased interactions with the aqueous solvent.

Cold-adapted enzymes from all of the six (Enzyme commission) classes have some potential for applications in LTOPB. It is anticipated that the earliest examples of LTOPB using cold-adapted enzymes will involve hydrolases.

7
Conclusions

There is no evidence that lower reaction temperatures lead to problems in OPB. On the contrary, low temperatures represent a levelling of the OPB "playing field" so that the unique properties of cold-adapted enzymes can be more fully exploited in this important area of biocatalysis. The wider use of LTOPB (with cold-adapted enzymes or mesophile derived enzymes) may require a re-evaluation of many OPB guidelines regarding solvent choice, medium engineering, product and substrate partitionin. [3-15] It seems that very stable enzymes, by virtue of their high rigidity require elevated temperatures for OPB. Cold-adapted enzymes offer exciting opportunities in the area of LTOPB.

8
References

1. Somero GN. Proteins and temperature. Ann Rev Physiol 1995; 57:43-68.
2. Tan S, Apenten RKO, Knapp J. Low temperature organic phase biocatalysis using cold-adapted lipase from psychrotrophic *Pseudomonas* P38. Food Chemistry 1996; 57:415-418.
3. Kunugi S, Koyasu A, Takahashi S, Oda K. Peptide condensation activity of a neutral protease from *Vibro* sp. T1800 (Vimelysin). Biotechnol Bioeng 1997; 53:386-390.
4. Cerrea G. Biocatalysis in water-organic solvent 2-phase systems. Trends Biotechnol 1984; 2:102-106.
5. Inada Y, Takahashi K, Yoshimoto T, Ajima A, Matsushima A, Saito Y. Trends Biotechnol 1986; 4:190-194.
6. Carrea G, Cremonesi P. Enzyme catalysed steroid transformation in water organic solvent 2-phase systems. Methods Enzymol 1987; 136:150-157.
7. Mattiasson B, Aldercreutz P. Tailoring the microenvironment of enzymes in water-poor systems. Trends Biotechnol 1991; 9:394-398.
8. Cowan, DA, Plant AR. Biocatalysis in organic media. ACS Symp Ser 1992; 498:86-207.
9. Halling PJ. Thermodynamic predictions for biocatalysis in nonconventional media-theory, tests, and recommendations for experimental design and analysis. Enzyme Microbiol Technol 1994; 16:178-206.
10. Wescott CR, Klibanov AM. The solvent dependence of enzyme specificity. Biochim Biophys Acta-Protein Str Mol Enzymol 1994; 1206:1-9.
11. Kvittingen L. Some aspects of biocatalysis in organic-solvents. Tetrahedron 1994; 50:8253-8274.
12. Lortie R. Enzyme catalysed esterification. Biotechnol Adv 1997; 15:1-15.
13. Fernandez-Mayoralas A. Synthesis and modification of carbohydrates using glucosidases and lipases. Topics Curr Chem 1991; 186:1-20.
14. Reetz MT. Entrapment of biocatalysts in hydrophobic sol-gel materials for use in organic chemistry. Advance Material 1997; 9:943.
15. Mabrouk PA. The use of poly(ethylene glycol) enzymes in nonaqueous enzymology. ACS Symp Ser 1997; 680:118-113.
16. Douzou P. Aqueous-organic solutions of enzymes at sub-zero temperatures. Biochimie 1971; 53:1135-1145.
17. Fink AL. Cryoenzymology: The use of sub-zero temperatures and fluid solutions in the study of enzyme mechanisms. J Theor Biol 1976; 61:419-445.
18. Makinen MW. Reactivity and cryoenzymology of enzymes in the crystalline state. Ann Rev Biophys Bioeng 1977; 6:301-343.

19. Fink AL, Geeves MA. Cryoenzymology: The study of enzyme catalysis at subzero temperatures. Methods Enzymol 1979; 63:336-370.
20. Douzou P. Cryoenzymology in aqueous media. Adv Enzymol 1980; 51:1-74.
21. Fink AL, Petsko GA. X-ray cryoenzymology. Adv Enzymol 1981; 52:177-246.
22. Singer SJ. The properties of proteins in nonaqueous solvents. Adv Protein Chem 1962; 17:1-68.
23. Franks F, Eagland D. The role of solvent intractions in protein conformation. CRC Crit Rev Biochem 1975; 3:165-219.
24. Douzou P, Hoa HBP, Maurel P, Travers F. Physical chemical data for mixed solvents used in low temperature biochemistry. In Faman ED, ed. Handbook of Biochemistry and Molecular Biology. 3rd ed. Cleveland:CRC Press, 1976:520-539.
25. Akerlof G. Dielectric constants of some organic solvent-water mixtures at various temperatures. J Am Chem Soc 1932; 54:4125-4139.
26. Aten WC. Solvent action and measurement. In Whim BP, Johnson PG, eds. Directory of Solvents. London:Blakie Academic & Professionals, 1996:11-47.
27. Schrier EE, Scheraga HA. The effect of aqueous alcohol solutions on the thermal transition of ribonuclease. Biochim Biophys Acta 1962; 64:406-408.
28. Schrier, EE, Ingwall RT, Scheraga HA. The effect of aqueous alcohol solutions on the thermal transition of ribonuclease. J Phys Chem 1965; 69:298-303.
29. Herkovits JT, Gadegbuku B, Jaillet H. Structural stability and solvent denaturation of proteins. Denaturation by alcohols and glycols. J Biol Chem 1970; 245:2588-2598.
30. Parodi RM, Bianchi E, Ciferri A. Thermodynamics of unfolding of lysozyme in aqueous alcohol solutions. J Biol Chem 1973; 248:4047-4051.
31. Owusu RK, Cowan DA. Thermostable microbial protein stability in aqueous: organic two solvent phase systems. Biochem Soc Trans 1989; 17:581-582.
32. Owusu RK, Cowan DA. A correlation between microbial protein thermostability and resistance to denaturation in aqueous-organic solvent two phase systems. Enzyme Microbiol Technol 1989; 11:468-474.
33. Zaks A, Klibanov AM. Enzymatic catalysis in organic media at 100°C. Science 1984; 224:1249-1251.
34. Slade L, Levine H. Beyond water activity: Recent advances based on an alternative approach to the assesment of food quality and safety. Crit Rev Food Sci Nutr 1991; 30:115-360.
35. Rupley JA, Careri G. Protein hydration and function. Adv Protein Chem 1991; 41:37-142.
36. Johnston DS, Castelli F. The influence of sugars on the properties of freeze-dried lysozyme and heamoglobin. Thermochim Acta 1989; 144:195-208.
37. Brandts JF, Hunt L. Thermodynamics of protein denaturation. III. The denaturation of ribonuclease in water and in aqueous urea and aqueous ethanol mixtures. J Am Chem Soc 1967; 89:4826-4838.
38. Fujita Y, Miyanaga A, Noda Y. Effect of alcohols on the thermal denaturation of lysozyme as measured by differential Scanning calorimetry. Bull Chem Soc Jpn 1979; 52:3659-3662.
39. Fujita Y, Izumiguchi S, Noda Y. Effec of dimethylsulfoxide and its homologues on the thermal denaturation of lysozyme as measured by differential scanning calorimetry. Int J Peptide Protein Res 1982; 19:25-31.
40. Fugita Y, Noda Y. The effect of organic solvents in the thermal denaturation of lysozyme as measured by differential scanning calorimetry. Bull Chem Soc Jpn 1983; 56:233-237.
41. Fink AL, Painter B. Characterization of the unfolding of ribonuclease A in aqueous methanol solvents. Biochemistry 1987; 26:1665-1671.
42. Velicelebi G, Sturtevant JM. Thermodynamics of the denaturation of lysozyme in alcohol-water mixtures. Biochemistry 1979; 18:1180-1186.
43. Jacobson AL, Turner CL. Specific solvent effects on the thermal denaturation of ribonuclease. Effect of dimethyl sulfoxide and p-dioxane on thermodynamics of denaturation. Biochemistry 1980; 19:4534-4538.
44. Fu L, Freire E. On the origin of the enthalpy and entropy convergence temperatures in protein folding. Proc Natl Acad Sci USA 1992; 89:9335-9338.

45. Creighton TE. Proteins: Structure and Molecular Properties. New York: WH Freeman & Co, 1984.

46. Privalov P, Gill SJ. Stability of protein structure and hydrophobic interaction. Adv Protein Chem 1988; 39:191-233.

47. Franks F. Protein destabilization at low temperatures. Adv Protein Chem 1995; 46:105-139.

48. Madan B, Sharp K. Heat capacity changes accompanying hydrophobic and ionic solvation. A Monte Carlo and random network study. J Phys Chem 1996; 100:7713-7721.

49. Madan B, Sharp K. Molecular origin of hydration heat capacity changes of hydrophobic solutes: Perturbation of water structure around alkanes. J Phys Chem B. 1997; 101:11237-11242.

50. Livingtone JR, Spolar RS, Record MT. Contribution to the thermodynamics of protein folding from the reduction in the water-assessible nonpolar surface area. Biochemistry 1991; 30:4237-4244.

51. Luisi PL, Henninger F, Joppich M, Dossena A, Casnati G. Solubilization and spectroscopic properties of α-chymotrypsin in cyclohexane. Biochem Biophys Res Comm 1977; 74:1384-1389.

52. Wolf R, Luissi PL. Micellar solubilization of enzymes in hydrocarbon solvents. Enzymatic activity and spectroscopic properties of ribonuclease in n-octane. Biochem Biophys Res Comm 1979; 89:209-217.

53. Barbaric S, Luisi PL. Micellar solubilization of biopolymers in organic solvents 5. Activity and conformation of a-chymotrypsin in isooctane-AOT reverse micelles. J Am Chem Soc 1981; 103:4239-4244.

54. Larsson KM, Pileni MP. Interactions of native and modified cytochrome-C with a negatively charged reverse micella liquid interface. Eur Biophys J 1993; 21:409-416.

55. Haber J, Maslakiewica P, Rodakiewicznwak J, Walde P. Activity and spectroscopic properties of bovine liver catalase in sodium bis (2-ethylhexyl)sulfosuccinate isooctane reverse micelles. Eur J Biochem 1993: 217:567-573.

56. Shasty MCR, Eftink MR. Reversible thermal unfolding of ribonuclease T-1 in reverse micelles. Biochemistry 1996; 35:4094-4101.

57. Laidler KJ, Peterman BF. Temperature effects in enzyme kinetics. Methods Enzymol 1979; 63:234-257.

58. Fruton JS. Proteinase-catalysed synthesis of peptide bonds. Adv Enzymol 1982; 53:239-306.

59. Morihara K. Using proteases in pepride synthesis. Trends Biotechnol 1987; 5:164-170.

60. Jonsson A, Aldercrueutz A, Mattiasson B. Effects of subzero temperatures on the kinetics of protease catalysed dipeptide synthesis in organic media. Biotechnol Bioeng 1995; 46:429-436.

61. Jonsson A, Sehtje E, Aldercreutz P, Mattiasson B. Temperature effects on protease catalysed acyl transfer reactions in organic media. J Molecular Catalysis B-Enzym 1996; 2:43-51.

62. Tongu V, Meos H, Hoga M, Aaviksaar A, Jakubke H-D. Peptide synthesis by chymotrypsin in frozen solutions. Febs Letts 1993; 329:40-42.

63. Jakubke HD, Eichhorn U, Hansler M, Ullmann D. Nonconventional enzyme catalysis: Applicatons of proteases and zymogens in biotransformations. Biol Chem 1996; 377:455-464.

64. Haensler M, Jakubke HD. Reverse action of hydrolases in frozen aqeous-solutions. Amino Acids 1996; 11:379-395.

65. Haensler M, Gerisch S, Rettelbush J, Jakubke HD. Application of immobilized alpha-chymotrypsin to peptide synthesis in frozen aqeous systems. J Chem Technol Biotechnol 1997; 68:202-208.

66. Haensler M, Wehofsky N, Gerisch S, Wissmann JD, Jakubke HD. Reverse catalysis of elastase from procine pancrease in frozen aqueous systems. Biol Chem1998; 379:71-74.

67. Herbert RA. A perspective on the biotechnological potential of extremophiles. Trends Biotechnol 1992; 10:395-402.

68. Teocjgraber P, Xache J, Knorr D. Enzymes from germinating seeds- potential applications in food processing. Trends Food Sci 1993; 4:145-149.

69. Margesin R, Schinner F. Properties of cold-adapted microorganisms and their potential role

in biotechnology. J Biotechnol 1994; 33:1-14.

70. Feller G, Narinz E, Arpigny JL, Aittaleb M, Baise E, Genicot S, Gerday C. Enymes from psychrophilic organisms. FEMS Microbiol Rev 1996; 18:189-202.

71. Brenchley JE. Psychrotrophic microorganisms and their cold-active enzymes. J Ind Microbiol Biotechnol 1996; 17:432-437.

72. Feller G, Gerday C. Psychrophilic enzymes: molecular basis for cold adaptation. CMLS Cell Mol Life Sci 1997; 53:830-841.

73. Marshall CJ. Cold-adapted enzymes. Trends Biotechnol 1997; 15:358-364.

74. Klibanov AM. Why are enzymes less active in organic solvents than water. Trends Biotechnol 1997; 15:97-100.

Lipases A and B from the yeast *Candida antarctica*

T. B. Nielsen[1]*, M. Ishii[1] and O. Kirk[2]

[1] Research and Development, Novo Nordisk Bioindustry Ltd., Makuhari Techno Garden CB-6, 3, Nakase, 1-chome, Mihama-ka, Chiba-shi 261-8501, Japan
[2] Bio-Organic Chemistry, Protein Discovery, Enzyme Research, Novo Nordisk A/S, Novo Allé, 2880 Bagsvaerd, Denmark

1
Introduction

The anamorphic basidiomyceteous yeast *Candida antarctica* was first isolated from sediment from the bottom of the antarctic lake Vanda, perennially covered with 3–5 m of ice.[1] The original name given to this yeast was *Sporobolomyces antarcticus* and since then the organism has been around under a number of different aliases. For instance, Centraalbureau voor Schimmelcultures (CBS) in Baarn-Delft, The Netherlands, now keeps Goto et al.'s original isolate as the filamentous fungus *Pseudozyma antarctica* CBS 214.83. However, when isolates belonging to this species in 1988 were identified as lipase producers, the generally accepted name was *Candida antarctica*. The numerous lipase-related publications always refer to the origin of these enzymes as *C. antarctica*, so for convenience this name will be used throughout this chapter.

Goto et al.[1] isolated only one strain of *C. antarctica* (AY-18-1) and found that it grew in the temperature range 5–32°C. It may therefore seem surprising that the sample it originated from was taken from the bottom of a permanently ice-covered lake, but the paper mentions, that temperatures well into the plus range have been measured in several lakes in Antarctica.

Nevertheless, it was quite surprising for us, when we conducted a large-scale screening for microbial lipases in the late 1980's, that several isolates of a brown-red, somewhat filamentous yeast from Japanese natural samples, producing a very thermostable lipase, turned out to be strains of *C. antarctica*. Intensive work has been done on one of these strains, LF058, and it was found that two quite different lipases are being produced. The lipases were named A and B, and cloning and expression in the filamentous fungus *Aspergillus oryzae* permitted their introduction as industrial catalysts. Since then, the character and potential applications of these two lipases have been described in hundreds of articles and patent applications from laboratories all over the world.

---

* Corresponding author

We will here review the characteristics of the unique biocatalysts Lipase A and Lipase B from *C. antarctica.*

2
Lipase screening

Lipases have for many years been regarded as interesting industrial catalysts for the hydrolysis of fatty acid esters, for ester synthesis, interesterifications, etc. One of the special features of lipases for these types of reactions is the enantiomeric selectivity, making it possible to obtain optically pure compounds under mild conditions on an industrial scale.

In the late 1980's, Novo Nordisk conducted a large-scale screening program for microbial lipases with potentials as industrial catalysts. Several different types of lipases were targeted and one of the important ones was a positional non-specific, thermostable lipase. Many lipases hydrolyze only the two outer positions in a triglyceride, but some also hydrolyze the secondary ester in the middle position. Novo Nordisk's Japanese laboratory had the task of finding such a lipase, and it turned out, that very few candidate strains showed up. The screening program was mostly based on microorganisms isolated from a wide variety of natural samples collected at many different locations in Japan. Good candidates were often brownish-red yeasts showing some degree of filamentous growth. They in the end all turned out to be isolates of *C. antarctica,* and the best lipase producer among the strains (LF058) was chosen for more detailed characterization and study. Despite *C. antarctica* being cold-tolerant, some isolates grow well even at 35°C. Actually LF058 was isolated from an enrichment culture at 35°C, pH 4.5.

As reported by Ishii et al.[2] and Heldt-Hansen et al.,[3] it was soon found that LF058 produces two different lipases, Lipase A, which is the very thermostable, positional non-specific lipase originally identified, and Lipase B, which also is thermostable, but more interestingly has a very unique ability to form and hydrolyze a variety of esters stereospecifically. For many of the potential applications, immobilized lipases are advantageous, so such preparations were prepared and characterized at a very early stage. However, LF058 produces only limited amounts of the two lipases, so obtaining pure components in larger amounts was quite tedious.

Patkar et al.[4] succeeded in purifying native lipases A and B in sufficient amounts for a detailed characterization. Using tributyrin as substrate in an assay system, where pH is kept constant, Patkar et al. showed that both lipases have a pH optimum around neutral, but keep more than half of the activity at pH 7 in the pH range 5–10. The very high thermostability of Lipase A was confirmed, as no decrease in activity was detected after 2 h at 60°C. The preparations of Lipase A as well as Lipase B were pure enough to allow the determination of the first ten terminal amino acids in both lipases. These sequences formed the basis for the cloning of the lipase genes.

3
Cloning of the two lipase genes, *LIPA* and *LIPB*

The cloning of *C. antarctica* Lipase B was reported by Uppenberg et al.[5] The N-terminal amino acid sequence was used for designing slightly degenerate oligonucleotide probes and the lipase gene *(LIPB)* was isolated from genomic DNA using standard molecular biology methods. The gene contains no introns and the full amino acid sequence could be deducted from the experimentally determined DNA sequence. The gene codes for a signal peptide of 18 amino acids, a short propeptide of 7 amino acids and a mature protein consisting of 317 amino acids. The deducted N-terminal amino acid sequence was in complete accordance with the sequence found experimentally in the native enzyme.

Lipases typically contain a catalytic triad composed of a serine, a histidine and an aspartic acid (rarely glutamic acid) residue. This turned out also to be the case for *C. antarctica* Lipase B, the active site residues being Ser105, His224 and Asp187 (Fig. 1). Another typical feature of lipases is the highly conserved sequence around the active site serine, Gly-X-Ser-Y-Gly. Unexpectedly, this motif is not found in Lipase B; instead the sequence is Thr-Trp-Ser-Gln-Gly. The significance of the threonine residue in position 103 was investigated by Patkar et al.[6] By site-directed mutagenesis they changed the threonine residue to the glycine normally found. The most striking difference found between the mutant and the wildtype lipase was that the mutant enzyme was significantly more thermostable. The half-life at 60°C increased from 6 to around 25 min. However, the specific activity as measured towards tributyrin at pH 7 was reduced from 500 units mg^{-1} for the wildtype lipase to 260 units mg^{-1} in the case of the glycine mutant. Both enzymes were immobilized and tested in a number of ester synthesis reactions, but no significant difference in performance was observed. The calculated molecular weight of the ma-ture protein is 33.0 kDa, which fits the experimental data well considering that the protein is glycosylated.

Lipase B shows only limited sequence homology to other lipases, however with one exception: the lipase cloned from *Hyphozyma* sp. LF132.[7] The *Hyphozyma* sp. LF132 lipase is similar in size (319 amino acids in the mature protein) and shows an overall identity of 73% with the mature Lipase B. Furthermore, the unusual Thr-Trp-Ser-Gln-Gly sequence around the active site serine (Ser107) is also found in the *Hyphozyma* lipase.

```
- 25  MKLLSLTGVA  GVLATCVAAT  PLVKRLPSGS  DPAFSQPKSV  LDAGLTCQGA
  26  SPSSVSKPIL  LVPGTGTTGP  QSFDSNWIPL  STQLGYTPCW  ISPPPFMLND
  76  TQVNTEYMVN  AITALYAGSG  NNKLPVLTWS  QGGLVAQWGL  TFFPSIRSKV
 126  DRLMAFAPDY  KGTVLAGPLD  ALAVSAPSVW  QQTTGSALTT  ALRNAGGLTQ
 176  IVPTTNLYSA  TDEIVQPQVS  NSPLDSSYLF  NGKNVQAQAV  CGPLFVIDHA
 226  GSLTSQFSYV  VGRSALRSTT  GQARSADYGI  TDCNPLPAND  LTPEQKVAAA
 276  ALLAPAAAAI  VAGPKQNCEP  DLMPYARPFA  VGKRTCSGIV  TP
```

Fig. 1. Full amino acid sequence of Lipase B. The signal peptide is shown by *single underlining;* the propeptide by *double underlining;* the consensus sequence around the active site serine by *bold underlining;* and the aspartic acid and histidine residues in the active site by *bold*

```
 - 31 MRVSLRSITS LLAAATAAVL AAPAAETLDR RAALPNPYDD PFYTTPSNIG
   20 TFAKGQVIQS RKVPTDIGNA NNAASFQLQY RTTNTQNEAV ADVATVWIPA
   70 KPASPPKIFS YQVYEDATAL DCAPSYSYLT GLDQPNKVTA VLDTPIIIGW
  120 ALQQGYYVVS SDHEGFKAAF IAGYEEGMAI LDGIRALKNY QNLPSDSKVA
  170 LEGYSGGAHA TVWATSLAES YAPELNIVGA SHGGTPVSAK DTFTFLNGGP
  220 FAGFALAGVS GLSLAHPDME SFIEARLNAK GQRTLKQIRG RGFCLPQVVL
  270 TYPFLNVFSL VNDTNLLNEA PIASILKQET VVQAEASYTV SVPKFPRFIW
  320 HAIPDEIVPY QPAATYVKEQ CAKGANINFS PYPIAEHLTA EIFGLVPSLW
  370 FIKQAFDGTT PKVICGTPIP AIAGITTPSA DQVLGSDLAN QLRSLDGKQS
  420 AFGKPFGPIT PP
```

Fig. 2. Full amino acid sequence of Lipase A. The signal peptide is shown by *single underlining;* the propeptide by *double underlining;* the two phenylalanine residues in the lipid contact zone, F135 and F139, by *bold;* and the consensus sequence around the active site serine by *bold underlining*

The cloning of *C. antarctica* Lipase A proceeded in parallel and was published in 1995 by Hoegh et al.[8] The cloning was again based on oligonucleotide probes designed using the knowledge obtained from the N-terminal amino acid sequence determination.[4] LIPA also contains no introns and the full amino acid sequence was deducted from the DNA sequence. The gene codes for a signal peptide of 21 amino acids, a propeptide of 10 amino acids and a mature lipase with 431 amino acid residues (Fig. 2). The deducted N-terminal amino acid sequence of the mature protein again matched the experimentally determined sequence perfectly. Lipase A contains the conserved sequence around the active site serine (Ser174), Gly-Tyr-Ser-Gly-Gly, but there is no general homology with any other lipase.

3.1
Recombinant expression making industrial uses feasible

Industrial applications set a number of requirements. The lipase must have the right character regarding substrate specificity, temperature requirements, pH optimum, etc., but, apart from this, the commercial product must be free of unwanted side activities (particularly other lipases and esterases) and, last but not least, the price must be at a level where it becomes economically interesting for the customer. The product economy has often been an important factor hindering the commercial success of an otherwise interesting lipase.

Cloning technology combined with high-level expression in suitable host organisms has now made it possible in many cases to put lipase products on the industrial market. The most successful example so far has been the lipase from the filamentous fungus *Thermomyces lanuginosus* (previously called *Humicola lanuginosa*), which is used in many standard detergents worldwide under the trade name Lipolase™.

The cloning of Lipases A and B from *C. antarctica* made it possible also to try these enzymes in the *Aspergillus oryzae* system for high expression and secretion of heterologous proteins.[9] As reported by Hoegh et al.,[8] both lipases are expressed in *A. oryzae* in amounts significantly higher than those obtained from LF058 itself. Both recombinant enzymes appear to be slightly more glycosylated than the native ones, but the catalytic properties are unchanged and all evidence suggests that the proteins are processed correctly in *A. oryzae*.

It therefore became possible to introduce Lipases A and B as industrial products practically free of contaminating side activities. As described in detail below, especially Lipase B has since then been the focus of intensive research in industrial as well as academic laboratories. In literature, the lipases are often referred to under various trade names. The by far most commonly seen are Novozym® 435 and SP435, both representing an immobilized version of Lipase B.

4
Lipase A

Despite the fact that *C. antarctica* in no way is a thermophilic microorganism (Goto et al. did not observe any growth of the original isolate even at 35°C[1]) Lipase A has been found to be probably the most thermostable lipolytic enzyme known today.

4.1
Stability

Svendsen et al.[10] investigated the thermostability at various pH values using highly purified, recombinant Lipase A. They used the method of differential scanning calorimetry (DSC), where the thermal denaturation temperature, T_d, is determined by slowly heating an enzyme solution at a constant rate until the protein denatures. The T_d values were as high as 96°C, 95°C and 93°C at the pH values 4.5, 5 and 7, respectively. The catalytic activity of Lipase A is highest around neutral pH, but, nevertheless, the enzyme was found also to be extremely stable at high temperatures in the acidic pH range.

4.2
Site-directed variants with surprising characters

From studies of other microbial lipases it was found that certain aromatic amino acids located in the so-called lipid contact zone of the lipase are important for the activity. In other lipases a tryptophan residue has been found to be a key amino acid in this respect, but Lipase A has instead two phenylalanine residues in positions 135 and 139 (Fig. 2). Svendsen et al.[10] substituted these two phenylalanines with tryptophan by site-directed mutagenesis and found that the variant lipase with tryptophan in position 135 had a roughly 4 times increased specific activity when measured with tributyrin as substrate in a pH-stat method at 30°C. The double mutant with two tryptophan substitutions had an even higher specific activity.

The denaturation temperature of the Trp135 variant was measured with the DSC method, and despite the variant enzyme being somewhat less thermostable than the wildtype, the denaturation temperature at pH 5 was still as high as 84°C.

4.3
Application possibilities at high temperatures

The exceptionally high temperature stability of Lipase A suggests it can be used in industrial processes, where other lipases quickly lose their activity due to thermal inactivation.

4.3.1
Pitch control

One such example of application is the pitch control concept in the paper and pulp industry, where the paper mills sometimes experience pitch troubles in paper-making processes based on mechanical pulp. Wood from coniferous trees contains variable amounts of so-called pitch components, fatty acid glycerides, resin acids, etc. This pitch can cause serious troubles by precipitating out on machinery and the paper itself. Lipase is already today used industrially to control the pitch problems, but the present products cannot be used at the hottest places in the paper-making process as the temperature often is well over 70°C. Matsukura et al.[11] tested Lipase A in mechanical pulp at high temperatures and found that even though the optimum temperature for the enzyme at these harsh conditions was around 70°C, Lipase A can completely hydrolyze the triglycerides in pulp even at 80°C.

4.3.2
Treatment of fabrics

Cotton naturally contains waxes and other lipophilic materials, but, as well as this, wax lubricants based on triglyceride esters are often applied to yarns in order to increase the speed of cotton weaving. In later parts of the process, it is desirable to remove these fatty materials, and lipases are one good method of doing this under mild and environmentally friendly conditions. However, some of these lubricants only melt at temperatures over 60–70°C and lipases generally show much better activity towards liquid than towards solid substrates. Lund et al.[12] showed that Lipase A performs well at 75°C under industrial conditions using 150 kg of denim jeans. Using a commercial wax lubricant, they also showed that temperatures over 60°C were necessary for efficient hydrolysis of this lubricant, and that Lipase A did an excellent job even at 70°C.

4.3.3
Esterification of tertiary alcohols – one more unique feature

Lipases are well known for their ability to hydrolyze and synthesize a large variety of esters, but the alcohols involved in these reactions are normally restricted to primary and secondary alcohols. Bosley et al.[13] tested Lipase A in various reactions involving tertiary alcohols and found that, for instance, it was possible to synthesize tertiary butyl oleate from a mixture of oleic acid and tertiary butanol. Fifteen other lipases were tested for comparison and, apart from a low level of product formation from *Candida rugosa* lipase, all other lipases tested showed no production of tertiary ester at all. So not only does the exceptional temperature stability make Lipase A something special, but it also performs reactions that no other lipase known today can match.

5
Lipase B

Lipase B is not only derived from an unusual cold-adapted organism, but also the enzyme in itself is a very unusual lipase. As outlined below, the unique characteristics of the enzyme have enabled many new applications in which Lipase B is a highly valuable catalyst.

5.1
Crystal structure

Lipase B has recently been crystallized and the three-dimensional structure resolved.[5,14] The structure is illustrated in Figure 3 with the three active site residues empha-

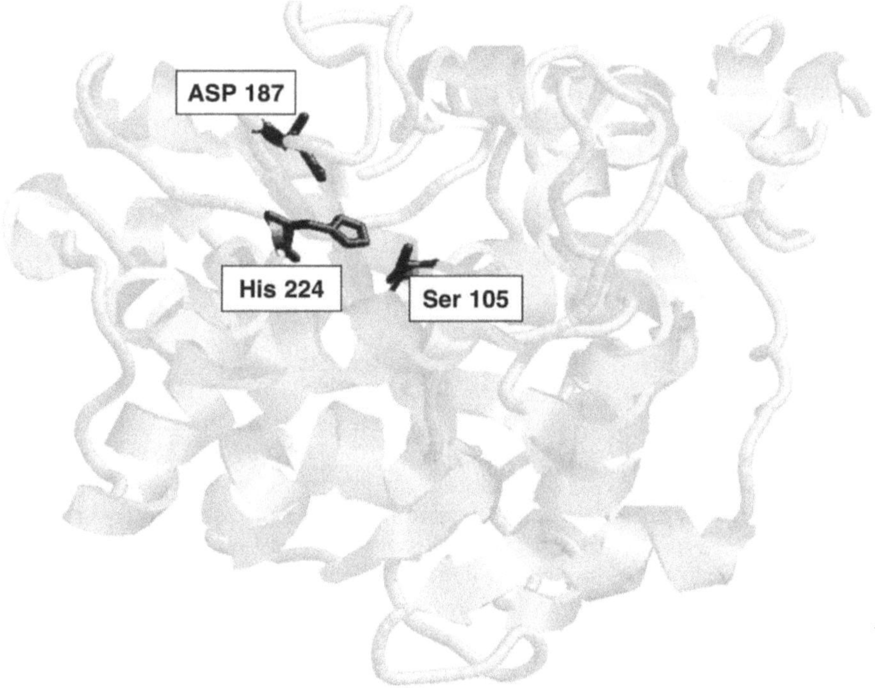

Fig. 3. Three-dimensional structure of Lipase B. The active site serine, histidine and aspartic acid residues are shown in *black*

sized. Even though the enzyme shares many features with other hydrolases, Lipase B has several unique characteristics. In general, lipases undergo conformational changes leading to activation when exposed to an interface between oil and water. In contrast, no interfacial activation is observed for Lipase B[15] and the enzyme can be regarded more as an intermediate between a lipase and an esterase. Furthermore, Lipase B has a very narrow active site pocket and can, accordingly, be expected to exhibit a very high degree of selectivity. Based on the structure, Lipase B is expected to exert a rather broad specificity towards acyl donors and a much higher degree of selectivity towards alcohol substrates,[16] and Uppenberg et al.[17] predicted that only the R-enantiomer of secondary alcohols should be able to form intermediates allowing for catalysis. As outlined below, all these predictions have been found to be in very close agreement with numerous experimental observations.

5.2
Stability

Also with respect to stability, Lipase B is a rather unique protein. Even though the optimal pH for catalysis is 7, the enzyme is stable in aqueous media in the pH range of 3.5–9.5. When immobilized, Lipase B is highly thermostable and can be used in continuous operation at 60–80°C without any significant loss in activity even after several thousand hours in use.[3,16] One very unique characteristic of Lipase B is its stability and activity in polar organic solvents. In general, enzymes are only stable and active in non-polar organic solvents.[18] However, even though Lipase B has both a high activity and stability in non-polar solvents such as hexane and iso-octane, it is also surprisingly stable and active in highly polar organic solvents such as acetone, acetonitrile and tert-butanol.[19–21] This feature has, as outlined below, enabled the use of Lipase B to catalyze conversion of polar substrates such as carbohydrates which are not compatible with the non-polar solvents traditionally used in lipase-catalyzed reactions.

5.3
Application possibilities

As predicted from the X-ray structure, Lipase B exhibits a very high degree of substrate selectivity. Combined with the robustness of the immobilized enzyme, this makes Lipase B a particularly useful catalyst. The enzyme has found use in a surprising diversity of reactions including both regio- and enantio-selective syntheses as evidenced by several hundred publications which have appeared over the last decade. The use of the lipase in organic synthesis has recently been reviewed.[22] Therefore, only a short overview of the different applications developed using Lipase B will be given below. Some examples of the wide variety of typical reactions catalyzed by Lipase B are shown in Figure 4.

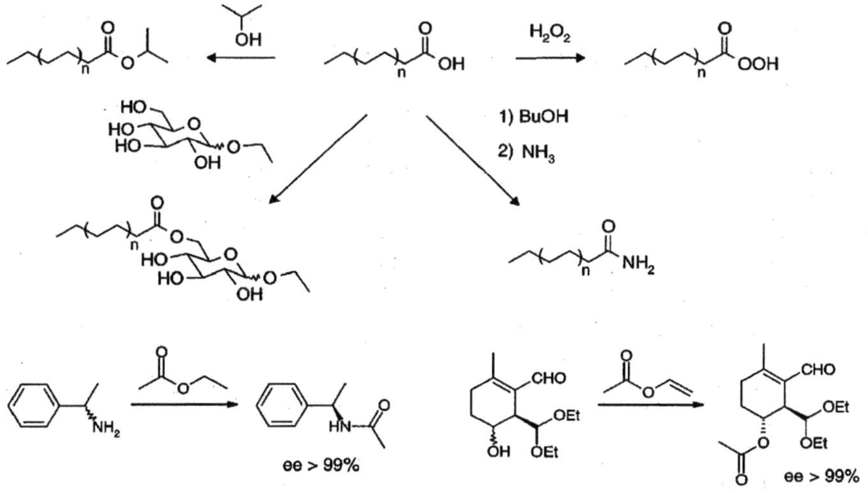

Fig. 4. Examples of typical reactions catalyzed by Lipase B

5.3.1
Fat and oil processing

As the enzyme is a non-selective lipase, it has been found to be especially useful for preparing homogeneous triglycerides either by interesterification or by direct esterification of glycerol with fatty acids. An example is the preparation of long-chain ω-3-type polyunsaturated fatty acid triglycerides characteristic of marine fat.[23,24] In this case the mild reaction conditions offered by the lipase make it possible to avoid chemical side reactions of the unsaturated acids, such as polymerization and migration or oxidation of double bonds. Another interesting application is the production of biodiesel as a substitute for mineral-based fuels. Short-chain alcohol esters of fatty acids are made from various fats and oils and Lipase B has been found to be particularly efficient as it is non-specific and tolerates the polar alcohols well.[25]

5.3.2
Applications within organic synthesis

Even surprisingly simple esters can advantageously be synthesized using Lipase B as catalyst. Examples include the production of isopropyl myristate[26] and myristyl myristate[27] which are both widely used in cosmetic formulations. The main advantage in using Lipase B in these simple reactions is again the mild reaction conditions which make it possible to avoid formation of un-desired by-products. The flavor and fragrance industries can, likewise, benefit from the mild conditions as exemplified by the esterification of geraniol and citronellol.[28]

5.3.3
Regioselective synthesis

As predicted from the X-ray structure, Lipase B can, in particular with respect to alcohol substrates, exhibit a very high degree of selectivity. In the field of regioselective synthesis the enzyme has proven to be a particularly useful catalyst for regioselective esterification of carbohydrates. Mono-esters of carbohydrates have very interesting surface-active properties and have potential applications in a number of areas. The earliest work with Lipase B in this area focused on the regioselective acylation at the 6-O-position of ethyl glucoside.[29] Mono-ester yields of more than 90% were obtained in a solvent-free process at 70°C and Lipase B was found to be much more selective compared to other lipases tested. The use of Lipase B as a regioselective catalyst has been successfully extended to other carbohydrates. Examples include glucose[19] and maltose,[30] of which mono-esters have been obtained in yields of more than 90% using solvents such as acetonitrile and t-butanol. Selective esterification of other polyhydroxy substrates has also recently been achieved. Examples include nucleosides[31] and different steroids[32] and steroid glucosides[33] which have been selectively acylated, providing pure compounds for biological and medical studies.

5.3.4
Enantioselective synthesis

For Lipase B, perhaps the most extensive area of study is in the resolution of racemic alcohols and acids, or the preparation of optically active compounds from meso-reactants. The resulting optically pure compounds are highly difficult to obtain by alternative routes and can be of great synthetic value, particularly in the pharmaceutical and agrochemical industries. Numerous examples exist in the literature, demonstrating the variety of substrates upon which Lipase B is active. As predicted, the enzyme has a highly pronounced catalytic preference towards the R-enantiomer of secondary alcohols. This selectivity can, depending on which product is isolated, be utilized to produce both the R- and the S-enantiomer. Products with ee values typically over 95% have been obtained using Lipase B. Examples of compounds which have been successfully resolved using Lipase B are secondary alcohols such as 1-phenylethanol[34] and 2-cyclopentenol,[35] diols such as 1,3-butandiol[36] and a complex intermediate of the anti-tumor agent Taxol.[37] The use of Lipase B is not restricted to alcohols as the enzyme has a remarkable activity also towards amines, amides, carbonates and thiols. Examples in this context comprise 1-phenylethylamine,[38] benzvinyl carbonate[39] and phenylethyl thiooctanoate.[40]

5.4
Future applications

Lipase B is now well recognized as a unique catalyst within organic synthesis, but work is well in progress to extend the use of the enzyme further. Some of the most interesting novel applications which are currently being explored using Lipase B as catalyst comprise the synthesis of aliphatic polyesters[41] and the preparation of amides,[41] peroxycarboxylic acids[42] and thioesters[40] involving non-natural nucleophiles such as amino, hydroperoxy and thiol substrates. The applicability of Lipase B in these very different reactions further highlights the unique properties of this enzyme derived from a unique cold-adapted organism.

6
Conclusions

Ten years have passed since *C. antarctica* LF058 was picked up from a screening plate as a good lipase producer. Since then, there has been tremendous interest from an academic as well as an industrial application point of view in the two lipases produced by this strain. Lipases A and B each have outstanding properties, unmatched by any of the hundreds of other lipases known today. The steady flow of articles and patent applications is clear evidence that also the coming years will show many new and interesting possibilities for research and industrial uses.

Most often, when psychrophilic (cold-tolerant) microorganisms are screened for useful enzyme activities, the application aimed at is one taking place at low temperatures. However, Lipases A and B are clear examples that one can find exceptionally thermostable enzymes even from an organism adapted to cold environments and first isolated from one of the coldest places on Earth — Antarctica.

7
References

1. Goto S, Sugiyama J, Iizuka H. A taxonomic study of Antarctic yeasts. Mycologia 1969; 61:748-774.
2. Ishii M, Suzuki E, Abo M, Nielsen TB, Heldt-Hansen HP. Thermostable, positional non-specific lipase from the yeast *Candida antarctica*. Hakkokogaku Kaishi 1988; 5:408-410 (in Japanese).
3. Heldt-Hansen HP, Ishii M, Patkar SA, Hansen TT, Eigtved P. A new immobilized positional non-specific lipase for fat modification and ester synthesis. In: Whitaker JR, Sonnet PE, eds. Biocatalysis in Agricultural Biotechnology. Washington DC: American Chemical Society, 1989:158-172.
4. Patkar SA, Björkling F, Zundel M, Schülein M, Svendsen, A, Heldt-Hansen P, Gormsen, E. Purification of two lipases from *Candida antarctica* and their inhibition by various inhibitors. Ind J Chem1993; 32B:76-80.
5. Uppenberg J, Hansen MT, Patkar, S, Jones, TA. The sequence, crystal structure determination and refinement of two crystal forms of lipase B from *Candida antarctica*. Structure 1994; 2:293-308.
6. Patkar SA, Svendsen A, Kirk O, Clausen IG, Borch K. Effect of mutation in non-consensus sequence Thr-X-Ser-X-Gly of *Candida antarctica* lipase B on lipase specificity, specific activity and thermostability. J Mol Catal B: Enzymatic 1997; 3:51-54.
7. Hashida M, Abo M, Takamura Y, Kirk O, Halkier T, Pedersen, S, Patkar SA, Hansen, MT. Lipases from *Hyphozyma*. International Patent Publication WO 93/24619, 1993.
8. Hoegh I, Patkar S, Halkier T, Hansen MT. Two lipases from *Candida antarctica*: cloning and expression in *Aspergillus oryzae*. Can J Bot 1995; 73 (Suppl. 1):S869-S875.

9. Christensen T, Wöldike H, Boel E, Mortensen SB, Hjortshoj K, Thim L, Hansen MT. High level expression of recombinant genes in *Aspergillus oryzae*. Biotechnology 1988; 6:1419-1422.

10. Svendsen A, Pathar SA, Egel-Mitani M, Borch K, Clausen IG, Hansen MT. *C. antarctica* lipase and lipase variants. International Patent Publication WO 94/01541, 1994.

11. Matsukura M, Fujita Y, Abo M, Itami R. Method for avoiding pitch troubles by use of thermostable lipase. International Patent Publication WO 92/13130, 1992.

12. Lund H, Nilsson TE, Pickard T. Treatment of fabrics. International Patent Publication WO 97/04160, 1997.

13. Bosley JA, Casey J, Macrae AR, Mycock G. Process for the esterification of carboxylic acids with tertiary alcohols. International Patent Publication WO 95/01450, 1995.

14. Uppenberg J, Patkar S, Bergfors T, Jones TA. Crystallization and preliminary X-ray studies of lipase B from *Candida antarctica*. J Mol Biol 1994; 235:790-792.

15. Martinelle M, Holmquist M, Hult, K. On the interfacial activation of *Candida antarctica* lipase A and B as compared with *Humicola lanuginosa* lipase. Biochim Biophys Acta 1995; 1258:272-276.

16. Arroyo M, Sinisterra JV. High enantioselective esterification of 2-arylpropionic acids catalyzed by immobilized lipase from *Candida antarctica*: A mechanistic approach. J Org Chem 1994; 59:4410-4417.

17. Uppenberg J, Ohrner N, Norin M, Hult K, Kleywegt GJ, Patkar S, Waagen V, Anthonsen T, Jones TA. Crystallographic and molecular-modeling studies of lipase B from *Candida antarctica* reveal a stereospecificity pocket for secondary alcohols. Biochemistry 1995; 34:16838-16851.

18. Reslow M, Adlercreutz P, Mattiasson B. Organic solvents for bioorganic synthesis 1. Optimization of parameters for a chymotrypsin catalyzed process. Appl Microbiol Biotechnol 1987; 26:1-8.

19. Ljunger G, Adlercreutz P, Mattiasson, B. Lipase catalyzed acylation of glucose. Biotechnol Lett 1994; 16:1167-1172.

20. Cao L, Bornscheuer UT, Schmid RD. Lipase-catalyzed solid phase synthesis of sugar esters. Fett/Lipid 1996; 98:332-335.

21. Córdova A, Hult K, Iversen T. Esterification of methyl glycoside mixtures by lipase catalysis. Biotechnol Lett 1997; 19:15-18.

22. Anderson EM, Larsson KM, Kirk O. One biocatalyst – many applications: The use of *Candida antarctica* B-lipase in organic synthesis. Biocatal Biotransform 1998; 16:181-204.

23. Haraldsson GG, Gudmundsson BÖ, Almarsson Ö. The synthesis of homogeneous triglycerides of eicosapentaenoic acid and docosahexaenoic acid by lipase. Tetrahedron 1995; 51:941-952.

24. Lee KM, Akoh CC. Immobilized lipase-catalyzed production of structured lipids with eicopentaenoic acid at specific positions. J Am Oil Chem Soc 1996; 73:611-615.

25. Nelson LA, Foglia TA, Marmer WN. Lipase-catalyzed production of biodiesel. J Am Oil Chem Soc 1996; 73:1191-1195.

26. Hills GA, Macreae AR, Poulina RR. Ester enzymatic preparation. European Patent Publication 0 383 405, 1990.

27. Garcia T, Martinez M, Aracil J. Enzymatic synthesis of myristyl myristate. Estimation of parameters and optimization of the process. Biocatal Biotransform 1996; 14:67-85.

28. Claon PA, Akoh CC. Enzymatic synthesis of geraniol and citronellol esters by direct esterification in n-hexane. Biotechnol Lett 1993; 15:1211-1216.

29. Adelhorst K, Björkling F, Godtfredsen SE, Kirk O. Enzyme catalysed preparation of 6-O-acyl-glucopyranosides. Synthesis 1990; 112-115.

30. Oosterom MW, Rantwijk F, Sheldon RA. Regioselective acylation of disaccharides in tert-butyl alcohol catalyzed by *Candida antarctica* lipase. Biotechnol Bioeng 1996; 49:328-333.

31. Morís F, Gotor V. A useful and versatile procedure for the acylation of nucleosides through an enzymatic reaction. J Org Chem 1993; 58:653-660.

32. Bertinotti A, Carrea G, Ottolina G, Riva S. Regioselective esterification of polyhydroxylated

steroids by *Candida antarctica* lipase B. Tetrahedron 1994; 50:13165-13172.

33. Danieli B, Luisetti M, Riva S, Bertinotti A, Ragg E, Scaglioni L, Bombardelli E. Regioselective enzyme-mediated acylation of polyhydroxy natural compounds. A remarkable, highly efficient preparation of 6'-O-acetyl and 6'-O-carboxyacetyl ginsenoside Rg$_1$. J Org Chem 1995; 60:3637-3642.

34. Frykman H, Öhrner N, Norin T, Hult K. S-ethyl thiooctanoate as acyl donor in lipase catalysed resolution of secondary alcohols. Tetrahedron Lett 1993; 34:1367-1370.

35. Johnson CR, Sakaguchi H. Enantioselective transesterifications using immobilized, recombinant *Candida antarctica* lipase B: resolution of 2-iodo-2-cycloalken-1-ols. Synlett 1992; 10:813-816.

36. Eguchi T, Mochida K. Lipase-catalyzed diacylation of 1,3-butanediol. Biotechnol Lett 1993; 15:955-960.

37. Johnson CR, Xu Y, Nicolaou KC, Yang Z, Guy RK, Dong JG. Enzymatic resolution of a key stereochemical intermediate for the synthesis of (-)-taxol. Tetrahedron Lett 1995; 36:3291-3294.

38. Reetz MT, Dreisbach C. Highly efficient lipase-catalyzed kinetic resolution of chiral amines. Chimia 1994; 48:570.

39. Pozo M, Pulido R, Gotor V. Vinyl carbonates as novel alkoxycarbonylation reagents in enzymatic synthesis of carbonates. Tetrahedron Lett 1992; 48:6477-6484.

40. Öhrner N, Orrenius C, Mattsson A, Norin T, Hult K. Kinetic resolutions of amine and thiol analogues of secondary alcohols catalyzed by the *Candida antarctica* lipase B. Enzyme Microbiol Technol 1996; 19:328-331.

41. Taylor A, Binns F. Enzymatic synthesis. International Patent Publication WO 94/12652, 1994.

42. Warwel S, Rüsch Gen Klaas M. Chemo-enzymatic epoxidation of unsaturated carboxylic acids. J Mol Catal B:Enzym 1995; 1:29-35.

Peptide hydrolases from antarctic krill – an important new tool with a promising medical potential

L. Hellgren[1], B. Karlstam[2], V. Mohr[3] and J. Vincent[2]*

[1] Department of Dermatology, University of Umeå, S-90187 Umeå, Sweden
[2] BioPhausia, AR 4, S - 741 74 Uppsala, Sweden
[3] The Research Council of Norway, St. Hanshaugen, Box 2700, N-0131 Oslo, Norway

1
Introduction

Krill, a group of reddish, thumb-length pelagic shrimp-like crustaceans which can reach a weight of about one gram, occupy a central position in the Southern Ocean food web.[1] The predominant species in this unique ecosystem is antarctic krill, *Euphausia superba*, which is widely distributed in dense swarms and represents the largest source of unutilized protein in the oceans. The potential catch has been estimated to be well in excess of the total annual harvest of fish in the world, suggesting that the biomass of this single species is the largest of any multi-cellular animal on the planet.

Although krill is the most extensively fished species in the Antarctic, the current commercial catch is only in the order of 500,000 tons per year. Krill is harvested mainly as feed supplement for fish and broiler farms, but also in limited quantities for human consumption. Further commercial expansion of the krill catch is restricted, however, for several reasons, such as environmental considerations, the short fishing season, high costs and distance, and by problems associated with the processing of the raw material. This is primarily because krill contains very potent digestive enzymes which break down the animal soon after death. As omnivores their digestive apparatus acts as a "vacuum cleaner" which sucks in and filters everything in its way, provided it can be physically retained. Krill's uniquely effective multi-enzyme system plays a key role for its survival in antarctic waters where the summer season when food is abundant is rather short. These digestive enzymes are specially adapted to ensure optimal digestion at ambient temperatures (seldom exceeding +2°C), including low activation energies and greater substrate affinity.[2]

---

* Corresponding author

2
Krill enzymes

The digestive enzymes in krill encompass a highly effective "battery" of proteolytic, carbohydrate-splitting, lipolytic and other enzymes assuring extensive breakdown of complex substrates such as proteins, carbohydrates and lipids to low molecular weight constituents exemplified in Table 1 by some of the enzymes identified so far.

Krill enzymes have been isolated by aqueous extraction of the animal followed by various purification procedures.[3–15] Beside more conventional methods, an autolytic procedure was also developed in which krill is homogenated and left to autolyze under defined conditions before final separation.[16,17]

The unique properties of krill enzymes offer novel and interesting possibilities both in the area of medicine and as a basis for enzyme technology. In an industrial context it is feasible to exploit the specific properties of krill enzymes such as their stability and high efficiency at low temperatures.[18,19] Furthermore, some of the krill enzymes (trypsin-like enzyme I, chymotrypsin-like enzyme, hyaluroni-dase or phospholipase) may serve as fine chemicals in biomedical research.

Table 1. A survey of important krill hydrolytic enzymes

Crude Extract	Purified Extract (Krillase™)
Proteases	Serine proteinases with
Serine proteinases with trypsin-like activity	trypsin-like activity
Serine proteinases with chymotrypsin-like activity	Enyzme I: Endo/exopeptidase
Carboxypeptidase A	Enzyme II: Endopeptidase
Carboxypeptidase B	Enzyme III: Endopeptidase
Aminopeptidase	
	Serine proteinase with
Carbohydrases	chymotrypsin-like activity
Amylases	
Carboxymethyl cellulase	
Chitinase	Carboxypeptidase A I
Endo/exoglucanases	Carboxypeptidase A II
Glucuronidase	Carboxypeptidase B I
Hyaluronidase	Carboxypeptidase B II
Glucosidases	
Galactosidases	
Fucosidases	
Other hydrolases	
Nucleases	
Phospholipase	

3
Proteolytic enzymes from krill

Krill peptide hydrolases (proteases), originating almost exclusively from the digestive tract of the animal, play an essential role in the breakdown of food.[20] They have been purified to varying degrees and their properties biochemically examined in detail.[4,8–12,15,21–26] Krill contains an array of endo- and exopeptidases, the major enzymes being: trypsin-like enzymes I–III, chymotrypsin-like enzyme; carboxypeptidases A1 and A2; carboxypeptidases B1 and B2; aminopeptidase. Most of these enzymes have also been extensively characterized immunochemically.[3,27,28]

Among the **endopeptidases** three serine proteinases with trypsin-like activity have been identified.[11] These enzymes have almost identical molecular weights around 30 kDa, display an alkaline pH optimum and show maximum activity at 50–54°C. They are inhibited by typical trypsin inhibitors such as phenyl methyl sulphonyl fluoride, tosyl lysyl chloromethyl ketone and soybean trypsin inhibitor. The krill trypsins are very stable at neutral pH. They become slowly inactivated at highly alkaline pH, contrary to their behaviour at low pH.

Trypsin I is an endo/exopeptidase whereas trypsin II and III are true endopeptidases.[11] Thus, trypsin I differs substantially from other trypsins by its broad specificity. These unique biochemical properties are further supported by results from immunochemical analysis using polyclonal antibodies on two-dimensional agarose gel electrophoresis and the Ouchterlony double immunodiffussion assay for characterization of individual enzyme components.[3] Moreover, it has been demonstrated that the krill trypsins have activation energies that are substantially lower than those of their mammalian counterparts with pronounced activities down to 0°C[11] mirroring their adaptation to efficient function at low temperatures. Finally, krill also possesses a unique chymotrypsin-like enzyme with collagenolytic activity that significantly contributes to the degradation of proteinaceous materials.[15]

The **exopeptidases** include two carboxypeptidases A and two carboxypeptidases B. pH-optima for the A-type are slightly on the acidic side, whereas for the B-type the optima occur at neutral pH. Furthermore, the carboxypeptidases A have higher temperature optima as well as greater thermostability than the carboxypeptidases B.[12] An aminopeptidase has also been described possessing temperature optimum and thermostability similar to those of carboxypeptidase A enzymes.[9,12]

Mohr and coworkers[11,24] and Osnes et al.[25] have demonstrated that krill peptide hydrolases are surprisingly resistant to breakdown and loss of activity when they are mixed. The simple digestive system of krill suggests that the individual enzymes are mutually protected against the degrading effect of each other. This property makes the krill peptide hydrolases unusually valuable as an enzyme composition for practical application in medicine.

4
Mode of action of the krill proteases

The mode of action of krill proteases was established by studying the effect of the purified enzymes both alone and in combination. Casein or synthetic peptides were selected as model substrates.[11,12,24]

Of the three trypsin-like enzymes only one (trypsin I) is able to release significant amounts of amino acids, in particular lysine and arginine, from casein, whereas the other two only enhance amino acid yield when used in combination with trypsin I.[11]

Carboxypeptidases A and B from krill yield relatively small amounts of amino acids from casein. However, when mixed with the trypsin-like enzymes, a synergistic effect was observed, in that the amino acid yield exceeded that of the sum of the amino acids liberated by the enzyme preparations alone. The aminopeptidase, on the other hand, exhibited broad specificity, although the individual amino acids were released in very limited amounts. Other studies on the hydrolysis of synthetic substrates support and extend the above data on the properties and specificity of the krill exopeptidases.[12]

Experiments on the combined effect of the purified krill enzymes further support the assumption that the digestive processes of *E. superba* depend on the synergistic action of several proteases of which the trypsin-like, serine-type of enzymes play an essential role. When combining the trypsin-like enzymes with purified preparations of carboxypeptidase A, carboxypeptidase B and aminopeptidase respectively, or in combination, virtually all of the amino acids in the model protein casein were liberated in varying proportions.[25]

The close correspondence between the pattern of amino acids liberated from casein by the crude extract of krill and by the mixture of purified peptide hydrolases seems to warrant the conclusion that these proteases are principally responsible for the digestive processes in krill.[25] The fact that the total proteolytic activity was somewhat lower for the combined, purified enzymes than for the crude extract in Osnes' study, may have been due to the absence of the chymotrypsin-like enzyme in the purified "pool". This enzyme was not known at the time but has later been isolated and characterized.[15]

5
Medical applications

5.1
Enzymatic debrider for treatment of necrotic ulcers

Wound healing is a complex physiological process that restores tissue defects in the skin. The first step in the treatment of almost all serious wounds is removal (debridement) of necrotic tissue (see ref. 29 and references therein). Debridement is a stepwise cleansing procedure in which necroses are removed either invasively (surgically) or non-invasively (enzymatically). Enzymatic debridement means that

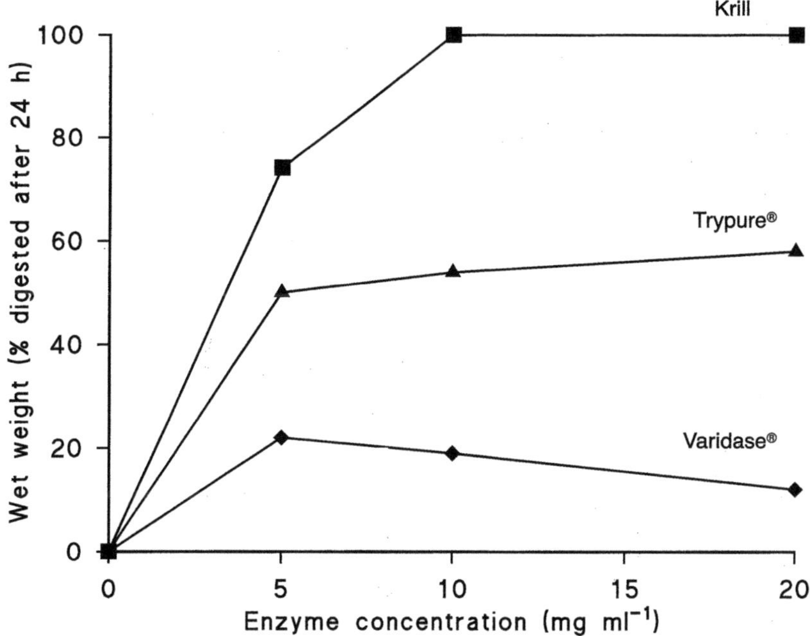

Fig. 1. Degradation of excised rat skin *in vitro* (37°C) by various enzymatic debriders

the necroses are selectively degraded (the viable tissues remain unaffected) by proteolytic enzymes. Although this kind of non-invasive debridement has been available for more than three decades with varying degrees of success, many of the products marketed have been discontinued because digestion of necrotic tissue was too slow or incomplete.

In contrast, experimental studies on the new multienzyme debrider (Krillase™), isolated from antarctic krill, indicate that it achieves both effective and rapid debridement, criteria which better meet the needs of modern wound care.[30–34] This is probably due to the complex multi-substrate nature of ulcer debris which is more easily degraded by a synergistic, multi-enzyme system (krill) than by single enzymes, such as trypsin (Trypure®), collagenase (Iruxol®), or combinations of two enzymes, such as streptokinase/streptodornase (Varidase®) or fibrinolysin/deoxyribonuclease (Elase®, Fibrolan®), (Fig. 1).

Krill enzymes appear to act in a two-step fashion when breaking down proteinaceous materials: the endopeptidases first attack the peptide bonds of the intrastructural parts of the polypeptide chains, and the resulting peptide fragments are subsequently cleaved by the exopeptidases into small peptides and free amino acids.

The major objective during the development of Krillase™ as a medical tool has been to maintain the natural composition of the main krill proteases intact throughout the purification procedures. The final product is a sterile, lyophilized powder which is reconstituted before topical application. Krillase™ is defined as a mixture of acidic endopeptidases (trypsin- and chymotrypsin-like enzymes) and exopeptidases (carboxypeptidase A and B). The main steps in commercial extraction include defatting of the raw material, extraction by two-phase equilibrium separation, ultrafiltration and size exclusion gel chromatography. The purified extract principally contains macromolecules in the molecular weight range of 20 to 40 kDa. Each processing step from raw material to the end product is thoroughly documented by advanced biochemical/immunochemical methods. Thus the product is well charracterized with respect to proteolytic activities, batch to batch variations and uniformity.

In both pre- and clinical comparative studies debridement with Krillase™ was fast and distinctly more effective than the currently marketed enzyme preparations. Data from these studies also indicate that krill enzymes possess a broad safety profile. The product was well tolerated and the treatment generally resulted in a clean granulating wound bed suitable for grafting.[35,36]

The impressive efficiency of krill enzymes, their synergistic interaction and the unique feature that they appear to co-exist without digesting each other, and thus may be mixed, constitute particularly important properties for distinctive medical applications, such as debridement of necrotic wounds. Krillase™ debridement appears to be efficient and safe, representing a new and important alternative to surgical debridement or to mammalian/microbial enzymatic debriders.[29,37]

The unique properties of krill enzymes can also be utilized for other indications. Some of the "blockbusters" are exemplified below.

5.2
Krill enzymes in odontology

Periodontitis and caries constitute the two most important causes of dental morbidity. Since both diseases are caused by dental plaque, efficient plaque removal by biocompatible and non-toxic agents has become a major goal in dental research.

At present no fully satisfactory antiplaque agent exists that offers effective antimicrobial, cleansing/debriding properties and at the same time is fully biocompatible. However, the effect of agents such as chlorhexidine or triclosan are limited and long term toxicology remains uncertain.[38,39]

A number of enzymes like pepsin, trypsin, dextranase and mutanase have been used with discouraging results. Their limited action may be due to their rather high specificity. As a result, they hydrolyze only a limited number of peptide and glucosidic bonds in proteins/polysaccharides, and hence lead only to limited digestion of the constituents of materials such as plaque. In contrast krill enzyme preparations, with both endo-and exopeptidases degrade proteins rapidly and extensively.[40]

A series of studies on the antiplaque effects of different krill enzyme compositions yielded very promising results. It has now been established *in vivo* that application of krill enzyme solutions on teeth that were initially clean significantly retards plaque formation on experimental buccal tooth surfaces compared to water placebo (S. Edwardsson et al., unpubl. data). It has also been established *in vivo* that krill enzyme solutions remove accumulated plaque more effectively than a commercially available denture cleaner containing Alcalase®. Other studies have confirmed that krill enzyme solutions remove surface cultures of *Candida albicans* in test systems, whereas no such effect was observed in controls using water (K. Hellgren et al., unpubl. data).

Recent data have further confirmed earlier findings that krill enzymes exert cleansing and prophylactic properties by directly inhibiting dental plaque formation. Krill was also more effective than papain (available as tooth paste), with clear implications for the prevention of periodontitis and caries. These new studies provide statistical evidence that krill enzymes are superior to other enzymes both in quantitative and qualitative terms, i.e. they are not only more effective but also more selective in their action. Furthermore, they neither affect viable cells nor do they have antibacterial effects (K. Hellgren and S. Kalfas, unpubl. data).

Krill enzymes seem to act more by selective removal rather than eradication of microbial flora, in contrast to chlorhexidine or antibiotics. This fits well with the modern therapeutic consensus in this field, namely that antiplaque agents should selectively reduce certain specific pathogens rather than remove or disturb the normal protective oral microflora. These recent studies also show that krill enzymes very effectively inhibit adhesion of one of the most important pathogenic bacteria, considered to be the major cause of enamel adhesion and subsequent dental plaque formation (K. Hellgren et al., unpubl. data).

5.3
Krill enzymes as a possible digestion promotor

The unique, concerted action of the krill multi-enzyme system offers particular promise in the treatment of diseases characterized by impaired digestion, such as chronic pancreatic insufficiency, indigestion caused by the presence of particular food ingredients and other short or long term disturbances of intestinal digestive capacity.[41]

Krill enzyme preparations, which consist of a natural mixture of endo- and exopeptidases, carbohydrate-splitting and lipolytic enzymes, may offer a number of advantages over existing digestion promotors. Firstly, the preparation contains a wide spectrum of highly active enzymes which induce extensive breakdown of the proteins, carbohydrates and lipids in ingested food. Secondly, the concept takes advantage of a particularly important property of krill digestive enzymes, namely the fact that they seem to be protected against mutual degradation. Thus, mixtures of krill enzymes maintain high activity over long periods of time. This contrasts with most of the existing digestion promotors, which frequently contain arbitrary mixtures of enzymes from different origins, such as animals, plants or fungi. The recognized drawback of such an approach is that, although each individual enzyme may be potent and efficient within the range of its specificity, such mixtures of enzymes from widely different sources are generally inefficient due to mutual degradation and loss of activity. This often means that very high levels of the individual enzymes are required. Krill enzymes, on the other hand, constitute a natural, multi-enzyme system ensuring rapid onset of action and extensive breakdown of complex substrates such as proteins, carbohydrates and lipids into low molecular, easy resorbable fragments. The projected composition should be administered in a form ensuring resistance to gastric juice, or formulated so as to gradually release the enzyme activity over a period of time.

5.4
Krill enzymes in chemonucleolysis

Chemonucleolysis involves the use of enzyme preparations to treat injured vertebral discs by direct injection into the affected disc. The main purpose of this approach is to diminish the disc height and subsequently stimulate the formation of a new matrix. Such non-invasive therapy has been shown to reduce the pressure on inflamed nerve roots caused by the bulging annulus fibrosus and eliminate or relieve sciatic pain.[42–44]

Some commercially available enzyme preparations have already been investigated as potential chemonucleolytic agents. These include Discase® and Chymodiatin®[45] both containing chymopapain from papaya as the active ingredient. In addition collagenase was earlier frequently studied but has been withdrawn. Preclinical data indicated that low doses of collagenase have a positive effect on the degradation of the nucleus pulposus and annulus fibrosus. Higher doses, however, were reported to have negative effects resulting in complete digestion of these structures and extensive vertebral haemorrhage.[46] These findings were further confirmed clinically resulting in withdrawal of collagenase from this market segment.[47]

Although enzyme preparations have received a somewhat mixed reception among clinicians engaged in spinal diseases, new approaches may offer hope. Recently, in co-operation with Ghosh et al., we designed a preclinical study on chemonucleolysis using two novel enzyme preparations isolated from antarctic krill.[48]

These krill preparations (krill proteases and krill trypsin-like enzyme I) were compared with each other and with chymopapain (Discase®) in two different models. The results showed that the mixture of endo- and exopeptidases (Krillase™) was much better in degrading proteoglycans than either the single highly purified krill trypsin-like enzyme (trypsin I) or Discase®. The former preparation degraded the proteoglycan polymers more completely.[48]

Based on these preliminary positive results Krillase™ was further evaluated on beagles by monitoring the disc height by lateral radiographs over a period of 32 weeks. After two weeks the disc height had decreased to half of the initial value. However, at the end of the observation period the disc height approached its pretreatment value, fully comparable to the data from the untreated control discs.[48]

It can thus be concluded that a relatively low dose of Krillase™ rapidly reduces disc height without side effects. The disc cells appear to be viable after the termination of chemonucleolytic treatment and can actively contribute to reconstitution of the nucleus pulposus to its original GAG content. It therefore seems fully realistic in the future to employ krill proteases as chemonucleolytic agents with considerable potential to heal and reconstruct different types of injured discs.

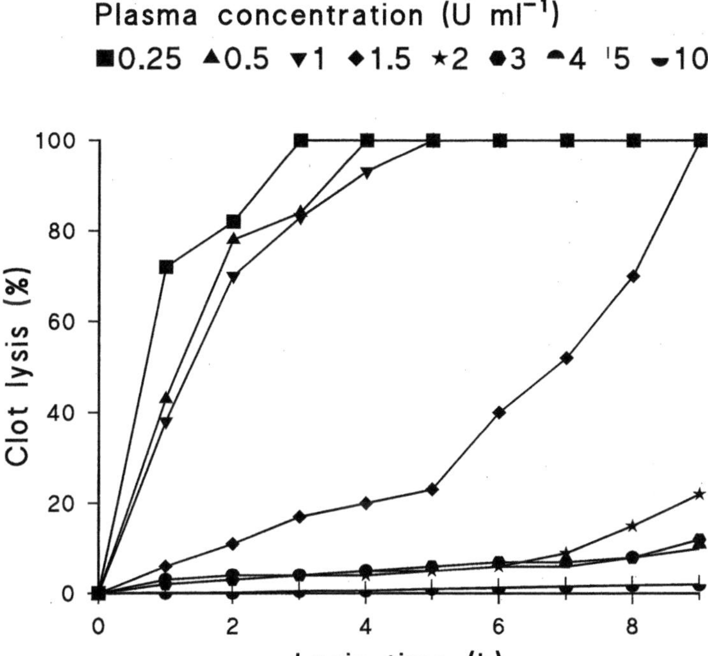

Fig. 2. Fibrinolytic activity of krill enzymes in human plasma as measured in a Chandler loop model

5.5
Krill enzymes as a potential thrombolytic agent

In vivo enzymatic lysis of intravascular thrombi offers promising potential in the management of stroke and myocardial infarction.[49] Commercially available preparations now in extensive use include streptokinase, urokinase and tissue plasminogen activator (tPA).

The thrombolytic/fibrinolytic potential of krill enzymes containing endo/exopeptidases has been evaluated in different *in vitro* and *in vivo* models previously used for testing and evaluation of other thrombolytic agents such as streptokinase or tPA. The fibrin plate assay showed that krill proteolytic enzymes possess strong fibrinolytic effect directly proportional to the enzyme concentration.

A particularly relevant model, the Chandler loop assay (using ^{125}I-labelled human fibrinogen activated by addition of Ca^{2+}-ions) confirmed the observations with krill proteases. The results obtained were very similar to data in parallel observations using the fibrin plate assay i. e. the higher enzyme concentration the better effect. Higher concentrations, doses of 4 to 10 U ml⁻¹, revealed the most rapid clot lysis ever observed in this particular model comparing tPA and streptokinase. Highly purified preparations representing a highly specific enzyme yielded even more dramatic effects (Fig. 2). To achieve maximum lysis in *in vivo* systems at enzyme concentrations as low as those in the *in vitro* studies, it is important to prevent inhibition by naturally occurring serum protease inhibitors like α_1-proteinase inhibitor, α_1-antichymotrypsin, α_2-macroglobulin and antithrombin III. This is nowadays conveniently achieved by employing intracoronary double balloon catheters with systemic heparinization to locally exclude serum from the thrombosed segment and thus mimic *in vitro* conditions. Another approach to avoid inactivation of enzyme(s) would be to modify the original enzyme(s) by protein engineering in order to eliminate the risk that serum related inhibitors recognize the enzyme(s). Both of these proposed modifications rely on the advantage that krill enzymes exhibit the more rapid onset in comparison with other proteolytic enzymes.[50] Further animal studies are required to strengthen this concept.

6
Conclusions

Despite the widespread occurrence of antarctic krill, its unique enzyme systems have been little exploited on a commercial basis. Krill has an exceptionally effective digestive apparatus designed to ensure swift breakdown of plankton at low temperatures, in order to build up sufficient lipid reserves during the short antarctic summer. The digestive apparatus contains a highly effective, wide spectrum multi-enzyme system which rapidly degrades complex substrates such as proteins, carbohydrates and lipids to low molecular weight constituents.

The principal enzymes involved in krill proteolysis have been purified and characterized. These proteases have been demonstrated to have highly interesting and important medical applications, particularly with respect to debridement of necrotic ulcerations, in addition to the exploratory areas such as odontotology, gastroenterology, chemonucleolysis and thrombolysis.

7
References

1. Laws R. Antarctica: a convergence of life. New Sci 1983; 1373:608-616.
2. Nicol S. Who's counting on krill? New Sci 1989; 1690:38-41.
3. Bucht A, Karlstam B. Isolation and immunological characterization of three highly purified serine proteinases from Antarctic krill *(Euphausia superba)*. Polar Biol 1991; 11:495-500.
4. Chen CS, Yan TR, Chen HY. Purification and properties of trypsin-like enzymes and a carboxypeptidase A from *Euphausia superba*. J Food Biochem 1978; 2:349-366.
5. Chen CS, Gau SW. Polysaccharidase and glycosidase activities of Antarctic krill *Euphausia superba*. J Food Biochem 1981; 5:63-68.
6. Chen CS, Lian KT. Purification and characterization of beta-D-glucosidases from *Euphausia superba*. Agric Biol Chem 1986; 50:1229-1238.
7. Karlstam B, Lunglöf A. Purification and partial characterization of a novel hyaluronic acid-degrading enzyme from Antarctic krill *(Euphausia superba)*. Polar Biol 1991; 11:501-507.
8. Kimoto K, Kusama S, Murakami K. Purification and characterization of serine proteinases from *Euphausia superba*. Agric Biol Chem 1983; 47:529-534.

9. Kimoto K, Murakami K. Purification and characterization of aminopeptidase from *Euphausia superba*. Agric Biol Chem 1984; 48:1819-1823.

10. Kimoto K, Yokoi T, Murakami K. Purification and characterization of chymotrypsin-like proteinase from *Euphausia superba*. Agric Biol Chem 1985; 49:1599-1603.

11. Osnes KK, Mohr V. On the purification and characterization of three anionic, serine-type peptide hydrolases from Antarctic krill, *Euphausia superba*. Comp Biochem Physiol 1985; 82B:607-619.

12. Osnes KK, Mohr V. On the purification and characterization of exopeptidases from Antarctic krill, *Euphausia superba*. Comp Biochem Physiol 1986; 83B:445-458.

13. Spindler KD, Buchholz F. Partial characterization of chitin-degrading enzymes from two euphausiids, *Euphausia superba* and *Meganyctiphanes norvegica*. Polar Biol 1988; 9:115-122.

14. Turkiewicz M, Galas E, Zielinska M. Purification and partial characterization of an endo-(1,3)-beta-D-glucanase from *Euphausia superba* Dana (Antarctic krill). Polar Biol 1985; 4:203-211.

15. Turkiewicz M, Galas E, Kalinowska H. Collagenolytic serine proteinase from *Euphausia superba* Dana (Antarctic krill). Comp Biochem Physiol 1991; 99B:359-371.

16. Ellingsen TE, Mohr V. Biochemistry of the autolysis process in Antarctic krill *post mortem*. Autoproteolysis. Biochem J 1987; 246:295-305.

17. Karlstam B, Vincent J, Johansson B, Brynö C. A simple purification method of squeezed krill for obtaining high levels of hydrolytic enzymes. Prep Biochem 1991; 21:237-256.

18. Kolakowski E. Seasonal variation of autoproteolytic activity in the Antarctic krill, *Euphausia superba* Dana. Pol Polar Res 1986; 7:275-282.

19. Kolakowski E. Proteolytic activity of Antarctic krill in relation to its feeding intensity in spring and summer. Pol Polar Res 1989; 10:141-150.

20. Ellingsen TE. Biokjemiske studier over Antarktisk krill (Biochemical studies on Antarctic krill). PhD thesis, Inst Techn Biochem NTH/University of Trondheim, Trondheim Norway, 1982.

21. Kimoto K, Thanh VV, Murakami K. Acid proteinases from Antarctic krill, *Euphausia superba*. Partial purification and some properties. J Food Sci 1981; 46:1881-1884.

22. Nishimura K, Kawamura Y, Matoba T, Yonezawa D. Classification of proteases in Antarctic krill. Agric Biol Chem 1983; 47:2577-2583.

23. Noguchi A, Yanagimoto M, Umeda K, Kimura S. Purification and some properties of protease of *Euphausia superba*. J Agric Chem Soc Jpn 1976;50:415-421.

24. Osnes KK, Mohr V Peptide hydrolases of Antarctic krill *Euphausia superba*. Comp Biochem Physiol 1985; 82B:599-606.

25. Osnes KK, Ellingsen TE, Mohr V. Hydrolysis of proteins by peptide hydrolases of Antarctic krill, *Euphausia superba*. Comp Biochem Physiol 1986; 83B:801-805.

26. Seki N, Sakaya H, Onozawa T. Studies on proteases from Antarctic krill. Bull Jpn Soc Sci Fish 1977; 43:955-962.

27. Karlstam B. Crossed immunoelectrophoretic analysis of proteins from Antarctic krill *(Euphausia superba)* with special reference to serine proteinases. Polar Biol 1991; 11:489-493.

28. Karlstam B, Johansson B, Brynö C. Identification of proteolytic isozymes from Antarctic krill *(Euphausia superba)* in an enzymatic debrider. Comp Biochem Physiol 1991; 100B:817-820.

29. Hellgren L, Vincent J. Debridement: an essential step in wound healing. In: Westerhof W, ed. Leg Ulcers: Diagnosis and Treatment. Amsterdam:Elsevier Science, 1993:305-312.

30. Anheller JE, Hellgren L, Karlstam B, Vincent J. Biochemical and biological profile of a new enzyme preparation from Antarctic krill *(Euphausia superba)* suitable for debridement of ulcerative lesions. Arch Dermatol Res 1989; 281:105-110.

31. Campbell D, Hellgren L, Karlstam B and Vincent J Debriding ability of a novel multi-enzyme preparation isolated from Antarctic krill *(Euphausia superba)*. Experientia 1987; 43:578-579.

32. Hellgren L, Mohr V, Vincent J. Proteases from Antarctic krill - a new system for effective enzymatic debridement of necrotic ulcerations. Experientia 1986; 42:403-404.

33. Hellgren L, Karlstam B, Mohr V, Vincent J. Krill enzymes: A new concept for efficient de-

bridement of necrotic ulcers. Int J Dermatol 1991; 30:102-103.

34. Mekkes JR, Le Poole IC, Das PK, Kammeyer A, Westerhof W *In vitro* tissue-digesting properties of krill enzymes compared with fibrinolysin/DNAse, papain and placebo. Int J Biochem Cell Biol 1997; 29:703-706.

35. Hellgren L, Vincent J. Débriding properties of krill enzymes in necrotic leg ulcers. Arch Dermatol 1989; 125:1006.

36. Westerhof W, van Ginkel CJW, Cohen EB, Mekkes JR. Prospective randomized study comparing the débriding effect of krill enzymes and non-enzymatic treatment in venous leg ulcers. Dermatologica 1990; 181:293-297.

37. Vanscheidt W, Weiss JM. Types of enzymes on the market. In: Westerhof W, Vanscheidt W, eds. Proteolytic Enzymes and Wound Healing. Amsterdam:Elsevier Science, 1994:59-73.

38. Hull PS. Chemical inhibition of plaque. J Clin Periodontol 1980; 7:431-442.

39. Marsh PD. The significance of maintaining the stability of the natural microflora of the mouth. Br Dent J 1991; 21:174-177.

40. Hellgren K, Hellgren L, Mohr V, Vincent J. Composition for dental use comprising krill enzyme. PCT patent WO 95/33470, 1995.

41. Hellgren L, Mohr V, Vincent J Enzyme composition acting as a digestion promoter on various levels in the alimentary tract, and a method for facilitating digestion. US patent 4.695.457, 1987.

42. Frymoyer JW. Back pain and sciatica. N Engl J Med 1988; 318:291-300.

43. Kato F, Mimatsu K, Kawakami N, Iwata H, Miura T. Serial changes observed by magnetic resonance imaging in the intervertebral disc after chemonucleolysis: a consideration of the mechanism of chemonucleolysis. Spine 1991; 17:934-939.

44. Nachemson AL, Rydevik B. Chemonucleolysis for sciatica. A critical review. Acta Orthop Scand 1988; 59:56-62.

45. Dabezies EJ, Langford K, Morris J, Shields CB, Wilkinson HA. Safety and effficacy of chymopapain (Discase) in the treatment of sciatica due to a herniated nucleus pulposus. Results of a randomized, double-blind study. Spine 1988; 13:561-565.

46. Garvin PJ. Toxicity of collagenase, the realtion to enzyme therapy of disc herniation. Clin Orthop 1974; 101:286.

47. Artigas J, Brock M, Mayer HM. Complications following chemonucleolysis with collagenase. J Neurosurg 1984; 61:679.

48. Melrose J, Hall A, Macpherson C, Bellenger CR, Ghosh P. Evaluation of digestive proteinases from the Antarctic krill *(Euphausia superba)* as potential chemonucleolytic agents. *In vitro* and *in vivo* studies. Arch Orthop Trauma Surg 1995; 114:145-152.

49. Sasahara AA, Loscalzo J. New Therapeutic Agents in Thrombosis and Thrombolysis. New York: M Dekker, 1997.

50. Hellgren L, Mohr V, Vincent J, Vincent J, Karlstam B. Intravasal thrombolysis. PCT patent WO 95/33471, 1995.

Heat-labile uracil-DNA glycosylase from a psychrophilic marine bacterium

H. Sobek

Boehringer Mannheim GmbH, Nonnenwald 2, D-82377 Penzberg, Germany

1
Introduction

1.1
Uracil-DNA glycosylases

Uracil DNA glycosylase (UDG, EC 3.2.2.3), also known as uracil-N glycosylase (UNG) is the first enzyme involved in the base repair pathway for removal of uracil from DNA. The base uracil is removed from mutagenic U/G mispairs resulting from the deamination of cytosine and from U:A pairs resulting from misincorporation of dUMP during DNA synthesis.[1,2] A mutagenic U/G mismatch leads to a C → T transition mutation in the next round of DNA synthesis. Although uracil is not mutagenic when base-pairing with adenine, uracil in place of thymine in regulatory DNA sequences can disrupt the binding of specific proteins.[3] Uracil-DNA glycosylases hydrolyse the N-glycosidic bond linking the base to the deoxyribose sugar, generating an abasic site, which is subsequently removed by an apurinic/apyrimidinic (AP)-endonuclease and a phosphodiesterase. The resultant gap is filled in by a DNA polymerase and sealed by a DNA ligase.[3] Uracil-DNA glycosylase shows high selectivity for the excision of uracil from DNA; it removes related bases such as isodialuric acid, 5-hydroxyuracil and alloxan at rates some orders of magnitude lower than with uracil.[4] No activity has been detected against any normal DNA base, nor against uracil in RNA. UNG excises uracil from single- and double-stranded DNA; for double-stranded DNA a preference for certain sequence contexts was shown.[5]

Uracil-DNA glycosylases have been identified in a variety of prokaryotic and eukaryotic organisms as well as in different viruses.[2] They consist of a single polypeptide chain and the sequences range between 199 and 359 amino acids with a conserved C-terminal component of 199–220 residues and variable N-terminal extensions.[6] The first UNG described and characterized was the enzyme from *Escherichia coli*.[7,8] Uracil-DNA glycosylases have also been characterized from different bacteria such as *Bacillus subtilis*,[9] *Bacillus stearothermophilus*,[10] *Micrococcus luteus*,[11] *Mycoplasma lactucae*[12] and *Thermothrix thiopara*.[13] Recently we described an uracil-DNA glycosylase from a psychrophilic marine bacterium.[14] Uracil-DNA glycosylases were also isolated from eukaryotic sources as e.g. human[15] and yeast.[16] The sequence alignement of several uracil-DNA glycosylases reveals a high degree of similarity between the genes and several clusters of residues that are highly conserved.[17]

The three-dimensional structures of the UNG from human and Herpes simplex virus type-I are very similar and consist of a single α/β domain containing a central four-stranded parallel β-sheet and eight α-helices.[6,17] Crystallographic and mutational studies have revealed the mechanisms for the selective binding and excision of uracil.[17] Single- and double-stranded DNA binds along a positively charged groove of the enzyme. The buried uracil-binding pocket is strictly conserved among UNGs and is characterized by extensive shape and electrostatic complementarity to uracil. The specific recognition of phosphate, deoxyribose and uracil occurs after a flipping-out of the damaged base from the DNA major groove.[18] A catalytic mechanism for the excision of uracil by hydrolytic cleavage of the N-glycosidic bond has been proposed.[6,17,18]

1.2
Applications

The enzyme uracil-DNA glycosylase is applied in several molecular biological techniques as e.g. the ligation and cloning of DNA,[19–22] the labeling of oligonucleotides,[23] the preparation of DNA for sequencing[24] and the examination of DNA-protein contacts.[3,25] Mainly UNG is used in the so-called carry-over prevention technique, which adresses the problem of the contamination of polymerase chain reaction (PCR) samples.[26,27] Amplification products generated by PCR can be substrates for subsequent PCR procedures. Furthermore, because the quantities of the amplification products can be large, the dispersal of even a small fraction of a PCR product into the laboratory area can lead to contamination of later amplifications and false positive results.[28] The carry-over prevention technique represents an improvement of PCR procedures by making amplification products distinguishable from naturally occuring DNA. It eliminates the products of previous amplifications from further amplifications by means of a treatment that leaves DNA from the sample unaffected in its ability to be amplified. In the carry-over prevention technique the deoxyribonucleotide triphosphate dUTP is incorporated into the PCR protocol resulting in an uracil containing PCR product. Discrimination between natural DNA and the uracil containing DNA (U-DNA) is obtained with the enzyme uracil-DNA glycosylase. Treatment of DNA samples containing uracil bases with UNG results in cleavage of the glycosidic bond between the deoxyribose and the uracil base. The loss of the uracil creates an apyrimidinic site in the DNA, which blocks DNA polymerases from using the strand as a template for the synthesis of a complementary DNA strand.

In the carry-over prevention protocol the PCR is modified in three ways: (1) dUTP is substituted for dTTP; (2) UNG is added to the PCR reaction mixture; and (3) an initial incubation is performed to allow UNG to destroy contaminating products of prior PCR reactions. The UNG itself is inactivated by high temperature prior to the first PCR cycle. This inactivation prevents UNG from destroying newly-synthesized PCR products.[26]

1.3
Approaches for improvement

The UNG from *E. coli* has been cloned and expressed in high amounts.[30,31] The recombinant enzyme is usually applied in molecular biological techniques. For many applications the enzyme shows a suitable performance. However, its application in the carry-over prevention technique is complicated as the enzyme is not completely inactivated by the heat denaturation step and the subsequent PCR. It has been shown that a sufficient amount of activity survives the inactivation to degrade the newly synthesized PCR product during extended incubations at either 4°C or 25°C. Also a reactivation of the inactivated uracil-DNA glycosylase after PCR is discussed.[29]

The PCR is an extreme sensitive technique to detect only a few molecules of a target DNA sequence. Therefore the carry-over prevention is applied in PCR techniques which assay and quantitate DNA samples as e.g. in medical diagnostics, forensic analysis or molecular biology. The possibility of a degradation of the newly synthesized PCR by residual activity represents a major complication as it may lead to false negative results. To overcome the residual activity it was suggested to add the uracil-DNA glycosylase inhibitor protein (Ugi) to the PCR after the last cycle.[29] Another possibility is the addition of denaturing reagents (e.g. SDS) or proteases to prevent UNG activity. However, for routineous applications any additional step in a PCR protocol is inconvenient. To overcome these additional manipulations we started a screening for uracil-DNA glycosylases that show less heat-stability than the UNG from *E. coli.*

2
Screening for heat-labile uracil-DNA glycosylases

Psychrophilic bacteria were considered as a possible source for heat-labile uracil-DNA glycosylases. Therefore a collection of psychrophilic bacterial strains was screened for the existence of heat-labile uracil-DNA glycosylases. The different strains were cultured in the appropriate media and small amounts of harvested cells (gramm scale) were lysed to release uracil-DNA glycosylase activity. As the crude extracts of the bacteria contained unspecific nucleases that influenced the identification of uracil-DNA glycosylase activity it was neccessary to remove these unspecific nuclease activities. By two chromatographic steps on hydroxyapatite and Q-Sepharose it was possible to separate the unspecific nucleases to an extent that allowed the measurement of uracil-DNA glycosylase activity. As an uracil containing substrate M13mp11 U-amber ss-DNA was used to assay the uracil-DNA glycosylases.[14]

Several strains were found to contain uracil-DNA glycosylase activity. The heat-lability of the uracil-DNA glycosylases was tested by incubating the enzymes in a

PCR buffer system at high temperatures. After different periods samples were removed and the residual activity was assayed. In the screening procedure the strain BMTU 3346 contained an uracil-DNA glycosylase activity that was rapidly inactivated by incubation at 50°C. The enzyme was found to be more thermolabile than the uracil-DNA glycosylase from *E. coli* that was tested as reference enzyme (Fig. 1).

Originally the psychrophilic strain BMTU 3346 containing this heat-labile uracil-DNA glycosylase was collected from marine samples. Taxonomic characterization identified the strain as an aerobe gram-positive coccus. The partial sequence of the 16S rRNA of the strain BMTU 3346 showed a high degree of homology to the 16S rRNA of *Arthrobacter globiformis* (96.2%) and *Micrococcus luteus* (96.6%).

3
Purification and characterization of heat-labile UNG

For detailed characterization the heat-labile uracil-DNA glycosylase from the strain BMTU 3346 was purified in a larger scale. Frozen cells (120 g) were lyzed by addition of lysozyme followed by disruption in a cell press. Polymin was added to precipitate cellular DNA and the precipitate was centrifuged. The enzyme was purified by chromatography on hydroxyapatite, Q-Sepharose fast flow and Phenyl-Sepharose fast flow. The obtained enzyme fraction showed a single band in sodium dodecyl sulphate (SDS) gel electrophoresis and was free of any detectable unspecific nucleases. From mobility in SDS gel electrophoresis the molecular weight of the uracil-DNA glycosylase was determined to be 23.4 kDa (+/- 1 kDa). In gel filtration experiments on a Superose 12 column the thermolabile UNG coelutes with chymotrypsinogen A (molecular weight 25 kDa) indicating a monomeric structure of the enzyme.[14] With respect to the monomeric character and the molecular weight the enzyme is similar to the uracil-DNA glycosylases from other prokaryotic organisms.[7,13] The molecular weight of 23.4 kDa closely resembles the molecular weights of the enzymes from *B. subtilis* and *E. coli* having a molecular weight of 24 kDa and 24.5 kDa respectively.[7–9] For the heat-labile UNG a specific activity of $5x10^4$ units mg^{-1} was determined. Like other uracil-DNA glycosylases the enzyme excises uracil from single- and double-stranded DNA.

The heat-labile uracil-DNA glycosylase shows stability for months when stored in complex storage buffers containing glycerol (50%) at -20°C. It is also stable in complex assay buffers (containing bovine serum albumin or glycerol) showing a linear reaction kinetics for 1 h at 37°C. However, it is rapidly inactivated in dilute buffers as usually applied in PCR technology. In PCR buffer containing 10 mM Tris/HCl, 50 mM KCl, 1.5 mM $MgCl_2$, pH 8.3 it showed a half-life time of 2 min (0.5 min) at 40°C (45°C).[14] As the uracil-DNA glycosylase showed excellent thermolability under these conditions it was further tested for its performance in the PCR carry-over prevention technique.

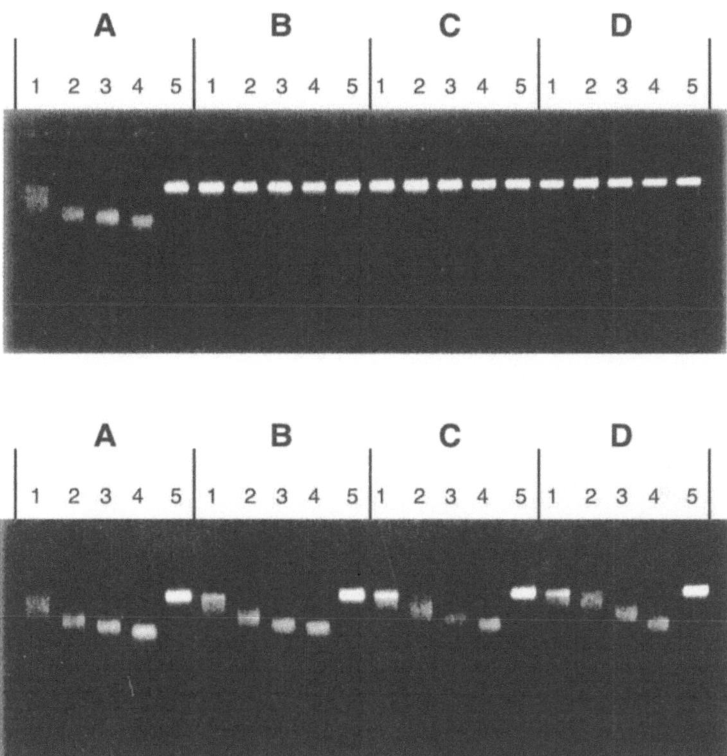

Fig. 1. Screening for heat-lability of uracil-DNA glycosylases
One unit of UNG from BMTU 3346 or *E.coli* were incubated in 100 µl PCR buffer at 50°C. After 2, 5 and 10 min samples were removed. 1, 3, 5 and 10 µl of the samples were incubated with 1 µg M13 mp11 U-DNA for 1 h at 37°C. After incubation aliquots were separated on agarose gels as described.[14] *Upper part:* UNG from BMTU 3346; *lower part:* UNG from *E.coli*. Lanes 1, 2, 3, 4 show the degradation of U-DNA by 1, 3, 5 and 10 µl of the samples. Lane 5: U-DNA without UNG treatment (negative control). Section A: without heat-inactivation. Section B, C and D: heat-inactivation of the UNGs for 2, 5 and 10 min, respectively.

4
Application in carry-over prevention technique

The thermolabile uracil-DNA glycosylase was tested for application in carry-over prevention using different experimental procedures. It was tested whether the application of uracil-DNA glycosylase leads to a degradation of an uracil containing DNA (U-DNA) and prevents the amplification of this uracil containing DNA by a subsequent PCR. As an artificial U-DNA a 264 bp fragment of the cloned human tissue plasminogen activator (tPA) gene was generated by PCR using the unnatural nucleotide dUTP (instead of dTTP). This U-DNA served as artificial

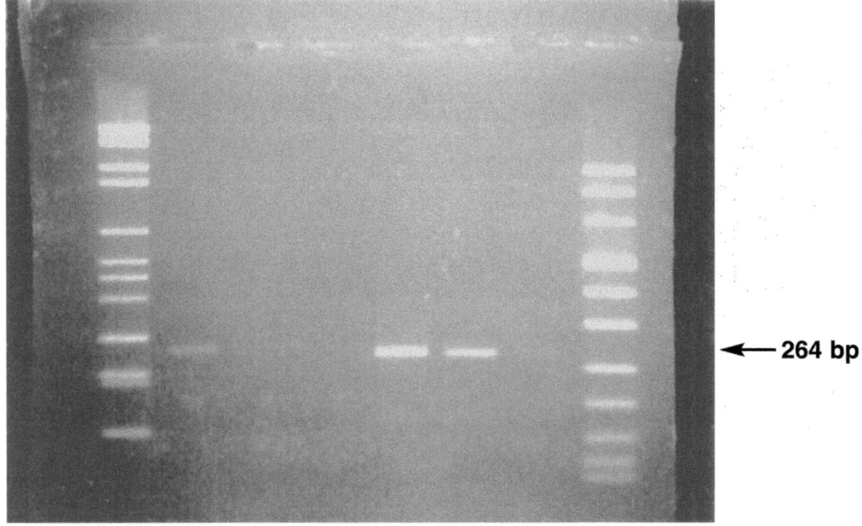

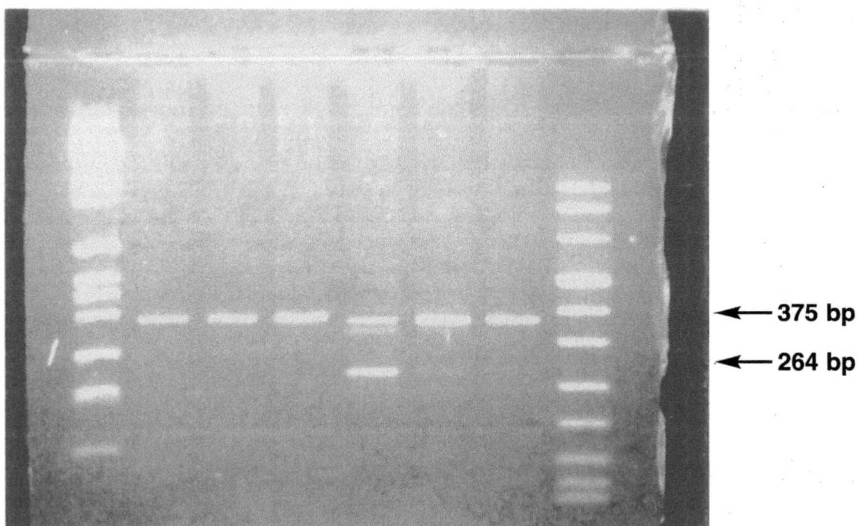

Fig. 2. Application of uracil-DNA glycosylase in carry-over prevention

Lanes 1–6: Amplification of 264 bp tPA fragment from an uracil containing template by PCR. 10 pg (lanes 1, 4), 1 pg (lanes 2, 5) and 0.1 pg (lanes 3, 6) of template were incubated with 2 units of UNG (lanes 1–3) or without UNG (lanes 4–6) for 10 min at 20°C. After inactivation of the UNG by incubation for 2 min at 94°C the fragment was amplified by PCR. **Lanes 7–12:** Simultaneous amplification of a 264 bp and a 375 bp fragment of tPA from uracil containing template (264 bp) and from thymidine containing template (375 bp). Template for the 375 bp fragment was human genomic DNA (500 ng). The protocol as described for lanes 1–6 was applied for lanes 7–12 respectively; with UNG: lanes 7-9, without UNG: lanes 10–12 contaminating DNA in the simulation of a carry-over prevention experiment consisting of incubation with uracil-DNA glycosylase followed by PCR. The incubation with UNG prevented the re-amplification of the uracil containing target (Fig. 2, lanes 2 and 3). High amounts (10 pg) of U-DNA were not completely degraded under the experimental conditions and resulted in a lower amount of PCR product (Fig. 2, lane 1) as compared to the control reaction (Fig. 2, lane 4). In control reactions (omitting UNG) PCR products were obtained with 10–0.1 pg U-DNA as template (lanes 4–6).

In another experiment it was tested whether the decontamination of U-DNA by treatment with uracil-DNA glycosylase influences the amplification of a thymidine containing DNA (T-DNA). The target of this amplification was a 375 bp fragment of the tPA gene which was amplified by the same primer pair as the shorter U-DNA (264 bp) and gave a longer PCR fragment due to the existence of intron sequences in the natural gene sequence. The 375 bp target was amplified from a constant amount of human genomic DNA spiked with varying amounts of the contaminating U-DNA. After incubation with uracil-DNA glycosylase a PCR was performed to amplify both tPA targets. As shown in Figure 2 the heat-labile UNG was successfully applied to prevent the amplification of the artificial contaminating U-DNA (lanes 7–9), whereas the T-DNA was successfully amplified. In control experiments (omitting UNG) both targets were simultanously amplified (lanes 10–12).

The results clearly demonstrate that incubation of uracil-containing template with the enzyme prevented amplification of the U-DNA by the following PCR whereas the amplification of the T-DNA was not affected by treatment with the heat-labile uracil-DNA glycosylase. A short incubation for 2 min at 94°C after digestion of the U-DNA was found to completely inactivate the UNG prior to PCR. Therefore the heat-labile uracil-DNA glycosylase is suitable for the application in the carry-over prevention technique; due to its short inactivation time it is more suitable than the UNG from *E.coli* for which an inactivation for 10 min at 94°C is usually applied.[26]

5
Characterization of residual activity

For the uracil-DNA glycosylase from *E. coli* it has been reported that sufficient activity survives the heat-inactivation (and following PCR) to degrade an uracil-containing PCR product during extended incubation after PCR. Attempts to precisely characterize the residual activity of the uracil-DNA glycosylase from *E. coli* were unsuccessful due to the insufficient sensitivity of the radioactive test system used.[29] Therefore a different experimental procedure was chosen to characterize the residual activity of the uracil-DNA glycosylases from *E. coli* and from strain BMTU 3346. The degradation of a digoxigenin-labeled (DIG-labeled) uracil containing PCR product was monitored to characterize residual uracil-DNA glycosylase activities. Using a 5′-DIG-labeled forward primer a 103 bp uracil containing DNA was synthesized by PCR. This PCR product was used to detect residual uracil-DNA glycosylase activity by monitoring the appearance of degradation products (Fig. 3).

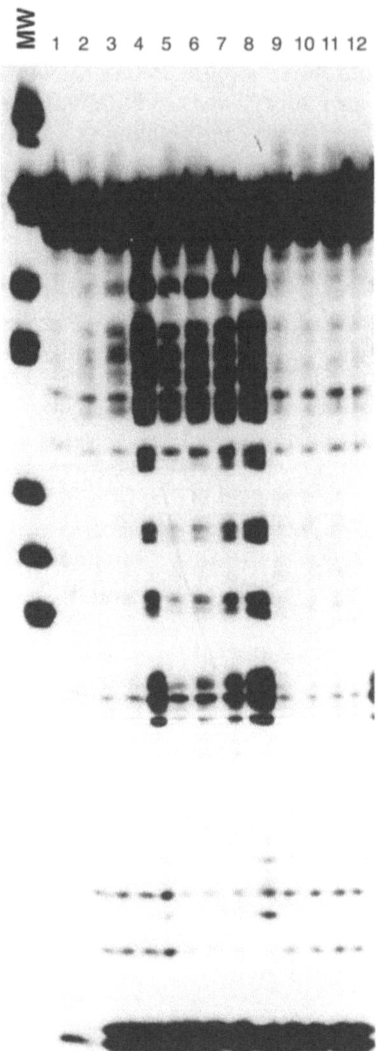

Fig. 3. Characterization of residual UNG activity

The degradation of a 103 bp DIG-labeled U-PCR product was monitored. 100 µl amplification reactions containing 10 mM Tris/HCl, pH 8.3, 50 mM KCl, 1.5 mM $MgCl_2$, 200 µM each of dATP, dCTP, dGTP, 600 µM dUTP, 2.5 units of Taq DNA polymerase, 1 ng pUC18 and 1 µm of each primer (DIG-labeled pUC18 sequencing forward primer, pUC18 sequencing reverse primer) were prepared. Samples were treated at 95°C for 2 min followed by PCR (25 cycles of 94°C, 1 min; 50°C, 1 min; 72°C, 3 min; followed by incubation at 4°C). Samples (20 µl) were removed at different times, treated with 5 µl of 0.6 M NaOH for 5 min at 37°C and 5 µl of 0.6 M HCl. After addition of formamide dye (4 µl) aliquots of each were analyzed by gel electrophoresis on a sequencing gel, blotting and chemoluminescent detection. Prior to PCR, 2 units of UNG from BMTU 3346 (lanes 1–4) or 2 units of UNG from *E. coli* (lanes 5–8) were added; negative control without UNG (lanes 9–12). Samples were removed at t = 0 (lanes 1, 5, 9), t = 1 h (lanes 2, 6, 10), t = 4 h (lanes 3, 7, 11) and t = 16 h (lanes 4, 8, 12)

Degraded PCR product were separated on sequencing gels and blotted on nylon membranes. The detection was performed by anti-DIG/peroxidase staining.[14] The results show that after inactivation of the UNG for 2 min at 95°C followed by PCR, degraded PCR products were detectable immediately after the last PCR cycle in the reaction mixture containing uracil-DNA glycosylase from *E. coli*. This strongly indicates that a minimal amount of uracil-DNA glycosylase from *E. coli* survives the inactivation and leads to degradation of the PCR product. After prolonged incubation at 4°C the amount of degraded PCR product increases, whereas the PCR samples containing uracil-DNA glycosylase from BMTU 3346 showed no de-graded PCR products after the last PCR cycle. This indicates a complete inactivation of the enzyme. The PCR samples treated with the heat-labile uracil-DNA gly-cosylase show a significant amount of degraded uracil containing PCR product only after a prolonged incubation, whereas samples containing the enzyme from *E. coli* showed a significantly higher degree of degradation.

The observed appearance of degradation products after prolonged incubation could be caused by (1) minimal amounts of uracil-DNA glycosylase activity that survive heat-inactivation, as a result of the residual activity degraded PCR pro-ducts are accumulated; and/or (2) uracil-DNA glycosylases are reactivated after PCR by refolding of the denatured enzymes. For the UNG from *E. coli* a reactiva-tion after PCR has been discussed previously.[29] However, under the experimental conditions used it is impossible to discriminate whether the accumulation of degradition products is the result of residual enzyme activity over time or/and is influenced by regain of activity by reactivation.

6
Conclusions

The results decribed in this chapter demonstrate that the recently isolated heat-labile uracil-DNA glycosylase can be successfully applied in the carry-over preven-tion technique to control the contamination of PCR samples by products from pre-vious amplifications. Due to its thermolability the uracil-DNA glycosylase can be inactivated by a shorter incubation step than the UNG from *E. coli*. Its less residual activity after the amplification reaction allows to extend the time frame for storage and manipulation of the obtained PCR products without additional inconvenient steps in the PCR protocol. Therefore the application of the heat-labile uracil-DNA glycosylase results in an improvement of the carryover prevention technique.

Recently the gene for the heat-labile uracil-DNA glycosylase has been cloned (S. Jaeger, unpubl. results) and mutagenesis studies to illuminate the nature of the thermolability of this enzyme are in progress.

7
References

1. Lindahl T. Instability and decay of the primary structure of DNA. Nature 1994; 362:709-715.
2. Krokan HE, Standal R, Slupphaug G. DNA glycosylases in the base excision repair of DNA. Biochem J 1997; 325:1-16.
3. Pu WT, Stuhl K. Uracil interference, a rapid and general method for defining protein-DNA interactions involving the 5-methyl group of thymines: the GCN4-DNA complex. Nucl Acids Res 1992; 20:771-775.
4. Dizdaroglu M, Karakaya A, Jaruga P, Slupphaug G, Krokan HE. Novel activities of human uracil DNA N-glycosylase for cytosine-derived products of oxidative DNA damage. Nucl Acids Res 1996; 24:418-422.
5. Eftedal I, Guddal PH, Slupphaug G, Volden G, Krokan HE. Consensus sequences for good and poor removal of uracil from double stranded DNA by uracil-DNA glycosylase. Nucl Acids Res 1993; 21:2095-2101.
6. Savva R, McAuley-Hecht K, Brown T, Pearl L. The structural basis of specific base-excison repair by uracil-DNA glycosylase. Nature 1995; 373:487-493.
7. Lindahl T. An N-glycosidase from *Escherichia coli* that releases free uracil from DNA containing deaminated cytosine residues. Proc Nat Acad Sci USA 1974; 71:3649-3653.
8. Lindahl T, Ljungquist S, Siegert W, Nyberg B, Sperens B. DNA N-glycosidases. Properties of uracil-DNA glycosidase from *Escherichia coli*. J Biol Chem 1977; 252:3286- 3294.
9. Cone R, Duncan J, Hamilton L, Friedberg E C. Partial purification and characterization of a uracil DNA N-glycosidase from *Bacillus subtilis*. Biochemistry 1977; 16:3194-3201.
10. Kaboev OK, Luchkina LA, Akhmedow AT , Bekker ML. Uracil-DNA glycosylase from *Bacillus stearothermophilus*. FEBS Lett 1981; 132:337-340.
11. Leblanc JP, Martin B, Cadet J, Laval J. Uracil-DNA glycosylase: purification and properties of uracil-DNA glycosylase from *Micrococcus luteus*. J Biol Chem 1982; 257:3477-3483.
12. Williams MV, Pollack JD. A mollicute (mycoplasma) DNA repair enzyme: purification and characterization of uracil-DNA glycosylase. J Bacteriol 1990; 172:2979-2985.
13. Kaboev OK., Luchkina LA, Kuziakina TI. Uracil-DNA glycosylase of thermophilic *Thermothrix thiopara*. J Bacteriol 1985; 164:421-424.
14. Sobek H, Schmidt M, Frey B, Kaluza K. Heat-labile uracil DNA glycosylase: purification and characterization. FEBS Lett 1996; 388:1-4.
15. Krokan H, Wittwer CU. Uracil DNA-glycosylase from HeLa cells: general properties, substrate specificity and effect of uracil analogs. Nucl Acids Res 1981; 9:2599-2613.
16. Crosby B, Prakash L, Davis H, Hinkle DC. Purification and characterization of a uracil-DNA glycosylase from the yeast, *Saccharomyces cerevisiae*. Nucl Acids Res 1981; 9:5797-5809.
17. Mol CD, Arvai AS, Slupphaug G, Kavil B, Alseth I, Krokan HE, Tainer J A. Crystal structure and mutational analysis of human uracil-DNA glycosylase: structural basis for specificity and catalysis. Cell 1995; 80:869-878.
18. Slupphaug G, Mol C D, Kavil B, Arvai A S, Krokan HE, Tainer JA. A nucleotide-flipping mechanism from the structure of human uracil-DNA glycosylase bound to DNA. Nature 1996; 384:87-92.
19. Andersson B, Povellini CM, Wentland MA, Shen Y, Muzny DM, Gibbs RA. Adaptor-based uracil DNA glycosylase cloning simplifies shotgun libraray construction for large-scale sequencing. Anal Biochem 1994; 218:300-308.
20. Rashtchian A, Buchman GW, Schuster DM, Berninger MS. Uracil DNA glycosylase-mediated cloning of polymerase chain reaction-amplified DNA: application to genomic and cDNA cloning. Anal Biochem 1992; 206:91-97.
21. Kumar NV, Varshney U. Excision of uracil from the ends of double stranded DNA by uracil DNA glycosylase and ist use in the high efficiency cloning of PCR products. Curr Sci 1994; 67:728-734.

22. Liu HS, Tzeng HC, Liang YJ, Chen CC. Ligation of multiple DNA fragments through uracil-DNA glycosylase generated ligation sites. Nucl Acids Res 1994; 22:4016-4017.
23. Craig AG, Nizetic D, Lehrach H. Labeling oligonucleotides to high specific activity. Nucl Acid Res 1989; 17:4605-4610.
24. Ball JK, Desselberger U. The use of uracil-N-glycosylase in the preparation of PCR products for direct sequencing. Nucl Acids Res 1992; 20:3255.
25. Devchand PR, McGhee JD, van de Sande JH. Uracil-DNA glycosylase as a probe for protein-DNA interactions. Nucl Acids Res 1993; 21:3437-3443.
26. Longo MC, Berninger MS, Hartley JL. Use of uracil DNA glycosylase to control carry-over contamination in polymerase chain reactions. Gene 1990; 93:125-128.
27. Udaycumar, Epstein JS, Hewlett IK. A novel method employing UNG to avoid carry-over contamination in RNA-PCR. Nucl Acids Res 1993; 21:3917-3918.
28. Kwok S, Higuchi R. Avoiding false positives with PCR. Nature 1989; 339:237-238.
29. Thornton CG, Hartley JL, Rashtchian A. Utilizing uracil DNA glycosylase to control carry-over contamination in PCR: characterization of residual UDG activity following thermal cycling. BioTechniques 1992; 13:180-183.
30. Duncan B, Chambers JA. The cloning and overproduction of *Escherichia coli* uracil-DNA glycosylase. Gene 1984; 28:211-219.
31. Varshney U, Hutcheon T, van de Sande JH. Sequence analysis, expression, and conservation of *Escherichia coli* uracil DNA glycosylase and its gene (ung). J Biol Chem 1988; 263:7776-7784.

Principle of cold-adaptation in the derivation of live attenuated respiratory virus vaccines

H. F. Maassab

School of Public Health, Department of Epidemology, University of Michigan, 109 Observatory Street, Ann Arbor, Michigan 48109, USA

1
Introduction

Adaptation of selected human respiratory viruses to grow at suboptimal temperatures has provided a valuable tool for use in the development of live respiratory virus vaccines in man. The early report that live influenza virus vaccines can be made suitable for administration to children and young adults by passage at 25°C, stimulated our interest in deterring whether cold-adaptation can be used regularly to obtain attenuated vaccines that were well tolerated by and immunogenic for humans of all ages.[1,2] This chapter describes the development, characterization and evaluation of the cold-adapted respiratory viruses in animals and also in limited testing in man.

2
Historical background – viral virulence

Virulence of a virus is a variable characteristic under polygenic control and is influenced by the virus strain selected, the clinical diagnosis, and the passage history of the isolated strain. A review of the literature provides numerous examples of factors modifying virulence, some are classified as host factors, while others are connected to the viral origin.[2,3] The properties of a virus that are important determinants of virulence are quality, e.g. inactive or defective virus which competes for cell receptor in the initiation of infection, the quantity or the dose of the virus, its genetic constitution and its ability to induce interferon. The mechanisms in the expression of virulence are also influenced by viruses that are poor inducers of interferon and those which tend to be resistant to interferon by their ability to spread and to exhibit a high rate of multiplication in the host cell. An additional property that also affects the course of infection is host temperature. *In vivo* observations have demonstrated that high temperature plays a defensive role and can suppress viral replication and limit spread whilst lower temperature allows a higher rate of viral replication and more severe expression of a disease.

The role of temperature in microbial disease has been the subject of numerous reports in the literature, a high normal body temperatures in certain species of animals being found to play a defensive role in immunity to infection. Enders and Shaffer presented definitive evidence that the capacity of Pneumococuss III to multiply at 41°C was a genetic property and thus gave the first hint of the effect of temperatures on virulence of different strains of the same microbe.[4] In general, it was stated that hyperthermia has no effect on highly virulent strains, whereas it has a marked effect on strains with low virulence. Burnet in 1936 demonstrated that an animal virus such as influenza produced lesions and death of eggs incubated at 36°C but not at 39°C.[5] The studies in animals, although qualitative and sometimes speculative in nature, have not offered any more insight into the mode of action: it is not clear whether the mechanism involves changes in the host or whether it affects the virus directly.

In vitro studies have shown that temperature alters viral development in tissue culture infected cells. Correlation was found between virulence and the ability of virulent strains to grow at temperatures above physiological range (ca. 40°C). From all points of view, virulence can be quantified and it is a quality that can be added to or lost experimentally from a virus strain.

3
Definition of cold adaptation

Adaptation of respiratory viruses to growth at sub-optimal temperature (25°C) implies:

1. A procedure used for attenuation of the virus by serial passage at 25°C,
2. conferring on the virus strain a "cold-adapted" (ca) phenotype which can be used as a marker to monitor attenuation,
3. the ca phenotype designation equates equivalence in infectivity titer at 25 and 33°C,
4. in general acquisition of the ca phenotype also confers the temperature-sensitive (ts) phenotype,
5. molecular characterization of the cold-adapted strain also resulted in mutational changes in every segment where compared to the original wild type strain,
6. the cold-adapted strains have been shown to be genetically stable and attenuated,
7. designation of the donor lines A/AA/6/60-H2N2-7PI and B/AA/1/66-7PI,
8. recombination-reassortment for antigenic updating of the live vaccine, and
9. designation of a 6/2 CR (cold reassortant) clone as a vaccine for trials in man.[2,3]

4
Adaptation, characterization, and derivation of live cold-adapted influenza virus

The study of epidemiology of influenza virus and illness has been carried out over many years employing different designs. When results differ there is always a question of how much of the variation is due to the effect of a changing virus and how much to different methods of study. There are certain constants that can be identified especially involving mortality. Other findings may be less constant, particularly with regard to transmission in families. Thus, these results are of particular relevance in designing and determining vaccination programs intended to protect and to interrupt transmission.

Prophylaxis of influenza has been a recurrent public health concern for many years with frequent re-examination of available vaccines and policies of their use. Recently research has been ongoing for development of an effective and safe attenuated live influenza virus vaccine for man.

It has also been recognized over a decade that live attenuated influenza virus vaccine offers a potential advantage over inactivated vaccine. These advantages offered are the possible use of a single dose, the administration by the natural route, wide range of antibody response, the induction of local immunity and the cost effectiveness.[3]

In recent years, in our laboratory, we have developed, characterized, and provided attenuated ca vaccine with two markers, ca and ts, and a stable attenuated phenotype. The procedure allowed us to designate two genetically stable attenuated donor lines. The passage history of the two influenza virus lines, type A, A/Ann Arbor/6/60-H2N2 and type B cold-adapted donor of B/Ann Arbor/1/66 are presented in Tables 1 and 2. The availability of these two lines provided a practical approach to the attenuation of new antigenic variants by genetic reassortment or reverse genetics using the donor strains to confer attenuation to the circulating wild type influenza virus of relevant surface antigens.[3,6] Thus, through genetics we were able to update and provide attenuated seed strains possessing the contemporary circulating surface antigens in 4 to 6 weeks for use in man. The live influenza vaccine is designated as CR or cold-reassortant. Thus, the ability of the two donor strains used in mixed infection to update the live vaccine has allowed us the introduction of the concept of 6/2 gene profile, the six internal genes from the donor lines and the two genes which code for the surface glycoproteins from the wild type parent. The CR vaccine has been proven to be genetically stable immunogenic and avirulent when administered intranasally to man at six clinical centers funded by the National Institute of Allergy and Infectious Disease. Data to this effect will be presented as well as data on the molecular basis of attenuation of these cold reassortant live influenza virus vaccines.[3]

Table 1. Passage history of cold variant A/AA/6/60 (H3N2)

Passage level	Procedure
PCK-1	Isolation in PCK cell cultures at 36°C
PCK-2	Passage in PCK at 36°C
PCK-3-9	Serial passage in PCK at 33°C
PCK-10-16	Serial passage in PCK at 30°C
PCK-17-23	Serial passage in PCK at 25°C
PCK-24-30	Serial cloning by plaque passages in PCK at 25°C
E-1-3	Serial passage in SPAFAS embryonated hens' eggs at 25°C
7 PI	Plaque-plaque purification seven times at 25°C

Table 2. Adaptation of strains of influenza virus type B to growth at 25°C

Strains of type B influenza	Prior passage history	No. of passages at various temperatures required for optimal growth			
		35°C	33°C	27°C	25°C
B/AA/1/66	CK1	2	2	1	1[a]

[a] Plaque-plaque purification seven times at 25°C.

4.1
Development of the cold-adapted "master strain" of type A and of type B influenza

The initial results in mice and ferrets upon intranasal administration of live cold-adapted influenza virus of type A and type B demonstrated that the procedure provided the basis for development of live influenza virus vaccine. The animals given the cold variants through the natural route showed no signs of clinical disease and protective antibodies were elicited. The animals were protected upon challenge with the homologoeus wild type parent.[3,7,8]

Thus, the protocol used to identify an acceptable donor of attenuated genes for both types of influenza virus was established on the following:

1. Passage, adaptation and growth at 25°C with infectious titer in primary chick kidney cells and embryonated eggs.
2. *In vitro* and *in vivo* characterization of the two master strain lines:
   - growth in tissue culture at 33, 37, 38 and 39°C (phenotypes),
   - marker designation as ca (cold-adapted) and ts (temperature sensitive) phenotypes,
   - plaque purification to established the genetic stability and molecular characteristic,
   - growth in the lungs and turbinates in ferrets (animal model),
   - reactogenicity in ferrets (clinical response) when compared to the wild type parent,
   - designation of type A (A/Ann Arbor/6/60 H2N2) and of type B (B/Ann Arbor/1/66) as attenuated donor of attenuated "genes" to the new antigenic variant.

4.1.1
Production of live influenza vaccine strains by the classical method of "cold reassortment"

As stated previously, considerable effort has been directed toward the development of live attenuated vaccines for the prevention of influenza. These preparations have a number of advantages over the traditional inactivated vaccines, especially with regard to duration and breadth of immunity induced, and the production of local antibody. Preparations of such live influenza vaccines pose a particular problem not encountered when dealing with other viruses. The surface antigens of influenza virus change over time, and the vaccine must change accordingly. Thus, not only must the vaccine be safe and effective, but it must also be possible to produce predictably a new effective vaccine variant quickly in response to changes in the circulating viruses.

Certain characteristics of influenza viruses can be transferred between different strains by genetic reassortment.[9] It seems likely therefore, that under proper conditions it should be possible to confer, by genetic interchange, the growth characteristics of a previously cold-adapted attenuated strain to a new antigenic variant. Such an accomplishment would considerably shorten the interval between virus isolation and delivery of a suitable seed virus for the production of live vaccine.

Earlier work on the transfer of the ca marker has shown that a 'cold hybrid' of influenza virus can be obtained in 5 weeks by reassortment at 25°C. The studies involved the cold variant A/PR8/34 and a wild type parent A/England/878/69 (H3N2). The cold reassortant contained the haemagglutinin of the current Hong Kong-like parent and the neuraminidase of the established cold mutant. It could grow and produce plaques at 25°C. It was attenuated for animals (mice) and highly immunogenic. However, the A/PR8/34 cold variant, because of its dubious passage history, cannot be used for the production of reassortants as live virus vaccine for man.[7] The availability of a donor line developed in our laboratory and characterized as being suitably attenuated, genetically stable, immunogenic, with defined markers has enabled us to create, in a short period of time, a cold adapted reassortant (these lines are designated CR) bearing the particular contemporary surface antigens desired. Two lines that have been designated as donors of attenuated genes (master strains) are the type A influenza virus A/AA/6/60 (H2N2)[10,11] and the type B influenza virus B/AA/1/66.[12,13] Both cold-adapted lines were cloned at 25°C and have been used since 1968 for obtaining cold reassortants for use as attenuated live vaccines.

To update the live cold reassortant vaccine with the new relevant surface antigens of the circulating strain (H3N2) and (H1N1) of influenza virus type A and of type B we have used two methods in the derivation of a 6/2 live attenuated lines with the two surface glycoprotein of the recent epidemic strain.[3]

The flow diagram of the classical method (method I) shown below illustrates the procedure followed:

Step 1   Mixedly infect PCK cells in 20 tubes with equal input of A/AA/6/60 cold mutant and wt parent viruses. Incubate at 25°C until CPE observed (multiplicity of infection=5 pfu⁻¹ for each parent).

Step 2   Harvest from each tube and repassage separately in PCK cells at 25°C in the presence of A/AA/6/60 antiserum.

Step 3   Repeat Step 2.

Step 4   Repeat Step 2, but omit immune serum.

Step 5   Prepare egg pools for each independent reassortant line.

Step 6   Identify HA and NA antigens for each egg pool.

Step 7   Grow each separate line or reassortant in PCK cells at 25°C under agar overlay, without immune serum. Pick one large well-defined plaque for each reassortant line.

Step 8   Repeat Step 7, using virus from plaques of Step 7.

Step 9   Repeat Step 7, using virus from plaques of Step 8.

Step 10   Verify cold adaptation and temperature sensitivity by plaque titration in PCK cells at 25, 33 and 39°C.
Step 11   Identify the gene constellation of each independent clone by appropriate electrophoresis techniques.
Step 12   On the basis of the above studies, select clones with the desired properties for evaluation of reactogenicity in the ferret animal model system.
Step 13   Select the clones with the desired *in vivo* and *in vitro* properties for the production and safety-testing of vaccine pools for human volunteer studies.
Step 14   Perform three times plaque to plaque purification of the 6/2 selected clones. Evaluation *in vitro* and *in vivo* of the selected vaccine clone.
Step 15   Production of the volunteers pool for human clinical trials.

In general, PCK cells are simultaneously infected at 25°C with approximately 5 plaque-forming-units (pfu) per cell of the two viruses, i.e. the attenuated donor line and the wild type virus with the relevant surface antigen, as shown above. The duration of the various manipulations from the time of co-infection of PCK cells until the derivation a cold-adapted reassortment clone(s) with the desired properties is ca. 5 weeks. The procedure employed can be relied upon to reproducibly yield suitable candidate vaccine strains if these major conditions are met: (i) vigorous growth with high plaquing efficiency at 25°C; and (ii) temperature sensitivity and loss of virulence in a ferret animal model system.

This procedure involves selective pressure applied to the production of independent, appropriately attenuated clones by (i) the temperature of incubation at 25°C, at which the wt parent cannot replicate efficiently; and (ii) the presence of specific antisera against the surface antigens of the attenuated donor strain. Thus, the isolation of cold reassortants, bearing the current haemagglutinin (HA) and neuraminidase (NA) of the virulent wild type parent, and the growth and other characteristics of the ca master strain, is favored.

4.1.2
Reverse genetics (method II)

Additionally, a novel process which combines the reverse genetics technology discovered by Dr. Peter Palese (Mt. Sinai Medical Center) with the classical method used to generate 6:2 reassortants has recently been developed.[6] It is now possible to selectively introduce the HA and NA gene segments of wild-type influenza A or B virus directly into the cold-adapted master donor strains. This process should significantly shorten the time required to prepare 6:2 reassortants for the annual vaccine, improve reliability of the methods used by eliminating gene dominance that occasionally occurs with traditional methods, and further reduces the risk of introduction of human advertitial agents in the overall manufacturing process.

4.2
Molecular basis of genetic stability of the two cold-adapted mutants of type A and type B influenza virus

Different biochemical techniques were used to reveal that each of the six internal genes of the ca /AA/6/60-H2N2 donor line displayed some difference from the corresponding wild type (wt) parent from which it was derived. Twentyfour nucleotide differences between the ca and wt were identified of which eleven were deduced to code for amino acid substitutions in the ca virus proteins. In general the procedures used detected differences in each of the six internal segments of the ca-variant compared to the wt parent.[10,11] Such genomic wide range changes were also detected in the ca-type B variant, B/aa/1/66. There were 105 sites of difference between the wt and ca sets of the six internal genes of the type B influenza virus. The differences resulted in 26 amino acid substitutions over the six proteins.[12,13]

For both of these donor lines, a catalogue of molecular changes exists between the wt and ca variants, that should provide a valuable basis for understanding the mechanisms by which the two ca-donors confer the ts, the attenuated phenotypes and genetic stability to new ca vaccine reassortants. The studies of Snyder et al. using single gene reassortant for type A live vaccine provided evidence that only four genes (those coding for PB1, PB2, PA and M) were required for the attenuation phenotype of the ca master strain and the genes coding for PB2 and PB1 could independently confer temperatures sensitivity. Other studies suggest that extragenic suppressions and gene constellation effects might have confounded the results of the single gene studies.

4.3
The role of immunity of cold-adapted live influenza virus vaccine – clinical studies

Protective immunity to influenza virus infection is dependent on the level of humoral antibodies and on the level of the cell-mediated immune response (CMI), with evidence that CMI plays the major role in efficacy and protection. Replicating virus is far more effective in generating cytotoxic T cells (CTL) memory cells than killed vaccine preparations. It has been demonstrated that infection with attenuated live virus vaccine generates an immune response which includes the induction of CTLs. In contrast, the subunit vaccine does not prime for CTL response; thus, it may produce an inferior vaccine response. However, yearly administration of killed vaccines in people at high risk is still important for protection. Although humoral antibody has shown to be protective, its role in recovery is still not known.[9,14,15]

In comparison to inactivated vaccines, the use of cold-adapted reassortant vaccine has provided the basis for the following:

1. Primary serum antibody response appears long lived.
2. In unprimed individual, the use of a low, twice-administered dose induced solid immmunity against homologous and a heterologous challenge.
3. The induction of local secretory IgA in nasal wash.
4. Bivalent and trivalent live cold-reassortant vaccines stimulated peripheral blood lymphocytes capable of secreting influenza-specific antibody.
5. In an animal model (ferret) natural cytotoxic (NC) activity and antibody dependent cellular cytotoxic (ADCC) activity of a alvelolar macrophages are directed against local infection with influenza virus.
6. Intranasal administration of cold-adapted virus can be correlated with pulmonary B cell responses rather than splenic responses, and the presence of anti-

body-secreting cells and their numbers in the lungs upon challenge can be correlated with protection. In contast, serum antibody levels and the number of antibody-secreting cells in stimulated splenic cells were shown not to be correlated with protection.

7. Peripheral blood lymphocytes in the upper respiratory tract are stimulated in children receiving a bivalent and trivalent cold-reassortant vaccine preparation. These cells also secrete both H3N2 and H1N1-specific antibody.

There is now general agreement that the presence of secretory IgA in human nasal epithalium plays a major role in preventing influenza infection by inhibiting its spread or otherwise modifying the disease process. In the immune system, neutralization of the s IgA response effectively eliminates post-infectious immunity.[3,9,14,16]

In summary, the rationale for the use of live virus vaccine is its ability to induce protective cell-mediated immunity with the generation of primary immune response and immunological memory.[16,17]

Recently the intranasal vaccine was well tolerated in children. In summary, the vaccine efficacy was 93% against culture confirmed influenza. Both the one dose regimen showed 89% efficacy, while the two dose regimen were 94% efficacious. The vaccinated children had significantly fewer febrile otitis media. Thus, live attenuated cold-adapted trivalent influenza virus that is given through the natural route (intranasally) may represent a convenient and effective approach to the prevention of influenza in children.[18,19]

5
Cold-adapted respiratory syncytial virus – pneumovirus

Respiratory syncytial virus (RSV), a member of the paramyxovirus family, is the most common cause of viral pneumonia and bronchiolitis in infants, young children and also in the older age group. It has been estimated that RSV is responsible for about 3 million cases of mild respiratory tract infection, one million cases of lower respiratory tract infectious and 95,000 cases of hospitalization and 4,500 deaths per year in the U.S. In addition, RSV is prevalent worldwide causing possibly 5 million deaths annually and is making RSV infection a major public health concern. These data suggest that it is worthwhile to develop RSV vaccine which will provide sufficient protective immunity against severe disease and death.[19]

The early well documented attempts at the prevention of RSV infection by vaccination used formulin-inactivated alum-precipitated whole virus. The results showed a high rate of seroconversion but no protection; vaccinees developed more serious respiratory disease during subsequent epidemics.[20] Several subunit of vaccine have been evaluated such as a chemeric F-G to initiate immunity with differing protective glycoproteins expressed in baculovirus and shown to induce immunity and partial efficacy in different hosts.[19,21] Overall there are several major features of RSV nonreplicating vaccines which make it contraindicated for human use: (i) These vaccines prove to be poorly immunogenic in seronegative chimpanzees and humans,[2] (ii) antibodies induced by these vaccines have a low level of functional activity *in vivo* and will probably not protect the vaccine, (iii) maternally-acquired serum antibodies can be expected to significantly suppress the immune response to the parenternally-administered vaccine. These features would limit the usefulness of nonreplicating RSV subunit vaccine in infants.

Development of live viral vaccines given intranasally offer advantages over non-replicating vaccines: (i) induction of more effective local mucosal immunity, (ii) greater duration of immunity, and (iii) during infection protective viral antigens are displayed in their mature configuration on the virus and in some instances in the surfaces of infected cells. These features of antigen presentation favor the induction of an effective, balanced response of neutralizing antibodies and cell-mediated immunity. Several types of live attenuated RSV vaccines have been evaluated. The host range vaccine, the cold-passaged vaccine, the temperature sensitive RSV vaccine and the rapid and prolonged passage of the virus in an acceptable substrate have found limited success in initiating the proper level immune response as in reverting to wild type parent.[19,22]

Our experiences in the attenuation of influenza virus by cold-adaptation have prompted us to evaluate the method in the development of live cold-adapted RSV vaccine for man. Recently Whitehead et al.[23] have shown that recombinant RSV vaccine with a defined mutation from cold-passaged RSV was attenuated in chimpanzees.

The development of an acceptable candidate of cold-adapted RSV vaccines was initiated by the isolation and serial passage of five strains of subgroup type A. The five clinical isolates obtained from a local pediatric hospital facility were inoculated into a human host system, diploid human embryonic lungs (MRC5) which offer the advantage of being an acceptable vaccine substrate and from our previous experience will not require a prolonged sequential passage in order to obtain vaccine candidates with the appropriate ca, ts and the attenuation phenotypes in an animal model.[24] The RSV virus was adapted at 25°C by passaging it 60 times in MRC5 diploid cells. After sequential passage the virus was purified by terminal dilution and the isolation of five distinct cold-passaged lines on the basis of early appearance of cytopathic effect at 25°C was performed as demonstrated in Figure 1.

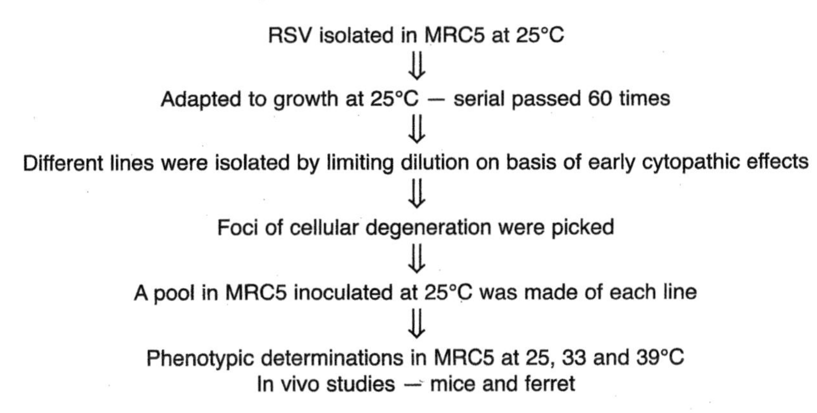

Fig. 1. Flow diagram of the cold-adaptation of respiratory syncytial virus (RSV)

The lines were designated as clones and titrated in MRC5 at 25, 33 and 39°C and compared to the wild type parented virus (Table 3). Genetic stability and immunogenicity were evaluated *in vivo* using the intranasal administration of each clone into three ferrets and measuring the shed virus for retention of the desired phenotypes and for immune response. Attenuation was based on the genetic stability after three consecutive passages from ferret to ferret and upon antibody response and the retention of the ca and ts phenotype. Table 4 illustrates the response in ferrets and the properties of shed virus. The ferrets did not exhibit any reactogenicity, they were used only for evaluating immunogenicity and growth of the virus in turbinates and lung.

It is evident that the five clones studied can be considered as live candidate vaccine for intranasal administration in a phase I clinical trial in adults, infants and children and also in children with natural immunity. At present we are proceeding with the cold-adaptation of subgroup B of RSV, in order to develop a live vaccine against both subgroups for intranasal administration in humans.

6
Development of cold-adapted human paramyxovirus parainfluenza type 1, 2, and 3 in two avian host systems

For the development of cold-adapted parainfluenza viruses to the three serotypes we were able to obtain, through proper channels, nasopharyngeal swabs from individuals with documented clinical illnesses. The three human serotypes were isolated, grown, and propagated in primary avian host-cells. Adaptation to growth at 25°C followed in primary chick kidney cells derived from 5 d-old SPAFAS chicks. The cell system is permissive for viral growth, cytopathic effect and quantitation by the plaque assay system. The plaque assay was used for selection and purification of clones to achieve the genetic stability of the cold-adapted vaccine candidates with the desired *in vitro* and *in vivo* characteristics. The derived clones will then be propagated in 10 d-old SPAFAS embryonated eggs which offer two advantages, (i) a high infectious yield, and (ii) production in large quantities in an economical substrate

Table 3. Characterization of five clones of cold-adapted subgroup A of RSV vaccine candidate

Clone designation	Infectivity titer TCID$_{50/ul}$[a]			Cytopathic effects[b]		
	25°C	33°C	39°C	25°C	33°C	39°C
Ia-CRSV-5	$10^{5.0}$	$10^{6.0}$	$<10^{3.0}$	+	+	−
Ia-CRSV-6	$10^{5.5}$	$10^{6.0}$	$<10^{3.0}$	+	+	−
30d-3 IaCRSV	$10^{6.0}$	$10^{7.5}$	$<10^{3.0}$	+	+	−
BC5-CRSV	$10^{5.0}$	$10^{5.3}$	$<10^{3.0}$	+	+	−
BC13-CRSV	$10^{6.5}$	$10^{6.5}$	$<10^{3.0}$	+	+	−
Wild type parent	no titer at					
	$10^{3.0}$	$10^{6.0}$	$10^{5.3}$	+	+	+

[a] All five lines were ca (cold-adapted) and ts (temperature-sensitive) when compared to the parental type.
[b] + implies 4+ CPE with syncytial formation.

Table 4. Ferrets response to five clones of cold-adapted respiratory syncitial virus[a]

Clone designation	Dose ml⁻¹ intranasal	Immune response[b] Neutralization titer[c]	Phenotype of shed virus	Comment
Ia-CRSV-5	$10^{5.0}$	180	ts, ca	immunogenic genetically stable
Ia-CRSV-6	$10^{5.5}$	90	ts, ca	immunogenic genetically stable
30d-3 Ia-CRSV	$10^{5.5}$	180	ts, ca	immunogenic genetically stable
BC5-CRSV	$10^{5.0}$	90	ts, ca	immunogenic genetically stable
BC13-CRSV	$10^{5.0}$	90	ts, ca	immunogenic genetically stable

ts temperature-sensitive, *ca* cold-adapted
[a] Nasopharyngeal swabs taken daily for 5 d; the phenotypic evaluation was on the virus recovered after 5 d post-infection.
[b] Ferrets bled 2 weeks post-infection and challenged with wild type parent.
[c] Dilution test to neutralize the virus.

for vaccine production and marketing. Figure 2 summarizes the protocol used.

Different approaches were used to develop vaccine candidates of parainfluenza viruses.[19] The procedures involved derivation of temperature sensitive mutants,[25] cold-adaptation,[26–29] and more recently a viable chimeric candidate was developed as possible method for developing live vaccine candidates.[28,30]

Virus isolated and grown in primary chick kidney cells
⇓
Series of serial passages (60) at 25°C
Each passage in tissue culture will be incubated for 10 d
⇓
Evidence of focal cellular degeneration using the plaque assay system at 25°C
⇓
Plaque titration — ten individual clones were picked
⇓
Three times plaque to plaque purification followed
⇓
Selection of the clone of each serotype which grew to a higher titer in a short period
at both 25 and 33°C for propagation in embryonated eggs
as a prototype cold-adapted candidate
⇓
A pool of each clone was made in embryonated eggs at 25°C for screening and
characterization in vitro and in vivo

Fig. 2. Diagramatic illustration for the derivation of cold variants to parainfluenza types 1, 2 and 3

Table 5. Characterization of live cold-adapted human parainfluenza types 1, 2 and 3

| Designation | Titer$_{\text{ID50/ul}}$ at | | | Ferret response[a] | Genetic stability |
	25°C	33°C	39°C	25°C	after three passages
Parainfluenza type 1					
Tissue (TC) line	$10^{6.0}$	$10^{7.3}$	$<10^{3.0}$	ND	Retention of ca, ts
Eggs (E) line	$10^{7.2}$	$10^{8.0}$	$<10^{3.0}$	1024	Retention of ca, ts
Parainfluenza type 2					
TC-line	$10^{5.5}$	$10^{6.5}$	$<10^{3.0}$	ND	Retention of ca, ts
E-line	$10^{6.0}$	$10^{7.0}$	$<10^{3.0}$	2048	Retention of ca, ts
Parainfluenza type 3					
TC-line	$10^{6.0}$	$10^{7.5}$	$<10^{3.0}$	ND	Retention of ca, ts
E-line	$10^{6.5}$	$10^{7.5}$	$<10^{3.0}$	1024	Retention of ca, ts

[a] Hemagglutination-inhibition antibody titer 2 weeks post-infection.

Data of the characterizations of these serotypes of human parainfluenza type 1, 2, and 3 which were isolated, grown and cold-adapted in an acceptable avian system, are presented in Table 5. It is evident that the potential live vaccine candidates were cold-adapted, temperature sensitive, immunogenic (in ferrets) and genetically stable. The results show the virus was recovered from the nasopharyngeal swabs up to 6 d post-infection and the infectivity of the virus recovered from the turbinates of the ferrets at the 4th d post-infection exhibited the ca and ts phenotypes throughout the period of observation. Passage of the three serotypes serially from ferrets to ferrets for three successive passages and the analysis of the recovered virus after the third passage also showed reduction of the attenuated phenotypes (ca, ts). The infected ferrets did not respond to the virus, no clinical signs were recorded. However the parental wild types of the three lines upon intranasal (IN) administration to ferrets were mildly reactogenic with 2 d of rhinitis, coryza and low-grade fever. Phase I clinical trials of this live cold-adapted parainfluenza virus vaccine candidate are being planned.

Acknowledgments. The author acknowledges with appreciation the hard work of Jean Humrich for her patience, typing and editing the manuscript.

7
References

1. Alexandrova GI, Polezhaev FI, Budilovsky GN, Garrueshova LV. Recombinant cold-adapted attenuated influenza A vaccines for use in children. Reactogenicity and antigenic activity of cold-adapted recombinants and analysis of isolates from the vaccine. Infect Immun 1984; 44:734-740.
2. Maassab HF, DeBorde DC. Development and characterization of cold-adapted virus for use as live virus vaccines. Vaccine 1985; 3:355-369.
3. Maassab HF, Herlocher ML, Bryant ML. Live influenza virus vaccine. In: Orenstein WA, Plotkin SM, eds. Vaccines, 3rd ed. Philadelphia, London, Toronto, Montreal, Sydney, Tokyo: WB Saunders Co, Division of Harourt Brace and Co., 1998 (in press).
4. Enders JF, Shaffa MF. Studies on natural immunity to pneumococcus type 14. Capacity of

strains of pneumococcus type 111 to grow at 4ic and their virulence for rabbits. J Exp Med 1936; 7:64-70.

5. Burnet FM. The Use of Developing Egg in Virus Research. London, HM Stationary Office, Great Britain Medical Research Council, Special Report Series No 220, 1936.

6. Li S, Mo D, Bilsel P, Bryant M. Recombinant inactivated and cold-adapted vaccines. Abstract: 10th International Conference on Negative Strand Viruses: Emergence and Re-emergence of negative strand viruses. Sept 21-26 Dublin, Ireland, 1997.

7. Davenport FM, Hennessy AV, Maassab HF, Minuse E, Clark F, Abrams GD, Mitchell JR. Pilot studies on recombinant cold-adapted live type A and type B influenza virus vaccines. J Infect Dis 1977; 136:17-23.

8. Maassab, HF. Biologic and immunologic characteristics of cold-adapted influenza viruses. 1969;102: 728-732.

9. Kilbourne ED. The control of influenza virus. In: Kilbourne ED, ed. Influenza. London: Plenum Medical Book Co, 1987:291-317.

10. Cox NJ, Kitame F, Kendal AP, Maassab HF, Naeve C. Identification of sequence changes in the cold-adapted, live attenuated influenza vaccine strains, A/Ann Arbor/6/60 (H2N2). Virology 1988; 167:554-562.

11. Herlocher ML, Maassab HF, Webster RG. Molecular and biological changes in the cold-adapter "master strain" A/AA/6/60-H2N2 influenza virus. Proc Nat Acad Sci USA 1993; 90:6032-6036.

12. Maassab HF, DeBorde DC, Donabedian AM, Smitka CW. Development of cold-adapted "master strain" for the type B influenza virus vaccines. In: Lerner RA, Chanock RM, Brown F, eds. Vaccines 85. Cold Spring Harbor, NY: Cold Spring Harbor Laboratory, 1985:327-332.

13. DeBorde DC, Donabedian AM, Herlocher ML, Naeve CW, Maassab HF. Sequence comparison of wild-type and cold-adapted B/Ann Arbor/1/66 influenza virus genes. Virology 1988; 163:429-443.

14. Clements ML, Betts RF, Murphy BR. Advantage of live attenuated cold-adapted influenza A virus vaccines in seronegative children. Lancet 1989; 2:705-708.

15. Wright PF, Okabe N, McKee JJ, Maassab HF, Karzon DT. Cold-adapted recombinant influenza A virus vaccines in seronegative children. J Infect Dis 1982; 146:71-79.

16. Wright PF, Ross KB, Thompson J, Karzon DT. Influenza infection in young children: Primary natural infection an protective efficacy of live-vaccine-induced or naturally acquired immunity. N Engl J Med 1977; 296:829-834.

17. Jones PD, Ada GL. Influenza-specific antibody secreting cells and B cell memory in the murine line after imunization with wild-types, cold-adapted variant and inactivated influenza viruses. Vaccine 1987; 5:244-248.

18. Belshe RB, Mendelman PM, Treaner J, King J, Gruber W, Piedia P, Bernstein D, Hayden F, Kotloff K, Zangwell K, Iacuzio D, Wolff M. The efficacy of live attenuated, cold-adapted, trivalent, intranasal influenza virus in children. N Engl J Med 1998; 338:1405-1412.

19. Crowe JJ, Collins PL, Chanock RM, Murphy BR. Vaccines against respiratory syncytial viruses and parainfluenza type 3 virus. In: Levine MM, Woodrow GC, Kaper JB, Cobon GS, eds. New Generation of Vaccines, 2nd ed, revised and expanded. New York, Basel, Hong Kong: Marcel Dekker Inc, 1997:711-725.

20. Connors M, Giesu N, Aulkarni A, Firestone C, Morse H III, Murphy BR. Enhanced pulmonary histopathology induced by respiratory syncytial virus (RSV) challange of formalin-inactivated RSV-immunized BALB/C mice is abrogated by depletion of interleukin (IL-4) and IL-10. J Virol 1994; 68:5321-5325.

21. Collins P, McIntosh K, Chanock RM. Respiratory syncytial virus. In: Fields DN, Knipe DN, Howley PM, Chanock RM, Melnick JL, Nonath TP, Roizman B, Strauss, SE, eds. Fields Virology, 3rd ed, vol 2. Philadelphia New York: Lippincott-Raven, 1996:1313-1352.

22. Herlocher ML, Ewasyshyn M, Sambhara S, Gharaee-Kermani M, Cho D, Lai J, Klein M, Maassab HF. Immunologic properties of plaque purified strains of live-attenuated respiratory syncytial virus (RSV) for human vaccine. Vaccine 1998 (in press).

23. Whithead S, Juhasz K, Firestone Cai-Yen, Collins P, Murphy BR. Recombinant respiratory syncytial virus (RSV) bearing a set mutations from cold-passaged RSV is attenuated in chimpanzees. J Virol 1998; 72:4467-4471.

24. Kim H, Arrobis JO, Brant C, Wright P, Jodes D, Chanock R, Parrott RH. Safety and immunogenicity of temperature-sensitive (ts) mutant of respiratory syncytial virus (RSV) in infants and children. Pediatrics 1973; 52:56-63.

25. Ray R, Galuiski M, Hemingway B, Newman F, Belshe RB. Temperature sensitive phenotype of the human parainfluenza virus type 3 candidate vaccine strains (cp45) correlates with a defect in the L gene. J Virol 1996; 70:580-584.

26. Belshe RB, Hissom FK. Cold adaptation of parainfluenza virus type 3: induction of three phenotypic markers. J Med Virol 1982;10:235-242.

27. Belshe RB, Hisson FK. Cold-adaptation of influenza virus type 3 induction of three phenotypic markers. J Med Virol 1982; 10:235-242.

28. Karron RA, Wright PF, Hall SL, Makhene M, Thompson J, Burns J, Tollefson S, Steinhoff MC, Wilson M, Harris D, Clements ML, Murphy BR. A live attenuated bovine parainfluenza virus type 3 vaccine is safe, infectious, immunogenic and phenotypically stable in infants and children. J Infect Dis 1995; 171:1107-1114.

29. Tao Tao, Durbin A, Whitehead S, Davoodi F, Collins PL, Murphy BR. Recovery of a fully viable chimeric human parainfluenza virus (PIV) type 3 in which the hemagglutinin - neuraminidase and fusion glycoproteins have been replaced by those of PIV type 1. J Virol 1998; 72:2955-2961.

30. Karron R, Wright P, Newman F, Makhene M, Thompson J, Smorodin R, Wilson M, Anderson E, Clements ML, Murphy BR. A live human parainfluenza type 3 virus vaccine is attenuated and immunogenic in health infants and children. J Infect Dis 1995; 172:1445-1450.

Cold-adapted microorganisms for use in food biotechnology

H. Okuyama[1,2*], N. Morita[1] and I. Yumoto[1]

[1] Bioscience and Chemistry Division, Hokkaido National Industrial Research Institute, Toyohira-ku, Sapporo 062-8517, Japan
[2] Laboratory of Environmental Molecular Biology, Graduate School of Environmental Earth Science, Hokkaido University, Kita-ku, Sapporo 060-0810, Japan

1
Introduction

Within the field of food biotechnology, microorganisms have been well utilized as bioreactors and source mechanisms for such useful biosubstances as enzymes, antibiotics, organic and amino acids, vitamins, and so on.[1,2] Most microorganisms thus applied have been mesophilic. Recently, the utilization of thermophilic organisms has increased because of their thermostable characteristics. However, the utilization of cold-adapted microorganisms in food biotechnology has been very limited.[3-5]

In the area of food biotechnology, cold-adapted microorganisms, that is, psychrophilic, psychrotrophic and psychrotolerant microorganisms, have generally been regarded as food-spoilage organisms rather than as potentially useful, an outlook that has grown especially since the introduction of refrigerators for food storage. Thus, research into the mechanisms of control of the growth of cold-adapted microorganisms and their enzyme activities has been very popular. Less attention has been paid to the fact that cold-adapted microorganisms and their enzyme systems can themselves be applied as potential biocatalysts at low temperature. Low-temperature reactions utilizing such biocatalysts have various advantages, e.g., low temperatures in the processing of foods prevent contamination by mesophilic organisms, and cold-adapted enzymes, due to their heat-lability, can be easily inactivated by heating when unneeded after use.[3,4]

In this chapter we focus on several aspects of the utilization of cold-adapted microorganisms, namely, on their use as a source organism of cold-adapted enzymes and useful biosubstances, as a biocatalyst in alcoholic production and food ripening, and as a tool for food processing.

---

* Corresponding author

2
Cold-adapted microorganisms as a potential source of cold-active enzymes

Several kinds of cold-active enzymes have been isolated from cold-adapted microorganisms. These enzymes include lipases,[6,7] proteases,[8-10] β-galactosidases,[11,12] alkaline phosphatases,[13] triosephosphate isomerase,[14] amylase,[15] and so on. Although some of these have potential use as biocatalysts in food biotechnology, virtually no applications have been carried out. For example, lipase produced by a psychrotroph, *Pseudomonas fluorescens* P38, was found to catalyze the synthesis of butyl caprylate in *n*-heptane at low temperatures. Since caprylic acid esters have a fruity flavor and are present as flavor ester components in several fruits, these principles might suggest an application of cold-adapted microorganisms in the enzymatic synthesis of heat-sensitive chemicals.[16]

3
Useful biosubstances produced by cold-adapted microorganisms

Ethanol and organic acids such as lactic acid, citric acid, and malic acid are formed as a primary product of microorganisms by fermentation. Because, at the manufacturing level, the fermentation is carried out using pure cultures of microorganisms at moderate or higher temperatures, manufacturers need pay less attention to the contamination of microbes during fermentation. Thus, fermentation using cold-adapted microorganisms at lower temperatures carries no advantage. Biosubstances produced by cold-adapted microorganisms and their practical applications are very limited. We know of no reports on biosubstances specifically produced by cold-adapted microorganisms. However, several microbes are known to produce some compounds when grown at low temperatures. *Brevibacterium* sp. P145, a psychrophilic bacterium isolated from soil, accumulated a much higher volume of L-glutamic acid and L-alanine at 5°C than at 28°C.[17]

3.1 Antibiotics

The production of antibiotics in marine bacteria has been reported.[18] An antibiotic having selective toxicity against several Gram-positive bacteria was extracellularly produced by facultatively psychrophilic *Streptomyces* sp. No. 81.[19] This strain has been shown to produce the antibiotic only at low temperature cultivations below 20°C. Microorganisms isolated from marine environments are frequently examined as source organisms of antibiotics. Although it was not a novel com-pound, 3-amino-3-deoxy-D-glucose was produced by a *Bacillus* strain isolated from a sediment collected at a depth of 4,310 m.[20] Since the amino sugar structure of 3-amino-3-deoxy-D-glucose is contained also in kanamycine, 3-trehalosamine, and hikizimycin, it has been suggested that the 3-amino-3-deoxy-D-glucose isolated from the bacteria of deepsea sediments is terrigenous in origin. Aplasmomy-cine, a boron-containing polyether ionophore, and istamycin, an aminoglycoside antibiotic, were produced by *Streptomyces griseus* and a new isolate of actinomycete, respectively,[21] although it is unclear whether or not these organisms are cold-adapted microorganisms.

3.2 Polyunsatured fatty acids

The most practical production of useful biosubstances by cold-adapted microorganisms is likely to be that of long chain polyunsaturated fatty acids (PUFAs). PUFAs, such as docosahexaenoic acid {DHA; 22:6(n-3)}, eicosapentaenoic acid {EPA; 20:5(n-3)}, arachidonic acid {AA; 20:4(n-6)}, and γ-linolenic acid {GLA; 18:3(n-6)} have recently attracted attention due to their physiological and pharmacological effects.[22–25] Also, PUFAs are economically invaluable as additives for the food, cosmetic, and health industries.[26] Accordingly, a large demand for PUFA production is anticipated. Currently, GLA is obtained from plant seeds[22,27] and PUFAs with 20 carbon atoms (C20-PUFAs) can be extracted from a number of animals, but are usually derived from marine fish oils,[28] since marine fish oils are rich in C20-PUFAs so that this makes their large-scale preparation more economic. However, several problems in PUFA preparation from fish oils remain to be solved. The undesirable fish-odor cannot be eliminated after the final process of PUFA extraction, which becomes a problem when using PUFAs as food additives. Also, it is very difficult and complicated to purify EPA or DHA from fish oils in large-scale preparation for industrial uses. Finally, we must consider marine pollution when obtaining PUFAs. To solve these problems, in the past decade much research has been conducted on the potentially economic derivation of PUFAs from a variety of microorganisms, such as microalgae, fungi, and bacteria. Among these sources, cold-adapted microorganisms might be the most likely candidates for PUFA production, since it is well-known that the proportion of unsaturated fatty acids in cold-adapted organisms is higher than that in mesophilic ones.[29–32]

It is generally accepted that phytoplanktons (microalgae) are the primary producers of PUFAs. They go up through the food chain step by step, and are eventually incorporated into fish oils and the fats of higher marine animals. Searches for microalgae as an alternative source of PUFAs to replace the commonly used fish oils have already been conducted. Several algae have been found as promising sources, including *Chlorella minutissima*,[33,34] *Porphyridium cruentum*,[35,36] and *Spirulina platensis*.[37] In *P. cruentum*, genetic manipulation techniques were used to increase productivity of EPA.[36] However, few studies have focused on searching for PUFA-producing cold-adapted microalgae.[38,39] It would therefore seem useful to conduct such a search.

Also, although it has been known for some time that some fungal cells contain PUFAs in their membrane lipids, very little recent attention has been paid to the formation of PUFAs by fungal cells, even though, at present, fungal cells are regarded as useful producers of oils that are difficult to obtain from animal or plant cells. The productions of GLA-containing oils from a filamentous fungus belonging to the genus *Mucor*[40,41] and of AA-containing oils from *Mortierella alpina*[42] are now possible at an industrial level.[43] These unique oils produced by such fungi are called "single cell oils". The psychrotrophic filamentous fungus *M. alpina* strain 1S-4 was reported as the first fungus to produce AA in relatively large quantities.[42] Dried cells (by heating) resistant to oxidant for a long time, are superior to fish oil because of its sensitivity to oxidation. Mass production of triacylglycerol containing high AA is also possible now. It is known that fungi belonging to the genus *Mortierella* cannot produce α-linolenic acid in the manner of animals, and, therefore, *Mortierella* cannot be used to synthesize EPA from glucose via the n-3 pathway. Shimizu and co-workers[44] studied the possibilities of producing EPA from AA via the n-6 pathway, as well as from precursors on the n-3 pathway, by adding them into growth medium. They found, first, that *M. alpina* 1S-4 synthesized EPA when grown at lower temperatures (6–12°C), and this phenomenon is generally found among fungi that produce AA as well.[44,45]

The production of EPA is increased by lowering growth temperatures. It should be noted that this low-temperature growth condition induces EPA production in fungi which are believed to synthesize no EPA, and that EPA is synthesized via the n-6 pathway. Induced EPA is esterified to phospholipids, in contrast to AA, which is mainly esterified to triacylglycerol, suggesting that EPA is likely to be synthesized for maintenance of membrane fluidity at cold environmental temperatures. Second, the authors found that *M. alpina* 1S-4 synthesized EPA from exogenously added α-linolenic acid at moderate temperatures (20–30°C).[46] This result indicates that *M. alpina* 1S-4 contains both the n-3 and the n-6 pathways. In this case the production of EPA was also increased by lowering growth temperatures.[47] Each of the cases mentioned above employed a relatively simple method: first, a screening for the fungal strain that accumulates a high volume of the desired PUFA esterified to triacylglycerol, and then the growth of that strain in a medium containing sugars. It is possible that other PUFAs are produced in *M. alpina* 1S-4 by genetic manipulation[48,49] or using inhibitors for desaturases involved in PUFA synthesis.[50,51]

Molecular biological elucidation of the biosynthetic pathway of PUFAs, as well as genetic engineering techniques, make possible the selective production of PUFAs. To date, several fungi capable of producing DHA have been found.[52,53] The psychrophilic snow mold fungus, *Microdochium nivale*, accumulates triacylglycerol (80% of total) when grown at low temperatures. Such a characteristic might be of substantial benefit for producing desired PUFAs through genetic manipulation.

Bacterial cells are thought to contain only saturated and monounsaturated fatty acids as membrane acyl components. At present, however, PUFA-producing bacteria are not considered rare. The PUFA producing-bacteria are isolated from marine animal intestine,[54-61] Antarctica, and various deep-sea locations.[31,32,62-71] The biosynthesis of PUFAs by bacteria appears to be limited to a minority of marine representatives from the genera *Flexibacter*, *Vibrio*, and *Shewanella*, or closely related organisms.

It is generally observed that either EPA or DHA is the sole PUFA in bacterial lipids. This characteristic is of considerable benefit for the easy fractionation of EPA or DHA from fatty acid components via HPLC, versus the more difficult fractionation using fish oil as a source. EPA and DHA are esterified to phospholipids, and their contents vary with growth temperature. Thus, these PUFAs could work to maintain membrane fluidity. Although it is presently unclear how EPA or DHA are synthesized in bacterial cells, the gene cluster involved in EPA biosynthesis has already been cloned from psychrotrophic EPA-producing *Shewanella*.[72] We can also consider the use of psychrophilic bacteria as a genetic resource. This gene cluster is expressed in mesophilic *Escherichia coli*, and the EPA production in *E. coli* has been observed to be dependent on growth temperature as it is in *Shewanella*.[54,72] If this gene is expressed in bacteria which accumulate triacylglycerol,[73] it could be quite practical for production of EPA. This gene cluster is also known to be expressed in mesophilic marine cyanobacteria.[74] Marine cyanobacteria cannot synthesize EPA, but they can be cultured at high densities and can grow faster than eucaryotic microalgae.[75] Psychrophilic cyanobacteria might be suitable as a host for the gene transformant. Large-scale cultivation of EPA-producing cyanobacteria is relatively economic because of their auxotrophic property, and aids in the reduction of CO_2 in the atmosphere.[75] The modification of the gene cluster itself through genetic engineering techniques may bring about the production of desired PUFAs that are thought to be intermediates of EPA biosynthesis. However, we have no information on the biosynthetic pathway of DHA, or on the gene(s) involved in DHA synthesis in bacteria. Therefore, the selective production of DHA in microorganisms including microalgae, fungi, and bacteria remains to be developed.

4

Application of cold-adapted microorganisms in the production of alcoholic beverages

Although industrial-scale alcohol fermentation is generally carried out at low temperatures, the practical application of cold-adapted microorganisms in this area of food biotechnology seems to be rare. One exception is the report of Bakoyianis et al.,[76] which showed that a kissiris-supported biocatalyst prepared by immobilization of the alcohol-resistant and psychrophilic strain of yeast on mineral kissiris was suitable for continuous wine production at low temperatures (5–19°C). *Saccharomyces cerevisiae*, a yeast used in the brewing of beer, is originally neither psychrophilic nor psychrotrophic, although fermentation at relatively low temperatures ranging from 8 to 12°C has been applied. The production of ethanol should be active at higher temperatures in yeasts, but the taste and flavor of the fermentation product at this low temperature range seem to be more desirable.

Several types of microorganisms are involved in the action of the microflora in sake starter mash. Organisms with a nitrate oxidation activity produce nitrites by which the appropriate conditions of microflora can be maintained, and by which the appearance of lactic bacteria can be followed. Konishi et al.[77-79] isolated a psychrophilic *Pseudomonas* spp. from sake starter mash. The addition of this strain to sake starter mash as a nitrite producer stabilized and enlarged the mash.

5

Utilization of cold-adapted microorganisms in the ripening of foods

Although the ripening of dairy products, meats, preserved vegetables, and so on entirely depends on such metabolic activities as the proteolytic and lipolytic activities of microorganisms, few microbiological and biochemical approaches to cold-adapted microorganisms have been carried out. Since in most cases food fermentation progresses largely during the winter months, it is highly likely that cold-adapted microorganisms are the main organisms involved in the fermentation. The best advantage in ripening foods by cold-adapted microorganisms under low temperatures is that it minimizes contaminating mesophilic microbes (and thereby prevents the production of undesirable tastes, colors, and odors) and the remaining activity is stopped by pasteurization.[3,5]

During the ripening of cheeses, various kinds of microorganisms are involved in the physicochemical characteristics of the final product.[80-82] Sablé et al.[83] presented a microbiological study concerning a mold-ripened soft cheese made from raw goat's milk. Psychrotrophic salt-tolerant bacteria were present in significant quantities only in the microflora of the rind during the ripening period. The main bacterial group was *Micrococcus,* the members of which were generally found to exhibit proteolysis and lipolysis. The psychrotrophic *Micrococcus* spp. would be involved in the modification of the texture and the development of taste. The *Micrococci* in the cheese were implicated in the metabolism of methionine, which leads to production of methanethiol and esters of methanethiol, which are precursors of numerous sulfur-containing compounds that participate in the hint of garlic of mold-ripened cheeses. In the rind of Camembert cheese, *Micrococcus* spp. played an important role in flavor development by producing significant amounts of methanethiol and other aromatic compounds.

Lactic acid bacteria are essential for the fermentation of sausages and hams. These bacteria lower the pH of the meat by producing lactic acid, which in turn prevents the growth of mesophilic microbes. Samelis et al.[84] characterized lactic acid bacteria isolated from naturally fermented, dried Greek salami. They isolated a total of 348 lactic acid bacteria. Among these strains, 169 were psychrotrophic and were identified as *Lactobacillus curvatus* (88 strains), *L. sake* (76 strains), and *L. sake/curvatus* (5 strains). *L. sake* and *L. curvatus* were suggested to be the best candidates to be used as starter cultures for dry sausage production, since they were proven to be highly competitive during fermentation. *L. sake* has often been shown to produce bacteriocins. Sakacin B, for example, is produced by *L. sake* 251. Based on these results, bacteriocin production has significant potential for use in meat preservation. Genetic manipulations will potentially enable the creation of a better starter and numerous protective cultures for meats. Since the predominance of commercially available lactic starters often results in poorer flavor and aroma than found in naturally fermented sausages, the use of microbes as lactic starters in the production of naturally fermented foods warrants further investigation.[84]

Haga et al.[85] have applied psychrotrophic lactic acid bacteria to the production of fermented hams. When the meat was cured and fermented with precultured lactic acid bacterium, *Lactobacillus* sp. SK-1001, at 5°C for 3 weeks, microflora in the products turned into lactic acid bacterial flora, and the growth of *Staphylococci* and coliforms was inhibited. The meat interior pH decreased from 5.5 to 4.97, and the desired color, cohesiveness, and consistency of the meat were obtained. This seems to be the only report in which a pure culture of psychrotrophic microorganisms has been applied to the meat fermentation. In the northern areas of Japan, a wide range of fermented vegetable products known as tsukemono are readily available. In the winter season, most of these vegetables are salted with koji, a culture of *Aspergillus oryzae* grown on steamed rice, and which partially metabolizes starch through the action of its amylases. Cuts of fish, such as herring and salmon, are sometimes salted with vegetables. Although *A. oryzae* and lactic acid-producing bacteria, which subsequently increase as fermenters of glucose, might originally be neither psychrotrophs nor psychrophiles, it is likely that a very slow ripening of vegetables (and fish meats) by these 'cold-active' microorganisms at low temperatures produces tsukemono with suitable texture and taste. At the industrial level, fermented vegetable production is performed under temperature-controlled conditions even in summer time. The principal reason for the cold-temperature production of tsukemono is that raw vegetable and fish materials are supplied in autumn, and fermentation at low temperatures minimizes the contaminating microbes. Unfortunately, however, few studies on cold-adapted microorganisms involved in Japanese tsukemono have been carried out.

6
Application of ice nucleation-active microorganisms in food biotechnology

It has been reported that bacteria such as *Pseudomonas syringae*, *Erwinia herbicola*, and *Xanthomonas campestris* are ice nucleation active (INA).[86-88] These bacteria are able to freeze water at a subzero temperature higher than -5°C and often cause frost injury in plants. At present, application of INA bacteria for food processing is limited, because these microorganisms are phytopathogenic. *X. campestris* INXC-1, which was isolated from tea shoots[89] and has been found to be non-phytopathogenic, has been used to freeze water at a subzero temperature higher than -5°C for egg processing.[90] High-pressure treatment at 300 MPa and 5°C for 5 min killed the cells without affecting their ice nucleation activity. This means that cells pressurized at 300 MPa can be applied to food processing procedures without any hygiene problems. Egg white containing the pressurized cells began to freeze with a slight degree of supercooling. Ice crystals with a dendritic structure were formed when the egg white froze in the presence of the killed bacterial cells. The procedure reduced the freezing and thawing time of the egg white. Furthermore, Watanabe and Arai[91] have reported that application of INA bacterial cells showed the advantages of reduced freezing time and energy conservation in the freeze-drying process as well.

Soy sauce has been concentrated by the following procedure:[92] freezing in the presence of ice nucleation-active *X. campestris* (permitted for food use by the Japan Ministry of Health and Welfare) cells at -25°C for 1 day, and the resulting frozen soy sauce being filtered through a 22-mesh screen to remove the ice and eutectic crystals of salt and water ($H_2O \times 2NaCl$). A gas chromatographic analysis of the concentrated soy sauce, performed to characterize its flavor showed that the number of volatile compounds was small and that each volatile compound was concentrated by about 1.6-fold. The free amino acid concentrations in the original soy sauce and in the concentrated product were compared using an amino acid analyzer. All amino acids except for tyrosine were concentrated 1.4- to 1.8-fold times in the product, with tyrosine concentration being 0.8 times the level in the original soy sauce because of its low solubility at -25°C. Thus, by applying bacterial ice nucleation activity to freeze concentration, a soy sauce with umami taste, taste from certain kinds of amino acids, and/or with low saltiness can be obtained. By a similar technique, it is possible to concentrate lemon juice without losing its original flavor.[93] The technique has also been applied successfully to the production of non-heated strawberry jam sample which is almost equal in texture, and which is superior in flavor and color, to conventional jam products.[94]

7
Other new potential applications

Although there are not many examples of industrial applications of cold-adapted microorganisms, these microorganisms have some potential. Cold-adapted microorganisms, for example, are good candidates for use in low-energy waste treatments in the food industry. Hydrogen peroxide is a powerful oxidant that is used in food processing and sterilization of food packages. In waste treatment, hydrogen peroxide should be removed because it is known to damage microorganisms in the activated sludge. If a bacterium that effectively decomposes hydrogen peroxide could be found, it could be utilized to decompose hydrogen peroxide in food industry waste. Although there have been several studies on optimizing the cultural conditions of microorganisms for the production of catalase,[95-98] as well as on the induction of active mutants that overproduce catalase,[99] there have been few studies using cold-adapted microorganisms isolated from environments that contain hydrogen peroxide.

A cold-adapted bacterium exhibiting extraordinarily high catalase activity was isolated from the drain pool of a fish-product processing plant in which hydrogen peroxide was used as a bleaching and microbicidal agent.[100] The bacterium was identified as *Vibrio* sp. according to phenotypic characteristics and 16S rRNA sequence analysis. A culture medium containing living cells and a cell-free extract prepared from 48-h cultured cells of the isolate exhibited catalase activity of 680 units mg^{-1} cells and 7,276 units mg^{-1} protein, respectively. The activity of the cell-free extract of the isolate was 2 orders greater than that of *E. coli*, *B. subtilis*, and *Vibrio parahaemolyticus*. Application of a newly isolated bacterium or its enzyme will provide an effective means of decomposing hydrogen peroxide contained in industrial waste produced in food industries.

Interactions between microorganisms are well-known phenomena. Substrate competition and antagonism play important roles in the selection of a microflora.[101] However, there are few reports on such interactions in food. *Pseudomonas* spp. are important spoilage microorganisms in chilled foods, such as milk,[102] chicken,[103] meat,[104] and fish.[105] This prominence is probably due to environmental distribution of *Pseudomonas* spp. and their rapid growth at cold temperatures. The antibacterial effects of 209 strains of *Pseudomonas* spp. isolated from spoiled iced fish and newly caught fish have been assessed.[106] Sixty-seven strains inhibited the growth of one or several of six target organisms (*E. coli, Shewanella putrefaciens, Aeromonas sobria, P. fluoresens, Listeria monocytogenes,* and *Staphylococcus aureus*). Two-thirds of the inhibitory strains possessed an ability to produce siderophores. Siderophore-mediated competition for iron[107] or due to the antibiotic activity of siderophores[108] may explain the inhibitory activity of these strains. All but nine of the inhibiting strains were found to inhibit the growth of 38 strains of the cold-adapted fish spoilage bacterium, *S. putrefaciens.* Siderophore-containing *Pseudomonas* culture supernatants inhibited the growth of *S. putrefaciens,* as did the addition of iron chelators (ethylenediamine dihydroxyphenylacetic acid). In particular, *Pseudomonas* strains isolated from newly caught and spoiled Nile perch (*Lates niloticus*) inhibited *S. putrefaciens.* This suggested that microbial interaction (e.g., competition or antagonism) may influence the selection of microflora for some chilled food products.

Cold-adapted ammonia-oxidizing bacteria could potentially be applied to eliminate the unpleasant odor caused by ammonia release in refrigerated fish. However, pure isolation of chemoautotrophic ammonia-oxidizing bacteria is quite difficult due to the small size of the colony on the agar plate and to the fact that other microorganisms grow faster than ammonia-oxidizing microorganisms, even when the culture contains no organic contaminants. Among ammonia-oxidizing bacteria, the genus *Nitrosomonas* is most commonly used for biochemical studies, and is believed to be the dominant genus. The optimum growth temperature for the ammonia-oxidizing bacteria is generally between 27 and 30°C, although thermophilic[109] and cold-adapted[110,111] strains have been isolated. However, there are not many examples of isolation of cold-adapted ammonia-oxidizing bacteria because little attention has been directed to cold-adapted chemoautotrophic bacteria. In the future, it is likely that further strains of cold-adapted ammonia-oxidizing bacteria will be isolated.

8
Conclusions

Practical applications of cold-adapted microorganisms in the area of food biotechnology are generally considered to be limited, and, indeed, the research thus far has consisted of only a very few laboratory (and no industrial) trials of wine and beer production using psychrophilic yeasts. In practice, however, cold-adapted microorganisms seem to be useful as biocatalysts and tools in the processing (though perhaps not in the biotechnology) of foods, and in the production of alcoholic beverages and soy sauces. Without exception, manufacturers of these foods told us that low-temperature fermentation and ripening of foods provides better taste, odor, and quality of the products when compared with products fermented at relatively high temperatures. Since isolation and cultivation of cold-adapted microorganisms can be performed easily enough compared to these processes for mesophilic microbes, detailed microbiological approaches to cold-adapted microorganisms would seem to provide great potential for enriching the production and quality of food.

Acknowledgements. We thank Prof. H. Obta (Kansai University), Dr. Y. Honda (Snow Brand Milk Products Co. Ltd.), Dr. T. Yokota (Sapporo Breweries Ltd.), and Dr. A. Yamane (Hyo-on Laboratories) for their helpful discussions on practical usage of cold-adapted microorganisms in food biotechnology. Thanks are also due to Ms. K. Hanada of the Hokkaido National Industrial Research Institute for her assistance in the preparation of the manuscript.

9
References

1. Adams MWW, Perler FB, Kelly RM. Extremozymes: expanding the limits of biocatalysis. Bio/Technology 1995; 13:662-668.
2. Burgess K, Shaw M. The application of enzymes in industry. Ind Enzymol 1983:260-283.
3. Brenchley JE. Psychrophilic microorganisms and their cold-active enzymes. J Ind Microbiol 1996; 17:432-437.
4. Herbert RA. A perspective on the biotechnological potential of extremophiles. Trends Biotechnol 1992; 10:395-402.
5. Gounot AM. Bacterial life at low temperature: physiological aspects and biotechnological implications. J Appl Bacteriol 1991; 71:386-397.
6. Arpigny JL, Feller G, Gerday C. Cloning, sequence and structural features of a lipase from the Antarctic facultative psychrophile *Psychrobacter immobilis* B10. Biochim Biophys Acta 1993; 1171:331-333.
7. Feller G, Thiry M, Arpigny JL, Mergeay M, Gerday C. Lipases from psychrotrophic Antarctic bacteria. FEMS Microbiol Lett 1990; 66:239-244.
8. Schinner F, Margesin R, P,mpel T. Extracellular protease-producing psychrotrophic bacteria from high alpine habitats. Arctic Alpine Res 1992; 24:88-92.
9. Margesin R. Schinner F. Characterization of a metalloprotease from psychrophilic *Xanthomonas maltophilia*. FEMS Microbiol Lett 1991; 79:257-262.
10. Gügi B, Orange N, Hellio F, Burini JF, Guillou C, Leriche F, Guespin-Michel JF. Effect of growth temperature on several exported enzyme activities in the psychrotrophic bacterium *Pseudomonas fluorescens*. J Bacteriol 1991; 173:3814-3820.
11. Trimbur DE, Gutshall KR, Prema P, Brenchley JE. Characterization of a psychrotrophic *Arthrobacter* gene and its cold-active β-galactosidase. Appl Environ Microbiol 1994; 60:4544-4552.
12. Loveland J, Gutshall K, Kasmir J, Prema P, Brenchley JE. Characterization of psychrotrophic microorganisms producing β-galactosidase activities. Appl Environ Microbiol 1994; 60:12-18.
13. Kobori H, Sullivan CW, Shizuya H. Heat-labile alkaline phosphatase from Antarctic bacteria: rapid 5' end-labeling of nucleic acids. Proc Natl Acad Sci USA 1984; 81:6691-6695.
14. Rentier-Delrue F, Mande SC, Moyens S, Mainfroid PTV, Goraj K, Lion M, Hol WGJ, Martial JA. Cloning and overexpression of the triosephosphate isomerase genes from psychrophilic and thermophilic bacteria. J Mol Biol 1993; 229:85-93.
15. Feller G, Lonhienne T, Deroanne C, Libioulle C, Beeumen JV, Gerday C. Purification, characterization, and nucleotide sequence of the thermolabile α-amylase from the Antarctic psychrotroph *Alteromonas haloplanctis* A23. J Biol Chem 1992; 267:5217-5221.
16. Tan S, Apenten RKO, Knapp J. Low temperature organic phase biocatalysis using cold-adapted lipase from psychrotrophic *Pseudomonas* P38. Food Chem 1996; 57:415-418.
17. Ogata K, Kato N, Ohsugi M, Tochikura T. Studies on the low temperature fermentation, Part II. Amino acid formation by facultative psychrophilic bacterium. Agric Biol Chem 1969; 33:711-717.
18. Faulkner DJ. Antibiotics from marine organisms. In: Sammes P, ed. Topics in Antibiotic Chemistry, vol. 2. Chichester: E. Horwood, 1978:13-58.
19. Ogata K, Yoshida N, Ohsugi M, Tani Y. Studies on antibiotics produced by psychrophilic microorganisms, Part I. Production of antibiotics by a psychrophile, *Streptomyces* sp. No. 81.

Agric Biol Chem 1971; 35:79-85.

20. Fusetani N, Ejima D, Matsunaga S, Hashimoto K, Itagaki K, Akagi Y, Taga N, Suzuki K. 3-Amino-3-deoxy-D-glucose: an antibiotic produced by a deep-sea bacterium. Experientia 1987; 43:464-471.

21. Okami Y. Potential use of marine microorganisms for antibiotics and enzyme production. Pure Appl Chem 1982; 54:1951-1962.

22. Wright S, Burton JL. Oral evening-primrose-seed oil improves atopic eczema. Lancet 1982; 2:1120-1122.

23. Dyerberg J. Linolenate-derived polyunsaturated fatty acids and prevention of artherosclerosis. Nutr Rev 1986; 44:125-134.

24. Harris WS. Fish oil, plasma lipids and lipoprotein metabolism in humans: a critical review. J Lipid Res 1989; 30:785-807.

25. Radwan SS. Sources of C20-polyunsaturated fatty acids for biotechnological use. Appl Microbiol Biotechnol 1991; 35:421-430.

26. Uauy-Dagach R, Valenzuela A. Marine oils as a source of omega-3 fatty acids in the diet: how to optimize the health benefits. Proc Natl Acad Sci 1992; 16:199-243.

27. Wolf BR, Kleiman R, England RE. New source of γ-linolenic acid (Boraginaceae, Scrophulariaceae, Onagraceae, Saxifragaceae). J Am Oil Chem Soc 1983; 60:1858-1860.

28. Ackman RG. Marine Biogenic Lipids, Fats and Oils, vols. I and II. Florida: CRC Press Inc., 1989.

29. Chan M, Himes RH, Akagi JM. Fatty acid composition of thermophilic, mesophilic, and psychrophilic *Clostridia*. J Bacteriol 1971; 106:876-881.

30. Nagy G, Kerekes R. Fatty acid composition of mesophilic and psychrophilic *Pseudomonas* species. Zbl Bakt II Abt 1980; 135:533-540.

31. DeLong EF, Yayanos AA. Biochemical function and ecological significance of novel bacterial lipids in deep-sea procaryotes. Appl Environ Microbiol 1986; 51:730-737.

32. Hamamoto T, Takada N, Kudo T, Horikoshi K. Characteristic presence of polyunsaturated fatty acids in marine psychrophilic vibrios. FEMS Microbiol Lett 1995; 129:51-56.

33. Seto A, Wong HL, Hesseltine CW. Culture conditions: effect on eicosapentaenoic acid content of *Chlorella minutissima*. J Am Oil Chem Soc 1984; 61:892-894.

34. Yongmanitchai W, Ward OP. Screening of algae for potential alternative sources of eicosapentaenoic acid. Phytochemistry 1991; 30:2963-2967.

35. Cohen Z. The production potential of eicosapentaenoic acid and arachidonic acid of the red algae *Porphyridium cruentum*. J Am Oil Chem Soc 1990; 67:916-920.

36. Cohen Z, Didi S, Heimer YM. Overproduction of γ-linolenic and eicosapentaenoic acids by algae. Plant Physiol 1992; 98:569-572.

37. Cohen Z, Vonshak A, Richmond A. Fatty acid composition of *Spirulina* strains grown under various environmental conditions. Phytochemistry 1987; 26:2255-2258.

38. Okuyama H, Morita N, Kogame K. Occurrence of octadecapentaenoic acid in lipids of a cold stenothermic alga, prymnesiophyte strain B. J Phycol 1992; 28:465-472.

39. Nagashima H, Matsumoto GI, Ohtani S, Momose H. Temperature acclimation and the fatty acid composition of an Antarctic green alga *Chlorella*. Proc NIPR Symp Polar Biol 1995; 8:194-199.

40. Nakahara T. Production of oil containing γ-linolenic acid. Nippon Nogeikagaku kaishi 1995; 69:708-710 (in Japanese).

41. Hiruta O, Kamisaka Y, Yokochi T, Futamura T, Takebe H, Satoh A, Nakahara T, Suzuki O. γ-Linolenic acid production by a low temperature-resistant mutant of *Mortierella ramanniana*. J Ferment Bioeng 1996; 82:119-123.

42. Shinmen Y, Shimizu S, Akimoto K, Kawashima H, Yamada H. Production of arachidonic acid by *Mortierella* fungi: selection of a potent producer and optimization of culture conditions for large-scale production. Appl Microbiol Biotechnol 1989; 31:11-16.

43. Akimoto K. A use development of single cell oils. Nippon Nogeikagaku kaishi 1995; 69:729-733 (in Japanese).

44. Shimizu S, Shinmen Y, Kawashima H, Akimoto K, Yamada H. Fungal mycelia as a novel source of eicosapentaenoic acid. Biochem Biophys Res Com 1988; 150:335-341.

45. Shimizu S, Kawashima H, Shinmen Y, Akimoto K, Yamada H. Production of eicosapentaenoic acid by *Mortierella* fungi. J Am Oil Chem Soc 1988; 65:1455-1459.

46. Shimizu S, Kawashima H, Akimoto K, Shinmen Y, Yamada H. Microbial conversion of an oil containing α-linolenic acid to an oil containing eicosapentaenoic acid. J Am Oil Chem Soc 1989; 66:342-347.

47. Shimizu S, Kawashima H, Akimoto K, Shinmen Y, Yamada H. Conversion of linseed oil to an eicosapentaenoic acid-containing oil by *Mortierella alpina* at low temperature. Appl Microbiol Biotechnol 1989; 32:1-4.

48. Jareonkitmongkol S, Kawashima H, Shirakawa N, Shimizu S, Yamada H. Production of dihomo γ-linolenic acid by a Δ5-desaturase-defective mutant of *Mortierella alpina* 1S-4. Appl Environ Microbiol 1992; 58:2196-2200.

49. Jareonkitmongkol S, Sakuradani E, Shimizu S. A novel Δ5-desaturase-defective mutant of *Mortierella alpina* 1S-4 and its dihomo-γ-linolenic acid productivity. Appl Environ Microbiol 1993; 59:4300-4304.

50. Shimizu S, Akimoto K, Kawashima H, Shinmen Y, Yamada H. Production of dihomo-γ-linolenic acid by *Mortierella alpina* 1S-4. J Am Oil Chem Soc 1989; 66:237-241.

51. Shimizu S, Akimoto K, Shinmen Y, Kawashima H, Sugano M, Yamada H. Sesamin is a potent and specific inhibitor of Δ5-desaturase in polyunsaturated fatty acid biosynthesis. Lipids 1991; 26:512-516.

52. Li ZY, Ward OP. Production of docosahexaenoic acid by *Thraustochytrium roseum*. J Ind Microbiol 1994; 13:238-241.

53. Nakahara T, Yokochi T, Higashihara T, Tanaka S, Yaguchi T, Honda D. Production of docosapentaenoic and docosahexaenoic acids by *Schizochytrium* sp. isolated from Yap islands. J Am Oil Chem Soc 1996; 73:1421-1426.

54. Yazawa K, Araki K, Okazaki N, Watanabe K, Ishikawa C, Inoue A, Numao N, Kondo K. Production of eicosapentaenoic acid by marine bacteria. J Biochem 1988; 103:5-7.

55. Ringo E, Sinclair PD, Birkbeck H, Barbour A. Production of eicosapentaenoic acid (20:5 n-3) by *Vibrio pelagius* isolated from turbot (*Scophthalmus maximus* (L.)) larvae. Appl Environ Microbiol 1992; 58:3777-3778.

56. Yano Y, Nakayama A, Saito H, Ishihara K. Production of docosahexaenoic acid by marine bacteria isolated from deep sea fish. Lipids 1994; 29:527-528.

57. Iwatani H, Yamaguchi T, Takeuchi M. Fatty acid metabolism in bacteria that produce eicosapentaenoic acid isolated from sea urchin *Strongylocentrotus nudus*. Nippon Suisan Gakkaishi 1995; 61:205-210 (in Japanese).

58. Jostensen JP, Landfald B. Influence of growth conditions on fatty acid composition of a polyunsaturated-fatty-acid-producing *Vibrio* species. Arch Microbiol 1996; 165:306-310.

59. Jostensen J-P, Landfald B. High prevalence of polyunsaturated-fatty-acid-producing bacteria in arctic invertebrates. FEMS Microbiol Lett 1997; 151:95-101.

60. Yano Y, Nakayama A, Yoshida K. Distribution of polyunsaturated fatty acids in bacteria present in intestines of deep-sea fish and shallow-sea poikilothermic animals. Appl Environ Microbiol 1997; 63:2572-2577.

61. Watanabe K, Ishikawa C, Ohtsuka, I, Kamata M, Tomita M, Yazawa K, Muramatsu H. Lipid and fatty acid compositions of a novel docosahexaenoic acid-producing marine bacterium. Lipids 1997; 32:975-978.

62. Oliver JD, Colwell RR. Extractable lipids of gram-negative marine bacteria: fatty-acid composition. Int J Syst Bacteriol 1973; 23:442-458.

63. John RB, Perry GJ. Lipids of the marine bacterium *Flexibacter polymorphus*. Arch Microbiol 1977; 114:267-271.

64. Wirsen CO, Jannasch HW, Wakeham SG, Canuel EA. Membrane lipids of a psychrophilic and barophilic deep-sea bacterium. Curr Microbiol 1987; 14:319-322.

65. Nichols DS, Nichols PD, McMeekin TA. Anaerobic production of polyunsaturated fatty

acids by *Shewanella putrefaciens* strain ACAM 342. FEMS Microbiol Lett 1992; 98:117-122.

66. Ringo E, Jostensen JP, Olsen RE. Production of eicosapentaenoic acid by freshwater *Vibrio*. Lipids 1992; 27:564-566.

67. Nichols DS, Nichols PD, McMeekin TA. Polyunsaturated fatty acids from Antarctic bacteria. Antarct Sci 1993; 5:149-160.

68. Henderson RJ, Millar RM, Sargent JR, Jostensen JP. *Trans*-monoenoic and polyunsaturated fatty acids in phospholipids of a *Vibrio* species of bacterium in relation to growth temperature. Lipids 1993; 28:389-396.

69. Hamamoto T, Tanaka N, Kudo T, Horikoshi K. Effect of temperature and growth phase on fatty acid composition of the psychrophilic *Vibrio* sp. strain no. 5710. FEMS Microbiol Lett 1994; 119:77-82.

70. Bowman JP, McCammon SA, Nichols DS, Skerratt JH, Rea SM, Nichols PD, McMeekin TA. *Shewanella gelidimarina* sp. nov. and *Shewanella frigidimarina* sp. nov., novel Antarctic species with the ability to produce eicosapentaenoic acid (20:5ω3) and grow anaerobically by dissimilatory Fe(III) reduction. Int J Syst Bacteriol 1997; 47:1040-1047.

71. Nichols DS, Brown JL, Nichols PD, McMeekin TA. Production of eicosapentaenoic and arachidonic acids by an Antarctic bacterium: response to growth temperature. FEMS Microbiol Lett 1997; 152:349-354.

72. Yazawa K. Production of eicosapentaenoic acid from marine bacterium. Lipids 1996; 31 Suppl: S297-S300.

73. Alvarez HM, Mayer F, Fabritius D, Steinb,chel A. Formation of intracytoplasmic lipid inclusions by *Rhodococcus opacus* strain PD630. Arch Microbiol 1996; 165: 377-386.

74. Takeyama H, Takeda D, Yazawa K, Yamada A, Matsunaga T. Expression of the eicosapentaenoic acid synthesis gene cluster from *Shewanella* sp. in a transgenic marine cyanobacterium, *Synechococcus* sp. Microbiology 1997; 143: 2725-2731.

75. Takano H, Takeyama H, Nakamura N, Soda K, Burgess JG, Manabe E, Mirano M, Matsunaga T. CO_2 removal by high density culture of a marine cyanobacterium *Synechococcus* sp. using an improved photobioreacter employing light-diffusing optical fibers. Appl Biochem Biotechnol 1992; 34/35: 449-458.

76. Bakoyianis V, Kanellaki M, Kaliafas A, Koutinas AA. Low-temperature wine making by immobilized cells on mineral kissiris. J Agric Food Chem 1992; 40:1293-1296.

77. Konishi Y, Tochikura K, Ogata K. Nitrite forming bacteria in *Yamahaimoto* (I). Isolation of specific bacteria. J Ferment Technol 1967; 45:795-802 (in Japanese).

78. Konishi Y, Tochikura K, Ogata K. Nitrite forming bacteria in *Yamahaimoto* (II). Selective medium for nitrite forming bacteria in *Yamahaimoto*. J Ferment Technol 1967; 45:803-808 (in Japanese).

79. Konishi Y, Tochikura K, Ogata K. Nitrite forming bacteria in *Yamahaimoto* (III). The change of microflora by increment of sugar concentration. J Ferment Technol 1967; 45:809-814 (in Japanese).

80. Molimard P, Spinnler HE. Compounds involved in the flavor of surface mold-ripened cheeses: origins and properties. J Dairy Sci 1996; 79:169-184.

81. Litopoulou Tzanetaki E, Tzanetakis N. Microbiological study of white-brined cheese made from raw goat milk. Food Microbiol 1992; 9:13-19.

82. Roostita R, Fleet GH. The occurrence and growth of yeasts in Camembert and blue-veined cheeses. Int J Food Microbiol 1996; 28:393-404.

83. Sablé S, Portrait V, Gautier V, Letellier F, Cottenceau G. Microbiological changes in a soft raw goat's milk cheese during ripening. Enzyme Microb Technol 1997; 21:212-220.

84. Samelis J, Maurogenakis F, Metaxopoulos J. Characterization of lactic acid bacteria isolated from naturally fermented Greek dry salami. Int J Food Microbiol 1994; 23:179-196.

85. Haga S, Kato T, Kotuka K. Effects of curing and fermenting time on the quality of hams fermented with psychrotrophic lactic acid bacteria. Nippon Shokuhin Kogyo Gakkaishi 1994; 41:797-802 (in Japanese).

86. Maki LR, Galyan EL, Chang-Chien MM, Caldwell DR. Ice nucleation induced by

Pseudomonas syringae. Appl Microbiol 1974; 28:456-460.

87. Lindow SE, Arny DC, Upper CD. Distribution of ice nucleation-active bacteria on plants in nature. Appl Environ Microbiol 1978; 36:831-838.

88. Zhao JL, Orser CS. Conserved repetition in the ice nucleation gene inaX from *Xanthomonas campestris* pv. *translucens*. Mol Gen Genet 1990; 223:163-166.

89. Watanabe M, Watanabe J, Makino T, Honma K, Kumeno K, Arai S. Isolation and cultivation of a novel ice nucleation-active strain of *Xanthomonas campestris*. Biosci Biotech Biochem 1993; 57:994-995.

90. Honma K, Makino T, Kumeno K, Watanabe M. High-pressure sterilization of ice nucleation-active *Xanthomonas campestris* and its application to egg process. Biosci Biotech Biochem 1993; 57:1091-1094.

91. Watanabe M, Arai S. Freezing of water in the presence of the ice nucleation active bacterium, *Erwinia ananas*, and its application for efficient freeze-drying of foods. Agric Biol Chem 1987; 51:557-563.

92. Watanabe M, Tesaki S, Arai S. Production of low-salt soy sauce with enriched flavor by freeze concentration using bacterial ice nucleation activity. Biosci Biotech Biochem 1996; 60:1519-1521.

93. Watanabe M, Watanabe J, Kumeno K, Nakahama N, Arai, S. Freeze concentration of some foodstuffs using ice nucleation-active bacterial cells entrapped in calcium alginate gel. Agric Biol Chem 1989; 53:2731-2735.

94. Watanabe M, Arai E, Kumeno K, Honma K. A new method for producing a non-heated jam sample: the use of freeze concentration and high-pressure sterilization. Agric Biol Chem 1991; 55:2175-2176.

95. Caridis KA, Christakopoulos P, Macris BJ. Control of catalase production and purity by altering nutritional factors of *Alternaria alternata* growth medium. Biotechnol Lett 1991; 13:35-38.

96. Caridis K-A, Christopoulos P, Macris B. Simultaneous production of glucose oxidase and catalase by *Alternaria alternata*. J Appl Microbiol Biotechnol 1991; 34:794-797.

97. Nishikawa Y, Kawata Y, Nagai J. Effect of Triton X-100 on catalase production by *Aspergillus terreus* IFO6123. J Ferment Bioeng 1993; 76:235-236.

98. Petruccioli M, Fenice M, Piccioni P, Federici F. Effect of stirrer speed and buffering agents on the production of glucose oxidase and catalase by *Penicillium variable* (P16) in bench top bioreactor. Enzyme Microb Technol 1995; 17:336-339.

99. Fiedureck J, Gromada A. Selection of biochemical mutants of *Aspergillus niger* with enhanced catalase production. Appl Microbiol Biotechnol 1997; 47:313-316.

100. Yumoto I, Yamazaki K, Kawasaki K, Ichise N, Morita N, Hoshino T, Okuyama H. Isolation of *Vibrio* sp. S-1 exhibiting extraordinarily high catalase activity. J Ferment Bioeng 1998; 85:113-116.

101. Fredrickson, AG, Stephanopoulos G. Microbial competition. Science 1981; 213:972-979.

102. Reddy MC, Bills DD, Lindsay RC. Ester production by *Pseudomonas fragi*. II Factors influencing ester levels in milk cultures. Appl Microbiol 1969; 17:779-782.

103. Pittard BT, Freeman LR, Later DW, Lee, ML. Identification of volatile organic compounds produced by fluorescent pseudomonads on chicken breast muscle. Appl Environ Microbiol 1982; 43:1504-1506.

104. Edwards RA, Dainty RH, Hibbard CM. Volatile compounds produced by meat pseudomonads and related reference strains during growth on beef stored in air and chill temperatures. J Appl Bacteriol 1987; 62:403-412.

105. Miller A III, Scanlan RA, Lee JS, Libbey LM. Volatile compounds produced in sterile fish muscle (*Sebastes melanops*) by *Pseudomonas putrefaciens*, *Pseudomonas fluorescens*, and *Achromobacter* species. Appl Microbiol 1973; 26:18-21.

106. Gram L. Inhibitory effect against pathogenic and spoilage bacteria of *Pseudo monas* strains isolated from spoiled and fresh fish. Appl Environ Microbiol 1993; 59:2197-2203.

107. Henry MB, Lynch JM, Femor TR. Role of siderophores in the biocontrol of *Pseudomonas tolaassi* by fluorescent pseudomonad antagonist. J Appl Bacteriol 1991; 70:104-106.
108. Neiland JB. Microbial iron compounds. Ann Rev Biochem 1981; 50:715-731.
109. Golovacheva RS. Thermophilic nitrifying bacteria from hot springs. Microbiology 1976; 45:377-379.
110. Jones RD, Morita RY, Koops HP, Watson SW A new marine ammonia oxidizing bacterium *Nitrosomonas cryotolerans.* Can J Microbiol 1988; 34:122-128.
111. Tokuyama T, Yoshida N, Matsuishi T, Takahashi N, Takahashi R, Kanehira T, Shinohara M. A new psychrotrophic ammonia-oxidizing bacterium, *Nitrosovibrio* sp. TYM9. J Ferment Bioeng 1997; 83:377-380.

Low temperature fermentation of wine and beer by cold-adapted and immobilized yeast cells

M. Kanellaki* and A. A. Koutinas

University of Patras, Department of Chemistry, Section of Analytical, Environmental and Applied Chemistry, GR-26500 Patras, Greece

1
Introduction

Over the last 20 years considerable research and development have been made in wine making and brewing technology with the aim of improving the productivity and quality, taste and aroma of wine and beer. Use of selected cultures and genetically modified yeasts with desired traits, use of immobilized cells on several supports, enzymatic treatments, addition of adjuncts to malt, modern development in fermentor design and low-temperature fermentation are some technological innovations in alcoholic fermentation. Among these technological innovations, lowtemperature fermentation by cold-adapted and immobilized yeast cells is reviewed here.

The genus *Saccharomyces* is the most important yeast to mankind. It is used as wine yeast, brewer's yeast, distiller's yeast and baker's yeast. Yeast is an important contributor to flavor development in fermented beverages. The major flavor compounds produced by yeast during fermentation are ethanol, higher alcohols, aldehydes, esters, sulphur compounds, vicinal diketones and others. Some of these compounds are highly desirable even in very small quantities while others are not. So, strains possessing valuable properties, such as low-temperature and ethanol tolerance, may give high fermentation rate, produce desirable aroma and taste, and may give hygienic composition of beverages, high cell viability and easy separation from alcoholic beverage at the end of fermentation. In addition, cell immobilization has advantages over a free cell system, such as (i) higher cell density inside a bioreactor, thereby offering a higher fermentation rate, (ii) immobilization of cells by entrapment within polymers or adsorption to solid carriers which protects the cells and so maintains their viability and activity more, with or without a limited increase in cell population, (iii) after the end of batch fermentation the alcoholic beverage is removed and the fermentor is refilled with substrate for a new batch. In the case of continuous fermentation the process can be monitored and optimized.

---

* Corresponding author

The capital cost is minimized. A basic problem in the cell immobilization is the nature of support, which must be hygienic for food production. Also the support must be stable, cheap, easily prepared or abundant in nature for probable industrial application.

1.1
Influence of temperature on alcoholic fermentation

The process of the alcoholic fermentation depends directly on the temperature. Temperature is a factor over which the winemaker and brewer have great control.[1,2] The upper, the lower and the optimum fermentation temperature differ depending on the different yeast strains as regards to the fermentation rate, the yeast growth, the yeast metabolism, the alcohol-tolerance of the yeast, fermentation by-products, etc.

When the fermentation is going to be carried out at the lower temperature limit, there is risk of yeast death. Also at low temperature, the fermentation start, process, cell growth and metabolism of yeast may be delayed, but the evaporation of ethanol and volatile by-products is minimized. It has been reported by Nagodawithana et al.[3] that when the fermentation temperature was increased between 15 and 30°C, the fermentation rate is also increased, but the cell viability decreased. The optimum fermentation temperature is higher than optimum temperature growth of yeast and the optimum process temperature must lie between the two above temperatures. The optimum fermentation temperature and the optimum temperature growth of yeast are decreased at high ethanol concentrations because the alcohol tolerance of the yeast is decreased at high temperatures.[3,4] Thermotolerant strains of *Saccharomyces* were isolated from samples collected at the end of the sugarcane harvest season in distilleries.[5] Their main characteristic was the maintenance of high cell viability, so they fermented sugar juice at 38–40°C in less than 10 h and without continuous aeration of the culture. Thermotolerant yeasts have been defined as those capable of growth at temperature ≥45°C and/or fermentation at 40°C. In general, *Kluyveromyces* strains were more thermotolerant than *Saccharomyces* and *Candida*, but *Saccharomyces* strains produced higher ethanol yields.[6] Research on the influence of temperature, dilution rate and sugar concentration on the establishment of steady-state in continuous ethanol fermentation of molasses was carried out in laboratory-scale vessels at 27, 32 and 37°C by Perego et al.[7] At 27°C the system attained a steady state. Steady states were never reached at 37°C, but at 32°C the system response depended on the values of the dilution rate and sugar concentration.

The effect of temperature on must fermentation in the laboratory and at an industrial scale (2.5×10^4 l) was studied by Caro et al.[8] Factors such as specific microbial (yeast) growth rate, conversion of substrate to other products except ethanol, ethanol evaporation and lower alcohol yield were investigated. Table 1 shows the ethanol amount evaporated during fermentation, the final ethanol concentration, the specific microbial growth rate, and the percentage fraction of substrate converted to other products at different fermentation temperatures for batch laboratory fermentors. The data in Table 1 show that there is important loss of ethanol

Table 1. Effect of temperature on some fermentation parameters (reprinted with permission from ref. 8)

Tempe-rature (°C)	Fermentation time (h)	Specific microbial growth rate (h^{-1})	Fraction of substrate converted to other products (%)	Ethanol amount evaporated (g l^{-1})	Final ethanol concentration (g l^{-1})
15	576	0.024±0.004	4.8	0.8	102
20	298	0.090±0.008	2.8	1.1	104
25	168	0.182±0.016	2.9	1.4	104
30	107	0.390±0.030	5.6	1.8	100
35	85	0.708±0.084	21.5	1.3	85

under usual working conditions (30°C) and this loss is decreased as the temperature is lowered. Besides ethanol evaporation, other factors, such as microbial growth and conversion of substrate to other products, lower ethanol yield. As the values of these factors are increased, the final ethanol concentration in the fermentor decreases. It is important to note that the specific microbial growth rate is dependent on the microorganism used.

Growth rate of yeast cells is particularly affected by temperature during the exponential growth phase. For example, cell division may occur about every 12 h at 10°C, every 5 h at 20°C and every 3 h at 30°C. The rate of cell division is also affected by °Brix value of juice and pH.[9] It has previously been reported that methanol is not formed by the action of yeast during the production of wine.[10] It is formed from pectin by a pectinolytic enzyme which splits the methoxyl group from the pectin present in crushed grapes. The concentration of methanol in wine is dependent on the grape varieties and processing techniques.[11,12]

Glycerol is the wine constituent that is derived from yeast fermentation. It is nonvolatile and does not contribute to wine aroma but contributes to the viscosity and smoothness of the wine. Gardner et al.[13] reported that many factors influence glycerol production by *Saccharomyces cerevisiae*, such as aeration, temperature and sulfide content. Combined effects of initial sulfite concentration (100–300 ppm), temperature (15–25°C), and agitation time (0–24 h) on production of glycerol in grape juice by *Saccharomyces cerevisiae* were studied. Under static conditions, the highest level of glycerol production was observed at 20°C, while incubation at 25°C gave the best results when the cultures were agitated for 24 h. In earlier studies by Ough et al.[14] and Rankine and Bridson[15] an increase in the temperature of fermentation increases the glycerol yield. The effect of must fermentation temperature on minor products formed by cryotolerant and noncryotolerant *Saccharomyces cerevisiae* strains has been studied by Castellari et al.[16] Other researchers reported that cryotolerant yeast strains produced higher amounts of glycerol, succinic acid and malic acid and lower amounts of acetic acid and ethanol than noncryotolerant yeast strains and they are considered interesting from an enological point of view.[17] Higher alcohols formed during alcoholic fermentation play an important role in the quality of wines. Low higher alcohol content in wines corresponds to an increase in the quality of the wines. The formation of higher alcohols during grape juice fermentation at various temperatures was studied by Ough et al.[18] and it was reported that the levels of fusel oil alcohols in wines were partially controlled by limiting the temperature of fermentation. Comparison of wines fermented at 10°C and at 21–27°C showed an increase of 6–131% (on average 47%) at the higher temperatures. The yeast strain used for fermentation is of great significance in determining the level of higher alcohol in beer. The formation of higher alcohols is highly dependent upon the fermentation temperature, with an increase in temperature resulting in increased concentrations in beer.[19] In wines, an increase of higher alcohol content was reported[20] when the fermentation temperature was higher than 20°C. Cryotolerant strains are reported to produce greater amounts of higher alcohols than noncryotolerant control strains, but the high amounts of these compounds are under the perception threshold and may not influence the wine quality.[21]

Many winemakers prefer low temperatures for wine making. Wines produced at low temperatures develop taste and aroma. The improved quality of wines produced by low temperature fermentations can be attributed to the reduction of higher alcohols and to the increase of the proportion of ethyl acetate in the total volatiles.[22,23] At low temperatures, fruit esters such as isobutyl, isoamyl and hexyl acetates are synthesized and retained to a greater degree.[24] Also, some volatile compounds (alcohols, esters and fatty acids) in wines from red, rose and carbonic maceration of Monastrell grapes were measured during fermentation.[25] The lower fermentation temperature and shorter skin contact time of rose vinification produced higher concentrations of esters and acids than in red wine.

Low temperature helps the sedimentation of some colloid compounds and so facilitates wine clarification.

2

Isolation, purification, identification and adaptation treatment of yeast strains

A great variety of microorganisms are found in grapes. Wine yeast *Saccharomyces cerevisiae* does not live in nature at all, but can be found only in the winery environment such as vineyard soil or on the surface of ripe grapes where they form small colonies.[26] As the grapes mature, the population of yeasts increases. In case of rain, several microbes are grown on the grape surfaces. In cool regions, the number of yeast strains is smaller than in warm regions.

Results on the diversity of strains of *Saccharomyces cerevisiae* isolated at different stages of fermentation of must formed from grapes collected from a single Malvasia grape vine have been reported by Polsinelli et al.[27] The spontaneous fermentation of a Malvasia must was dominated by different strains of *Saccharomyces cerevisiae* at different stages of fermentation. Genetic diversity and geographical distribution of wild *Saccharomyces cerevisiae* strains from the wine-producing area of Charentes, France, were reported by Versavaund et al.[28]

In the maturation phase of grapes, the yeast population consists mainly of strains of *Saccharomyces cerevisiae*. Also, various species of other yeasts like *Kloeckera, Torulaspora, Kluyveromyces, Zygosaccharomyces* can be found.[29] In diseased or damaged grapes, acetic acid bacteria and filamentous fungi, such as *Aspergillus* and

Botrytis, are found on the grapes. Must composition favors the viability and fermentative role of *Saccharomyces cerevisiae* that generally predominates during must fermentation. During the first days of spontaneous wine fermentation, *S. cerevisiae* species and other yeast species grow in the must simultaneously.[30] *Saccharomyces cerevisiae* dominates its environment and keeps down the growth of the other types of microorganisms, which pass into a declining phase and are dead because they are not alcohol-tolerant.[31,32] Also, the effects of temperature and pH on the growth of yeast species during the fermentation of grape juice have been studied.[33]

During their growth phase the yeasts produce several compounds which may contribute to wine aroma, flavor and quality. For commercial reasons, the aroma, flavor and quality of alcoholic beverages must be retained as stable. Therefore, pure or mixed cultures of selected yeasts are used as starters for alcoholic fermentation for the production of alcoholic beverages with stable flavor and with particular but balanced features.

2.1
Isolation

A way of isolating the yeast strain is the fermentation method described by Koutinas and Pefanis.[34] The grape variety, climatic conditions and the altimeter reading of the vineyard from which the grapes are going to be obtained for the isolation of yeast strains must be taken into account. For alcohol-tolerant strains, maturated grapes must be collected, crushed and the must density must be at least 13.5°Be at 17–18°C. The alcohol concentration (% v/v) is numerically a little more than the must density Baumé (°Be); we take 1 ml of the fermenting must and add it to 5 ml of synthetic medium containing (figures in g l^{-1}): $(NH_4)_2SO_4$ 1, KH_2PO_4 1, $MgSO_4$ 5, yeast extract 0.4, and glucose 20. The synthetic medium has been previously sterilized at 130°C for 15 min and has been cooled at room temperature. The yeast strain and/or strains are incubated at an appropriate temperature, and after 40–48 h the culture is examined under the microscope and inoculated by a ring onto an agar plate. The yeast and/or yeasts form colonies in 2–3 d at about 25–30°C. Some yeasts are psychrophilic and require lower incubation temperature. If we wish to isolate alcohol-tolerant yeast strains, we repeat the liquid cultivation of yeast strains every day in alcohol concentrations higher than 12% (v/v).

2.2
Purification

By the above process, we produce a mixed culture of alcohol-resistant yeast strains, not a pure culture. There are five methods for the isolation of single yeast strains:[34]

1. Streaked plate method: with the ring, we take the mixed culture of yeast strains and draw successive parallel lines so that the mixed culture is diluted continuously. In the final two lines or one line, different colonies are formed, each consisting of one different strain.
2. Poured plate method: we dilute successively the liquid mixed culture (dilution

steps method), reataining a sample from each dilution. After that we transfer a little portion of each dilution with a sterilized pipette onto respective agar plates containing semisolid agar with nutrients. We incubate the culture to form colonies. We proceed in this way until only one kind of cells forms colonies. If the colonies are well separated, we can inoculate from each kind of colonies into liquid nutrient medium.

3. Method of successive dilutions: a liquid mixed culture is diluted successively and is incubated until only one cell is in the final nutrient liquid.
4. Enrichment method: we enrich the nutrient medium with compounds suitable for the growth only of certain microorganisms and for the prevention of others.
5. Method of environmental conditions: changing several factors such as pH, temperature, osmosis, ethanol tolerance etc. may help to isolate a certain yeast strain. Alcohol-resistant *Saccharomyces cerevisiae* strains were isolated from grapes from Greek agricultural areas.[35,36]

2.3
Characterization and identification

The species of the isolated yeast strain can be characterized[34] by the morphology of the cell and its colony, the characteristics of cell growth, certain physiological characteristics, ethanol tolerance, high fermenting ability in solution with high sugar concentration (high osmotic pressure), its behavior with temperature, production of by-products and fermentation rate.

Concerning the cell morphology, we examine the cell shape and size and the colony shape by microscope. As regards the physiological characteristics, we examine the ability of the strain to ferment sugars. *Saccharomyces carlsbergenesis* secretes melibiase and so it ferments melibiose, but not the related *Saccharomyces cerevisiae*. Also, we examine the ability of yeast to degrade proteins. Some yeasts secrete proteases in appreciable amounts but *Saccharomyces* species have only a limited proteolytic activity. Osmotolerant yeasts have been isolated from Tokaj wines by Miklos et al.[37] Another criterion for strain identification is the production level of higher alcohols, as studied by Polsinelli et al.[27] Growth temperatures and electrophoretic karyotype were used as tools for practical discrimination of *Saccharomyces bayanus* and *Saccharomyces cerevisiae*. Growth temperatures of 1 and 2°C accompanied by 35°C clearly distinguished 16 strains of *Saccharomyces bayanus* and nine strains of *Saccharomyces cerevisiae*. Fermentation characteristics such as fermentation rate at a low temperature (7°C) and ethanol yield for fermentation at an intermediate temperature (28°C) supported this distinction.[38] Castellari et al.[39] isolated cryotolerant *Saccharomyces cerevisiae* strains capable of growing well at 6°C from grape musts using the enrichment method. These strains were capable of vigor fermentation (12–30°C), alcohol tolerant and produced H_2S or not. Also, some yeast strains can assimilate inorganic salts because they have the suitable enzymatic system, others cannot. The test of yeast strain for alcohol tolerance is made by inoculation of the yeast strain in nutrient medium containing alcohol in several known concentrations. Specifically, we can streak a culture of yeast strain onto some agar plates containing sterile solidified growth medium and alcohol in percentages of 12, 14 and 18 (% v/v). After some days of incubation, we examine whether colonies have been formed and so we decide the alcohol tolerance of yeast strain. In addition, the fermenting ability of the yeast strain at high sugar solutions (12–20°Be) is tested. The promotion of the fermentation in high sugar concentrations concludes that the used strain is an alcohol-resistant one. The selected culture influences the composition of beverage and its organoleptic characteristics. Another factor which is tested is the temperature tolerance of the yeast strain. Thermotolerant strains are desirable for rapid fermentations in the distilleries, particularly in summer when it is difficult to cool the broth below 33°C. Also, yeast strains able to ferment at low temperature are desirable. The retention of viability during storage in the refrigerator (4°C) or in the freezing state (-7°C) must be studied. The isolated yeasts should be examined from time to time to ensure that they are the correct organisms.

Identification of the alcohol-resistant yeast strains AXAZ-1 and Visanto isolated from grapes,[36] was made by using an API 20C AUX yeast identification system. To verify the identification, protein profiles by SDS-PAGE according to Hantula et al.[40] and measurements of alcohol dehydrogenase (ADH) activity according to Nikolova and Ward[41] were used. Fermentation kinetics, mainly at low temperatures, was used to differentiate the strains. ADH activity was measured by a spectrophotometric method according to Giolfi.[42]

2.4
Adaptation

Adaptation treatment concerns yeast growth at low temperatures before low-temperature wine making. This procedure consists of gradually establishing feeding and temperature conditions during yeast production, similar to those that the strain encounters during wine fermentations at low temperatures. In this case, yeasts were grown in a culture media that contained gradually increasing amounts of kerino must, a sugar component, rather than glucose. In addition, the culture temperature was kept relatively low, at not more than 20°C.[36]

3
Methods of yeast preservation

Some preservation methods are applied to yeasts: subculturing, suspension in distilled water, freeze-drying and cryopreservation in liquid nitrogen.[43]

3.1
Subculturing

Subculturing preservation is a simple, safe method and suitable for a less well-equipped laboratory. We can streak a culture of yeast onto two petri dishes containing sterile solidified growth medium. After the colonies have grown, usually at 25°C, they are stored in a refrigerator at 4°C. Also, the yeast culture may be kept on slopes of solidified growth medium in a refrigerator. Screw-capped bottles give a good surface when sloped. Periodically, at about 6-monthly intervals, we must inoculate and incubate two containers of fresh culture. One culture is used as a source of yeast strain for research or other work and the other one is stored at about 4°C for the two inoculations after 6 months.[44]

3.2
Suspension in distilled water

The yeast is grown in a nutrient liquid medium for about 2 d, centrifuged at low temperature and suspended in sterile distilled water. Portions of this suspension are stored in stoppered bottles at room temperature. Storage in most breweries usually involves maintaining the yeast in beer or water in varying concentrations from collection to collection. These conditions are often far from ideal for growth or even maintenance, since limited assimilative carbon and soluble nitrogen are present, together with a relatively high concentration of ethanol. Under these conditions the yeast must survive for an indeterminate period of time and to do this it requires a basal level of metabolic energy. Good yeast storage should avoid inclusion of oxygen in the slurry, should include cooling of the yeast slurry to 4–6°C as soon as possible after collection and recognition of the yeast prior to pitching.[19]

3.3
Freeze-drying (lyophilization)

Freeze-drying is a method for long-term preservation for viability and stability of yeast. Yeast collections usually supply yeast strains in freeze-dried state with instructions for their reactivation. Cells are suspended in a solution containing one or more protecting media as cryoprotectant, e.g., trehaloze, glycerol or skim milk. Cooling at a rate of usually -1°C min[-1] follows, and thereafter freeze-drying of the culture at very low temperature and pressure. The freeze-dried culture is stored in vacuum-sealed ampoules in a refrigerator. Cooling rate and protecting media are two important parameters for the viability and stability of yeast cells during freeze-drying, as discussed by Berny and Hennebert.[45] In general, freeze-drying is not a very successful method of maintaining brewing yeasts. The viability tends to be very low and those cells surviving may well be mutants.[46] The possibility of supplying wineries with ready-to-use immobilized cells directed Loureiro[47] to perform freeze-drying experiments of cells immobilized by gel occlusion.

3.4
Cryopreservation in liquid nitrogen

Cryopreservation is considered the best method of long-term preservation of viability and stability of fungi.[43,48] Slow cooling, usually at 1°C min[-1], of the yeast in the presence of a cryoprotectant, e.g., dimethyl sulfoxide (DMSO) or glycerol, and storage in vials under nitrogen vapors or in liquid nitrogen at ultra low temperature is the cryopreservation method in brief. For yeast recovery (reactivation) a

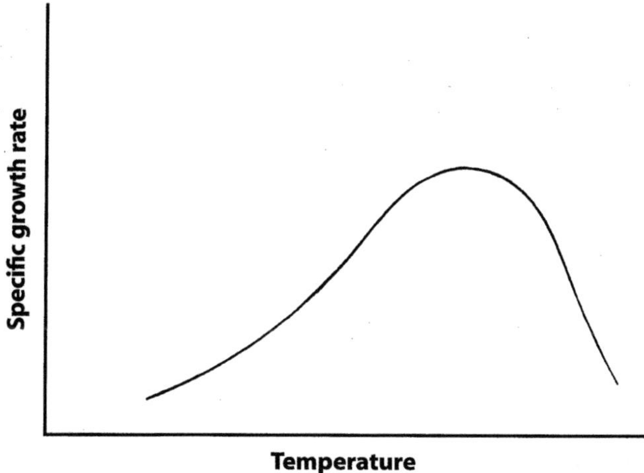

Fig. 1. Effect of temperature on growth rate of microorganisms (reprinted with permission from ref. 51)

quick warming at about 37°C is required. The growth method, the cryoprotectant used, the rate of cooling, the storage temperature and the rate of warming are factors affecting the viability and stability of yeast strain and its recovery.[48] Psychrotolerant yeast strains grown at low temperature are better adapted to the cryopreservation process, resulting in greater recovery.[49] Successive preservations at 0°C of cells of the yeast strain Visanto immobilized on mineral kissiris increased the biocatalytic stability of the cells and the ethanol and wine productivities.[49] For brewing yeasts, considerable success has been reported in maintaining yeasts in sealed vials containing a glycerol-serum medium and submerged in liquid nitrogen (-196°C).[46]

4
Cryotolerant yeast strains

All microorganisms have a characteristic optimal growth temperature and an upper growth temperature. At optimal growth temperature they demonstrate their highest growth and reproduction rates. Above the upper growth temperature and below the minimal growth temperature they do not grow. According to their optimal growth temperature, the microorganisms are characterized as psychrophiles (0–20°C), mesophiles (20–45°C) and thermophiles (45–90°C) by Atlas and Bartha.[50] The optimal growth temperature results from a balance between the growth rate and death rate of the organism, according to the equation:

$$\frac{dM}{dt} = M(A\,e^{-E/RT} - A'e^{-E'/RT}) \tag{1}$$

where M is the cell mass; A and A′ are constants for growth and death; E and E′ are activation energies for growth and death; R is the gas constant, and T is the absolute temperature. A typical plot of the overall growth rate is shown in Figure 1.[51]

Manufacturers know that alcoholic beverages produced at low temperatures develop taste and aroma and so cryotolerant yeast strains are desirable. Therefore, important products such as the Greek semi-sweet vinosanto, the semi-sparkling zitsa and the famous French sparkling champagne are produced by a secondary fermentation during the winter. The improved quality of wines produced by low temperature fermentation can be attributed to the reduction of higher alcohols and to an increase in the proportion of ethyl acetate in the total volatiles.[22,23] As a result, some researchers have focused their interest on low temperature wine making. Fermentations using free cells and performed in the temperature range 15–30°C have been reported by Mauricio et al.[52] Cryophilic yeast strains were selected and the correlation between their physiological activity and fermentation ability was studied by Kusewicz.[53] The cryophiles with high physiological activity at low temperatures showed high capacity for must fermentation.

Two cryophilic wine yeast strains, YM-84 and YM-126, were selected from 100 strains of the wine yeast *Saccharomyces cerevisiae*.[54] Their viability (specific growth rate) and fermentability at low temperatures (7 and 13°C) were superior to those of mesophilic wine yeast strains W_3 and OC-2. Also, isolation and initial characterization of cryotolerant *Saccharomyces* strains were reported by Castellari et al.[39] They studied 51 cryotolerant strains isolated from grape musts using the enrichment method and capable of growing well at 6°C.

However, there are very few publications concerning wine making using free cells at temperatures lower than 10°C. Wine making at temperatures lower than 15°C is not usual on an industrial scale since the productivity is very low. In order to decrease the temperature of fermentation below 10°C while retaining productivity, the psychrotolerant *Saccharomyces cerevisiae* strains Visanto[55] and AXAZ-1[56,57] yeast strains were immobilized on mineral kissiris, delignified cellulosic (DC) material and gluten respectively, so that the solid supported biocatalysts to reduce the activation energy. Various *Saccharomyces cerevisiae* strains have been isolated and are suitable for wine making. These strains came from the Greek agricultural area and specifically from the Achaia province and the Aegean islands Santorini, Samos, Thasos and Halkidiki. The strains were improved to ferment at lower temperatures by adaptation treatment.[36]

Over the last few years, the wine and brewing industries have started to be interested in yeast genetic engineering in order to obtain modified yeasts suitable for specific functions. Genetic engineering can involve standard procedures such as hybridization, backcrossing and mutagenesis, or newer techniques such as transformation and somatic fusion.[58] Through the application of hybridization technique, six hybrids were obtained between the cryophilic wine yeast *Saccharomyces bayanus* and the mesophilic wine yeast *Saccharomyces cerevisiae* by Kishimoto.[59] The fermentabilities of these hybrids at the low temperature of 7°C were superior to the mesophilic wine yeast and the same as the cryophilic wine yeast. The hybrids also produced higher amounts of malic acid and flavor compounds such as higher alcohols, isoamyl acetate, β-phenylethyl alcohol, β-phenylethyl acetate and lower amounts of acetic acid than those of mesophilic wine yeast. Through genetic engineering, improvement of wine yeasts has been reported for certain properties such as killer activity,[60] higher yields of glycerol,[61] higher fermentation efficiency and tolerance to SO_2,[62] and higher cell flocculation at the end of fermentation.[63] Perhaps, yeast genetic engineering gives optimistic results for the improvement of certain yeast properties and consequently for the quality of alcoholic beverages, but there are some problems: does a desirable new yeast feature arise without loss of others desirable characters that previously existed? Is it always possible to predict the results and consequences of the intervention in the process of the alcoholic fermentation and other yeast metabolic pathways? Intervening in a regular metabolic pathway, we may have unexpected results, as reported by Guerzoni et al.[64] and by Lee et al.[65]

5
Immobilization of cold-adapted yeast strains

5.1
Immobilization

Differences in specific metabolic activity between free and immobilized yeast have been reported by numerous authors[66–69] in the past; the cause of these changes in cellular activity is still unclear.

5.1.1
Wine making

Continuous wine making by immobilized yeast cells in contrast to potable and grade-fuel alcohol production has additional prerequisites; i. e., a final alcohol content of at least 11.5% (v/v) and a support of food grade purity. Therefore, the biocatalyst prepared for wine making by immobilization of yeast cells on solid supports has to be alcohol resistant. White wine with 11% (v/v) alcohol content was obtained by yeast cells fixed in sodium alginate gel through which grape must was passed for fermentation.[70] Likewise, white wine was produced in a batch fermentation by immobilized *Saccharomyces cerevisiae* in calcium alginate gel, κ-carrageenan, agar and pectic acid.[71] Yeast immobilized on alginates has also been used to produce fruit[72] and sparkling wines.[73] *Saccharomyces cerevisiae* immobilized on glass beads[74] and minerals such as polygorskite, montmorilonite and hydromica according to Ageeva et al.[75] produced wine in a rapid fermentation. Mineral kissiris was used for yeast immobilization to obtain continuous low-temperature wine making.[55] To immobilize yeast cells on cellulose for continuous wine making, researchers covered it with sodium alginate gel.[76] Also, for continuous fermentation, diethylaminoethyl (DEAE) cellulose has been used as support.[77]

Further research is needed to obtain cells immobilized on a support that is more hygienic for food production, cheap, abundant in nature and which gives an alcohol-resistant biocatalyst by immobilization. To satisfy the above prerequisites for the industrialization of continuous wine making by immobilized cells, the delignified cellulosic (DC) material[56] and gluten pellets[57] were used for immobilization of yeast cells. DC material was prepared using sawdust as starting material. To delignify sawdust, it was treated with 1% solution of sodium hydroxide. The soluble sodium phenolic salt of lignin was formed, removed and a porous DC material was prepared. Its pores accommodate cells going through. Therefore, this immobilization was performed by entrapment of cells in pores and by formation of hydrogen bonds between amino, hydroxyl and carboxyl groups of cell wall with the hydroxyl groups of the DC material. Gluten pellets were prepared using wheat flour as starting material. They were prepared as follows: wheat flour was mixed with tapwater, kneaded and the dough was washed extensively with tapwater for removal of starch. The gluten remaining was shaped by hand into pellets which were dried, cooled to room temperature and used for cell immobilization. Likewise, the attachment of cells on this support was made by van der Waals forces (adhesion of cells on gluten pellets) and by entrapment in its pores.

5.1.2
Brewing

A number of studies have been published on the production of beer by batch fermentation with immobilized cells on polyethelene film,[78] on ceramic or polyethylene ring,[79] in alginate gel[80] and in hollow polyvinylacetate (PVA) gel beads.[81] Also, a number of studies have been published on brewing by continuous fermentation with yeast cells immobilized on diatoms, polyvinylchloride (PVC) and plastic,[82] on pieces of brick,[83] in Ca-alginate,[84–86] and in ceramics.[87] Also, DC material[88] and gluten pellets[89] were used as supports for immobilization of yeast cells for brewing.

5.1.3
Activation energy

Low-temperature fermentations and hence low temperature wine making and brewing is a new scientific area in the field of food biotechnology. The cause of recently undertaking research on this subject was the promotion effect of the alcoholic fermentation with free cells using solids such as DC material, mineral kissiris and γ-alumina pellets. Furthermore, it was known that fermentations performed at relatively lower temperature than usual resulted in an improvement of the quality of the wine, beer and distillates. The promotion effect observed in the fermentation of must by DC material-supported biocatalyst was the reason to undertake experiments to investigate its effect on the activation energy (Ea) in the case of alcoholic fermentation. Thus, fermentations of must at 15 and 30°C were performed in the presence of DC material-supported biocatalyst and separately in the presence of free cells in an equal concentration with those of immobilized cells. Figure 2 illustrates that the DC material-supported biocatalyst causes an increase in the fermentation time of 158% and free cells of 331% as the temperature of the fermentation drops from 30 to 15°C. This means that the proportion of reaction speed constants at 30 and 15°C (K_{30}/K_{15}) becomes lower in the case of the presence of DC material-supported biocatalyst compared with those of free cells. Taking into account the Arrhenius equation:

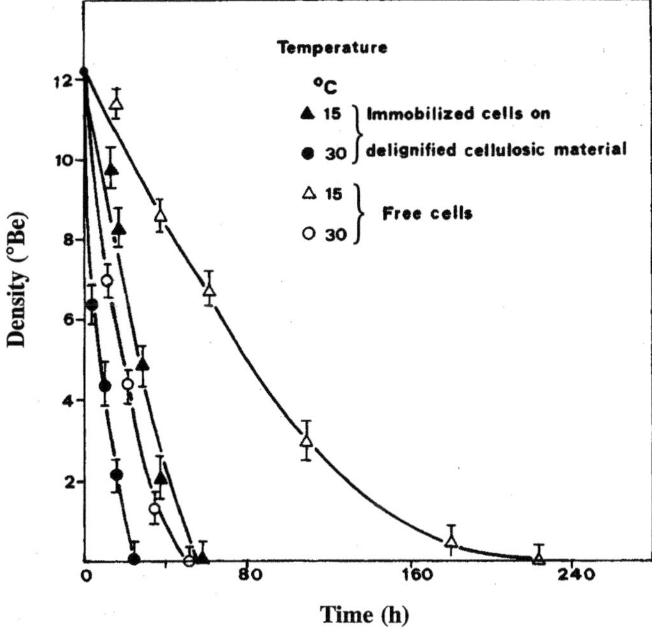

Fig. 2. Fermentation kinetics of must observed by delignified cellulosic material-supported biocatalyst and free cells at temperatures of 15 and 30°C (reprinted with permission from ref. 56)

$$\log \frac{K_2}{K_1} = \frac{E_a}{2.303R} \cdot \frac{T_2 - T_1}{T_1 \cdot T_2} \tag{2}$$

the activation energy (E_a) is reduced in the case of DC material supported biocatalyst. The latter shows that this biocatalyst behaves as a biocatalyst or as a promoter of the catalytic activity of the enzymes involved in the process.

5.2
Preservation of immobilized cells

Kissiris, a volcanic mineral material, was proposed as a useful cell support by Kana et al.[90] The low cost and natural abundance of kissiris prompted further research with kissiris-supported biocatalysts for the study of continuous potable alcohol production[91] and low-temperature wine making.[55] Furthermore, the immobilized biocatalyst must be preserved even during periods when alcohol production and wine making are halted. Repeated batch fermentations of grape must were carried out at 30°C, every time after successive preservations at 0°C, for 900 d.[49] These preservations increased the biocatalytic stability of the cells and the ethanol and wine productivities. Specifically, the results with preservation showed higher conversion of sugars and ethanol concentration than those in fermentations with kissiris-supported biocatalyst without successive preservations at 0°C.[49] Also, the ethanol productivity and conversion achieved were higher than continuous fermentations with immobilized *Zymomonas mobilis* on γ-alumina without preservation at 0°C.[92] The long-term stability is a prerequisite for a solid-supported biocatalyst to be employed in an industrial process. Moreover, kissiris-supported or any other biocatalyst will have to be preserved throughout the year for days during which the installa-tion needs maintainance (e.g. vacation periods). Therefore, the possibility of preservation of kissiris-supported biocatalyst at a low temperature during these periods is of technological importance and makes this abundant and low-cost support able to fulfill all the prerequisites for cost-effective immobilization and industrial application. Furthermore, the large improvement of ethanol and wine productivity, after preservation of immobilized cells at 0°C, is also of technological importance. In the cases of cells immobilized on gluten,[57,89] γ-alumina[23] and DC material[56,88] for wine making and brewing, the immobilized biocatalyst was preserved in distilled water or sugar solution in a refrigerator (4°C) during the halt of the fermentation process. No infection was reported of the yeast strains.

6
Low-temperature wine making

6.1
Effect of temperature

Reduction of the activation energy in alcoholic fermentation by the promoters was the reason to undertake research on low-temperature wine making and brewing by immobilized cells on the above supports. To ensure the possibility of low-temperature fermentations, cryotolerant yeast strains were used. Therefore, the combination of the aforementioned two factors led to fermentations carried out at the low temperatures of 0–13°C.[55–57] Table 2 shows the ethanol productivities obtained in wine making at low temperatures, using immobilized cells on various supports.[57,93–95] Immobilized cells resulted in an important increase of productivity at

Table 2. Ethanol productivities in low-temperature fermentation of must by free and immobilized cells on various supports

Temperature (°C)	Ethanol productivity (g l⁻¹ d⁻¹)										
	Free cells	Immobilized cells									
		Batch fermentation					Continuous fermentation				
		Kissiris	γ-Alumina	Alginates	DC	Gluten	Kissiris	γ-Alumina	Alginates	DC	Gluten
27	17	49	49	60	80	66	73	69	81	49	47
13	6	14	14	16	71	30	23	20	30	–	46
7	2	5	5	6	20	15	17	14	23	–	38
5	0.6	–	–	–	11	6	–	–	–	–	33
0	0.3	–	–	–	2	4	–	–	–	–	–

DC delignified cellulose

all temperatures as compared with free cells. At 7°C the increase was 3- to 10-fold. Continuous fermentations gave an increased productivity compared with the batch process, which becomes higher at lower temperatures. For example, at 7°C, kissiris resulted in a 2.5-fold increase in batch fermentation as compared to free cells while in continuous fermentation the increase was 8-fold. The organic supports were more effective than the inorganic ones. Gluten-supported biocatalysts at 5°C in continuous process resulted in a 55-fold increase in productivity as compared with free cells.

The above discussion shows the possibility of cost-effective wine making at 5°C. Continuous fermentation at 5°C led to a 2-fold higher productivity compared with free cells at 27°C. Analogous results were obtained by the other supports.

6.2
Characteristics of wine

The volatile acidities of the wines prepared from repeated fermentations of must by cells immobilized on gluten[57] were lower than those found in dry wines and those of wines obtained from repeated fermentations of must using DC material-supported biocatalyst[56] at the same temperature. This may be due to the amphoteric nature of proteins and the fact that small molecules such as acetic acid can react with the amino acid groups of protein molecules. In the case of DC material-supported biocatalyst,[56] total acidity and volatile acidity were in the ranges of those found in wines. In wine produced by cells immobilized on mineral kissiris,[55] the total acidity was 47% of that of must at 5°C, which was found to be equal to 6.75 g tartaric acid l^{-1}. It is estimated that this reduction is higher than that obtained in wines produced by natural fermentation. This is an advantage in the case of must with high total acidity which must be reduced to obtain better organoleptic character. Total and volatile acidities became lower as the temperature dropped.

6.3
Volatile by-products

To test the quality of the wine produced and to examine the effect of low-temperature fermentations on volatile by-products, samples were analyzed for methanol, amyl alcohols, ethyl acetate, isobutyl alcohol, propanol-1 and acetaldehyde.[96] Volatile compounds are responsible for aroma because they have greater vapor pressure. Among the carbonyl compounds of wine, the aldehydes are the most important as aroma compounds[97] and acetaldehyde accounts for more than 90% of total aldehyde content.[98] Esters are also significant components of wines, ethyl acetate being the most important.[99,100] Isoamyl acetate and ethyl caproate are especially important for a pleasant fruity aroma of wines.[101] Other major flavor compounds produced by yeast during fermentation are higher alcohols.[100] Small amounts of higher alcohols contribute positively to wine flavor, but excessive amounts do not.[102] Higher alcohols participate in the formation of esters during aging of wines.[103,104]

Amyl alcohols, ethyl acetate, isobutyl alcohol, propanol-1 and acetaldehyde were determined by gas chromatography according to Cabezudo et al.[105] using a stainless steel column, packed with Escarto-5905 consisting of 5% Squalene, 90% Carbowax-300 and 5% (v/v) Di-2-ethyl-hexyl sebacate. Nitrogen carrier gas at 20 ml min^{-1}, an injection port temperature of 210°C and a detector temperature of 220°C were used. Samples of 1 ml wine were injected directly into the column and the concentrations of the above compounds were determined using standard curves. Methanol was determined by gas chromatography according to Lee et al.[106] using Porapac S as column material, nitrogen carrier gas at 40 ml min^{-1}, an injector temperature of 210°C, a column temperature programming of 120–153°C and a detector temperature of 220°C. In both cases, pentanol-3 was used as internal standard at a concentration of 1% (v/v).

Table 3 shows the concentrations of amyl alcohols and ethyl acetate determined for inorganic and organic support compared with free cells.[96] Immobilized cells resulted in wines with less amyl alcohols than free cells. Furthermore, wines produced by immobilized cells contained more ethyl acetate compared with those produced by free cells. When the temperature was lowered, amyl alcohols decreased. Similar results were obtained for the other higher alcohols determined in wines. Kissiris and DC material resulted in higher ethyl acetate concentration as well as γ-alumina and DC material led to less amyl alcohols at low temperatures than the other supports.

6.4
Nutritional value and quality

According to the above explanation for low-temperature wine making by immobilized cells, wine contains less amyl alcohols and other higher alcohols. Due to their toxicity this reduction improves the nutritional value of this alcoholic beverage. Likewise, less higher alcohols and less diminution of the formation of ethyl acetate contributes to an increase in the percentage of total volatiles. This explains the improvement of the aroma and taste of wines produced by low-temperature fermentations using immobilized cells.

6.5
Effect on distillates

Wines produced by γ-alumina-supported biocatalyst[23] had low volatile and total acidities, fine aroma better than the other wines produced by immobilized cells on DC, gluten and kissiris. However, these wines were not taste tested due to their higher aluminium content which characterize these wines as unsuitable for drinking. Furthermore, these wines would be an excellent raw material for distillates production.

Table 3. Amyl alcohols and ethyl acetate formed by repeated batch and continuous fermentations of must by cells immobilized on different supports

Temperature (°C)	Amyl alcohols (ppm)						Ethyl acetate (ppm)					
	Free cells	Immobilized cells					Free cells	Immobilized cells				
		Kissi-ris[a]	γ-Alu-mina[a]	Algi-nates[a]	DC[b]	Gluten[b]		Kissi-ris[a]	γ-Alu-mina[a]	Algi-nates[a]	DC[b]	Gluten[b]
30	143[b]				152	136(143)	38[b]				126	74(110)
27	146[a]	(113)	(140)	(132)			67[a]	(101)	(115)	(111)		
20		(103)	(130)	(129)					(98)	(103)		
16		(91)	(114)	(117)	124			(83)	(89)	(93)	105	
15	160[b]					145(115)	31[b]	(76)				59(94)
13		(75)	(95)	(99)					(85)	(76)		
10	125[b]	(63)	(73)	(88)	72	159(137)	18[b]	(72)	(68)	(72)	83	75(65)
7	104[a]	(45)	(53)	(75)			22[a]	(63)	(62)	(56)		
5						165		(43)				54
0	123[b]				50	129	14[b]				69	22

Values in parentheses are referred to as continuous processes; *DC* delignified cellulose
[a] The *Saccharomyces cerevisiae* strain used was Visanto.
[b] The *Saccharomyces cerevisiae* strain used was AXAZ-1.

7
Low-temperature brewing

7.1
Effect of temperature

DC material-supported biocatalyst was used for repeated batch fermentations of wort at temperatures in the range of 30–0°C and for continuous fermentations at temperatures between 15 and 3°C, as studied by Bardi et al.[88] Also, gluten pellets-supported biocatalyst was used for repeated batch fermentations of wort at temperatures in the range of 30–0°C and for continuous ones at temperatures in the range of 15–5°C.[89]

AXAZ-1, a cryotolerant and alcohol-resistant *Saccharomyces cerevisiae* strain isolated from grapes[35] was immobilized on the above supports. In order to compare fermentation times and other parameters obtained with free cells and with gluten or DC material-supported cells, fermentations were carried out simultaneously with the same cell concentration. As shown in Table 4 the fermentation times are shorter for immobilized cells than for free ones at each temperature. The glucose uptake rate was probably greater in immobilized cells than in free cells.[107]

7.2
Characteristics of beer

Beer produced by repeated batch fermentations contains a high alcohol content (Table 4). Also, beer produced by primary fermentation does not differ in its common characteristics such as real extract, apparent extract, refractive index or pH from beer produced by coventional batch fermentation. An important group of flavor compounds is the diketones diacetyl (2,3 butane dione) mainly and 2,3 pentane dione. In commercial beer, diacetyl is usually less than 1 ppm, with an optimum value less than 0.15 ppm. High levels of diacetyl give an off-flavor (butterscotch) to beer. During yeast metabolism, the precusors of the diketones (α-acetolactate and α-ketobutyrate) are secreted by the yeast into the fermenting wort and then these precursors are converted to diketones. After that the yeast is able to convert the diacetyl to a tasteless substance, 2,3 butane diol, during beer maturation mainly.[108] Beer maturation is the most consuming process in conventional brewing, but little investigation has been carried out for a rapid removal of diacetyl. Immobilized yeast cells have been used for this purpose.[109,110] Analysis of young beers for diacetyl was performed just after their preparation; at all temperatures studied diacetyl concentrations were in the range of 0.08–0.14 ppm for immobilized yeast cells (Table 5). In comparison with free cells (Table 5), the diacetyl concentrations were lower as with immobilized yeast cells, while the fermentation time was much longer with free cells, providing the opportunity for diacetyl to be reduced. It is possible that immobilized cells showed an increased metabolic activity for the reduction of diacetyl to produce 2,3 butane diol which is almost flavorless.

Commercial products produced by the primary fermentation of the wort contain relatively high polyphenol contents in the range of 190–250 ppm and further treat-

Table 4. Some kinetic parameters obtained in brewing by free and immobilized yeast cells on two supports

Temperature (°C)	Repeated batch fermentation						Continuous fermentation			
	Free cells		Immobilized cells				Immobilized cells			
	Fermentation time (h)	Ethanol concentration (% v/v)	Fermentation time (h)		Ethanol concentration (% v/v)		Fermentation time (h)		Ethanol concentration (% v/v)	
			DC	Gluten	DC	Gluten	DC	Gluten	DC	Gluten
30	144	5.5	12	12	5.0	4.9				
20				26		5.4				
15			25	26	5.4	5.5	24	24	5.9	5.7
10	624	5.4	63	72	5.4	5.4	24	24	6.5	5.6
7			136	164	5.8	5.9				
5			216	240	6.0	6.3		24	6.4	4.8
3							24		5.3	
0	1320	5.4	288	282	6.4	6.3				

DC delignified cellulose

Table 5. Some characteristics of beer obtained in brewing by free and immobilized yeast cells on two supports

Temperature (°C)	Repeated batch fermentation								Continuous fermentation				
	Free cells		Immobilized cells						Immobilized cells			
	Diacetyl (ppm)	Polyphenols (ppm)	Diacetyl (ppm)		Polyphenols (ppm)		Diacetyl (ppm)		Polyphenols (ppm)	
			DC	Gluten	DC	Gluten	DC	Gluten	DC	Gluten
30	0.14	308	0.09	0.08	150	144				
20				0.10		116				
15			0.09	0.07	157	152	0.11	0.09	53	77
10	0.20	338	0.10	0.10	158	154	0.24	0.19	115	80
7			0.11	0.09	101	126				
5			0.11	0.09	114	139	0.36	0.18	145	146
3							0.31		147	
0	0.21	306	0.10	0.14	94	84				

DC delignified cellulose

Table 6. Ethanol productivities obtained in brewing by free and immobilized cells on two supports

Temperature (°C)	Ethanol productivity (g l^{-1} d^{-1})				
	Free cells	Immobilized cells			
		Repeated batch fermentation		Continuous fermentation	
		DC	Gluten	DC	Gluten
30	2.9	54.4	58.1		
20			26.6		
15		28.8	27.9	42.2	40.3
10	0.7	13.5	11.3	46.4	38.4
7		5.7	5.3		
5		3.7	3.4	45.7	18.2
3				26.2	
0	0.3	3.0	2.9		

DC delignified cellulose

ment is required to decrease and remove hazes. At all temperatures studied, the polyphenol concentrations were less than 158 ppm and about 30–50% of those contained in beers brewed by using free cells (Table 5). It is probable that the supports (gluten, DC material) adsorbed some polyphenols. The beers had very good clarity after the end of the primary fermentation, especially at temperatures below 10°C, with low concentrations of polyphenols and little cell growth. No filtration was therefore necessary for beer clarification. Young beers had low bitterness, a golden-yellow color which remained stable for 10 months and an improved aroma and taste.

Continuous brewing followed using yeast cells immobilized on DC material and gluten pellets separately, and the bioreactors were operated continuously for 3 months without contamination and loss of ethanol yield. Productivities were higher than in repeated batch fermentations, especially at temperatures lower than 10°C (Table 6). During the 3-month period of the continuous operation of the bioreactor, no increase in free cell concentration was observed. For young beers produced by the continuous process, the values of the original extract, real extract, apparent extract, refractive index, color and pH lie in the range of the most commercial beers. Ethanol concentration was similar to that of batch fermentations.

The concentration of diacetyl was determined in samples just after collection and remained at low levels (Table 5). At 10°C and below, the diacetyl content was increased but remained at acceptable levels. This may be due to the continuous process as well as to entrance of aerated wort. Since diacetyl is formed earlier in the fermentation, batch fermentation accommodates the diacetyl removal by the yeast.[19,86]

The concentrations of polyphenols and bitterness increased as the temperature was decreased but were in the range of the most commercial beers (Table 5). In the batch system, the values of these parameters decreased as the temperature was reduced. It is possible that in the continuous system smaller concentrations of these substances may be adsorbed onto supports (DC material or gluten) than in batch fermentations because of the shorter fermentation time in the bioreactor. At temperatures below 10°C, beer productivities were much higher than those obtained at repeated batch fermentations (Table 6). The preliminary taste characterized the young beers as sweet, with pleasant soft aroma and taste, fruity aroma and taste in the case of gluten, with body, aftertaste and distinct from commercial beers. The aroma and taste of young beers produced by batch and continuous systems were similar and more intense in beers produced at low temperatures. Taste and aroma remained stable for about 10 months of storage, but after this time they began to weaken.

7.3
Volatile by-products

Research concerning beer flavor has been intensively carried out during the last 30 years. The flavor of beer is the result of a complex combination of components, each of them giving its distinctive personality. Furthermore, many factors can modify the balance and hence the flavor of the product, e. g., yeast strain, incubation temperature, wort pH, wort gravity, etc.[111] However, an important part of this flavor is due to the result of yeast metabolism.[112] According to Bardi et al.[113] higher alcohols were reduced as the temperature was decreased in all experiments by immobilized and free cells (Table 7). However, ethyl acetate, determined as a percentage of total

Table 7. By-products of beers produced by repeated batch fermentations and continuous one of wort using cells immobilized on some supports and free cells

Temperature (°C)	Acetaldehyde (ppm)	Ethyl acetate (ppm)	Propanol-1 (ppm)	Isobutanol-1 (ppm)	Amyl alcohols (ppm)	Methanol (ppm)	
30	21(16)	19(77)	21(32)	21(46)	77(178)	45(26)	Repeated batch
15	20(23)	9(34)	13(28)	6(43)	54(158)	40(25)	fermentations by cells
10	10(25)	17(28)	13(27)	5(41)	38(139)	20(25)	immobilized on DC
7	19(20)	23(29)	10(19)	3(38)	38(97)	41(34)	material and on
5	9(17)	18(29)	14(18)	4(36)	36(92)	39(28)	gluten (in parantheses)
0	12(17)	16(21)	8(15)	3(32)	30(78)	42(28)	
30	11	8	8	45	59	64	Free cells
15	16	6	8	15	61	42	
10	22	7	11	12	41	43	
7	17	7	8	9	41	17	
5	22	9	6	4	42	52	
0	12	11	6	4	44	32	
15	24(24)	19(40)	20(33)	9(40)	66(165)	29(33)	Continuous
10	17(17)	14(26)	15(25)	8(46)	41(153)	34(28)	fermentation by cells
5	18(23)	25(26)	13(24)	8(33)	31(91)	33(31)	immobilized on DC
3	19	24	7	3	24	24	material and on gluten (in parantheses)

Above values are average values of at least two measurements.
DC delignified cellulose

volatiles minus methanol, was higher for immobilized cells at temperatures below 15°C compared with free cells, and was increased as the temperature was decreased down to 15°C. These two results lead to better aroma and less toxicity of beers. More specifically, results have also shown that the beers produced from batch fermentations of wort by cells immobilized on DC material contained, in general, lower concentrations of higher alcohols, acetaldehyde and methanol. Ethyl acetate was higher than that of free cells. Beers produced from fermentations of wort by cells immobilized on gluten pellets contained higher alcohols at concentrations higher than those of free cells. The concentrations of acetaldehyde were similar, methanol was lower and ethyl acetate was 2- to 8-fold higher than those of free cells.

The increased amounts of ethyl acetate determined as a percentage of total volatiles and the reduction of higher alcohols may give the fine aroma and taste to the beers produced. Continuous process by DC material-supported biocatalyst seems to give a higher percentage of ethyl acetate determined on total volatiles than gluten- supported biocatalyst and free cells at temperatures lower than 15°C.

7.4
Nutritional value and quality

According to the above explanation for low-temperature brewing by immobilized cells, beer contains less amyl alcohols and other higher alcohols. Due to their toxicity this reduction improves the nutritional value of this alcoholic beverage. Likewise, less higher alcohols and less diminution of the formation of ethyl acetate contribute to an increase in percentage of total volatiles. This explains the improvement of the aroma and taste of beer produced by low-temperature fermentations using immobilized cells. Furthermore, the reduction of polyphenols and diacetyl of beer contributes to a product with improved quality.

7.5
Effect on distillates

By distillation of the above prepared beers, we could take distillate with fine aroma and low content of higher alcohols and diacetyl. Probably, this distillate could be use for the production of whiskey.

8
Cost of production and capacity of factories

It is well known that fermentation technology is related to the production of wine, beer, single-cell protein (SCP), baker's yeast, potable alcohol and distillates. Therefore, fermentation technology is related to food biotechnological processing. The research objectives in industrial food biotechnological processing will have to aim at:

1. decreasing the size of the factory,
2. reducing production cost,

3. improving quality of food and
4. developing environmentally friendly technologies.

All these factors are related to the production cost, size of factory and therefore construction cost. More specifically, the decrease of the size of the factory and the reduction of production cost are related to the increase of productivity. The reduction of production cost is also related to reduction of the energy demand. Both of these are affected by the promotion of fermentation and immobilized cells. To improve quality of food we have to industrialize low-temperature fermentation, to isolate suitable yeast strains and to achieve the production of a protein-rich livestock feed using agro-industrial wastes. The development of environmentally friendly technologies needs:

1. to increase final alcohol concentration in industrial fermentations (the reduction of the energy demand is obtained by the increase of the final alcohol concentration),
2. to industrialize immobilized cells and
3. to increase methane fermentation rate in anaerobic treatment of relative liquid wastes.

According to this, the reduction of the size of the factories is affected by the improvement of kinetics parameters. This contributes to reduction of the construction cost. Likewise, this affects labor cost and cost of the area needed to install the plant. The increase of productivity also reduces labor cost and consequently cost of production. The reduction of the energy demand is related to oil consumption and to high alcohol concentration. It has been proved that some solids increase the fermentation rate and reduce the activation energy, and therefore they are suitable supports of cells for low-temperature fermentations by immobilized cells. This contributes to low-temperature wine making, distillates production and cost-effective brewing at very low temperatures of 0–10°C. Furthermore, low-temperature fermentations decrease amyl alcohols and other higher alcohols of the wine and beer. Likewise, low-temperature brewing produces beer containing less polyphenols, less diacetyl, less bitterness and lower pH compared with the traditional product. The above composition of wine and beer contributes to new products which have improved nutritional value (due to the toxicity of amyl and other higher alcohols) as well as improved quality. These improved new products create the possibility for the development of a new technology and create new capital value.

9
Conclusions

Alcoholic beverages produced at low temperatures develop more taste and aroma, therefore cold-adapted yeast strains are desirable. Such strains can be isolated from grape must by using various methods, and their fermentation at low temperatures can be improved by adaption methods. Cell immobilization on supports suitable for food production has advantages over a free cell system; immobilized biocatalysts can be preserved without loss of stability or viability. Wine making and brewing by cold-adapted and immobilized yeast cells result in an important increase of productivity for batch and continous systems as compared with free cells, and products are characterized by a fine aroma taste and nutritional value.

10
References

1. Ough CS, Amerine MA. Controlled fermentation. VI. Effects of temperature and handling on rates, composition, and quality of wines. Am J Enol Viticult 1961; 12:117-128.
2. Ough CS. Fermentation rates of grape juice. I. Effects of temperature and composition on white juice fermentation rates. Am J Enol Viticult 1964; 15:167-177.
3. Nagodawithana TW, Castellano C, Steinkraus KH. Effect of dissolved oxygen, temperature, initial cell count and sugar concentration on the viability of *Saccharomyces cerevisiae* in rapid fermentation. Appl Microbiol 1974; 28:383-391.
4. Van Uden N, Da Cruz Duarte H. Effects of ethanol on the temperature profile of *Saccharomyces cerevisiae*. Z Allg Mikrobiol 1981; 21:743-750.
5. Laluce C, Palmieri MC, Lopes da Cruz RC. Growth and fermentation characteristics of new selected strains of *Saccharomyces* at high temperatures and high cell densities. Biotechnol Bioeng 1991; 37:528-536.
6. Hacking AJ, Taylor IWF, Hanas CM. Selection of yeast able to produce ethanol from glucose at 40°C. Appl Microbiol Biotechnol 1984; 19:361-363.
7. Perego L Jr, Cabral de S Dias JM, Koshimizu LH, De Melo Cruz MR, Borzani W, Vairo MLR. Influence of temperature, dilution rate and sugar concentration on the establishment of steady-state in continuous ethanol fermentation of molasses. Biomass 1985; 6:247-256.
8. Caro I, Perez L, Cantero D. Development of a kinetic model for the alcoholic fermentation of must. Biotechnol Bioeng 1991; 38:742-748.
9. Ough CS. Fermentation rates of grape juice. II. Effects of initial °Brix, pH, and fermentation temperature. Am J Enol Viticult 1966; 17:20-26.
10. Bertrand G, Silberstein L. Does fermentation of sugar normally produce methanol? Compt Rend 1950; 230:800-803.
11. Lee CH, Robinson WB, Van Buren JP, Acree TE, Stoewsand GS. Methanol in wines in relation to processing and variety. Am J Enol Viticult 1975; 26:184-187.
12. Stella C, Testa F, Viviani C, Sabatelli MP. Heat treatment of grapes: effect on wines from different varieties. Vignevini 1991; 18:47-50.
13. Gardner N, Rodrigue N, Champagne CP. Combined effects of sulfites, temperature, and agitation time on production of glycerol in grape juice by *Saccharomyces cerevisiae*. Appl Environ Microbiol 1993; 59:2022-2028.
14. Ough CS, Fong D, Amerine MA. Glycerol in wine: determination and some factors affecting formation. Am J Enol Viticult 1972; 23:1-5.
15. Rankine BC, Bridson DA. Glycerol in Australian wines and factors influencing its formation. Am J Enol Viticult 1971; 22:6-12.
16. Castellari L, Magrini A, Passarelli P, Zambonelli C. Effect of must fermentation temperature on minor products formed by cryotolerant and non-cryotolerant *Saccharomyces cerevisiae* strains. Ital J Food Sci 1995; 7:125-132.
17. Giudici P, Zambonelli C, Passarelli P, Castellari L. Improvement of wine composition with cryotolerant *Saccharomyces* strains. Am J Enol Vitic 1995; 46:143-147.
18. Ough CS, Guymon JF, Crowell EA. Formation of higher alcohols during grape juice fermentations at various temperatures. J Food Sci 1966; 31:620-625.
19. Stewart GG, Russell I. Centenary review. One hundred years of yeast research and development in the brewing industry. J Inst Brew 1986; 92:537-558.
20. Ruzic N. Effect of temperature on yeast activity and chemical composition of wine. Zbomik

Radova 1991; 22:35-44.

21. Bertolini L, Zambonelli C, Giudici P, Castellari L. Higher alcohols production by cryotolerant *Saccharomyces* strains. Am J Enol Viticult 1996; 47:343-345.

22. Bakoyianis V, Kana K, Kaliafas A, Koutinas AA. Low temperature wine making by kissiris-supported biocatalyst: Volatile by-products. J Agric Food Chem 1993; 41:465-468.

23. Bakoyianis V. Formation of volatile by-products in the continuous wine making by immobilized cells on mineral kissiris, γ-alumina and calcium alginates and an application in industrial scale pilot plant of a fermentation in the presence of kissiris. PhD Thesis, University of Patras, Greece, 1995:140-150.

24. Killian E, Ough CS. Fermentation esters — formation and retention as affected by fermentation temperature. Am J Enol Viticult 1979; 30:301-305.

25. Gomez E, Laencina J, Martinez A. Vinification effects on changes in volatile compounds in wine. J Food Sci 1994; 59:406-409.

26. Martini A. Origin and domestication of the wine yeast *Saccharomyces cerevisiae*. J Wine Res 1993; 4:165-176.

27. Polsinelli M, Romano P, Suzzi G, Mortimer R. Multiple strains of *Saccharomyces cerevisiae* on a single vine. Lett Appl Microbiol 1996; 23:110-114.

28. Versavaund A, Courcoux P, Roulland C, Dulau L, Hallet J-N. Genetic diversity and geographical distribution of wild *Saccharomyces cerevisiae* strains from the wine-producing area of Charentes, France. Appl Environ Microbiol 1995; 61:3521-3529.

29. Brown CM, Campbell I, Priest FG. Introduction to Biotechnology. Oxford: Blackwell Scientific Publications, 1987:93.

30. Heard GM, Fleet GH. Growth of natural yeast flora during the fermentation of inoculated wines. Appl Environ Microbiol 1985; 50:727-728.

31. Gao C, Fleet GH. The effects of temperature and pH on the ethanol tolerance of the wine yeasts, *Saccharomyces cerevisiae*, *Candida stellata* and *Kloeckera apiculata*. J Appl Bacteriol 1988; 65:405-409.

32. Kunkee RE. Selection and modification of yeasts and lactic acid bacteria for wine fermentation. Food Microbiol 1984; 1:315-332.

33. Heard GM, Fleet GH. The effects of temperature and pH on the growth of yeast species during the fermentation of grape juice. J Appl Bacteriol 1988; 65:23-28..

34. Koutinas AA, Pefanis ST. Biotechnology of Foods and Drinks. University of Patras, Greece, 1992:149-153.

35. Argiriou T, Kalliafas A, Psarianos K, Kana K, Kanellaki M, Koutinas AA. New alcohol resistant strains of *Saccharomyces cerevisiae* species for potable alcohol production using molases. Appl Biochem Biotechnol 1992; 36:153-161.

36. Argiriou T, Kaliafas A, Psarianos K, Kanellaki M, Voliotis S, Koutinas AA. Psychrotolerant *Saccharomyces cerevisiae* strains after an adaptation treatment for low temperature wine making. Process Biochem 1996; 31:639-643.

37. Miklos I, Sipiczki M, Benko Z. Osmotolerant yeasts isolated from Tokaj wines. J Basic Microbiol 1994; 34:379-385.

38. Kishimoto M, Goto S. Growth temperatures and electrophoretic karyotyping as tools for practical discrimination of *Saccharomyces bayanus* and *Saccharomyces cerevisiae*. J Gen Appl Microbiol 1995; 41:239-247.

39. Castellari L, Pacchioli G, Zambonelli C, Tini V, Grazia L. Isolation and initial characterization of cryotolerant *Saccharomyces* strains. Ital J Food Sci 1992; 4:179-186.

40. Hantula J, Kurki A, Vuoriranta P, Bamford DH. Rapid classification of bacterial strains by SDS-polyacrylamide gel electrophoresis: population dynamics of the dominant dispersed phase bacteria of activated sludge. Appl Microbiol Biotechnol 1991; 34:551-555.

41. Nikolova P, Ward OP. Production of L-phenyl-acetyl carbinol by biotransformation product and by-product formation and activities of the key enzymes in wild-type and ADH isoenzyme mutants of *Saccharomyces cerevisiae*. Biotechnol Bioeng 1991; 38:493-498.

42. Giolfi G. Acetaldehyde formation in relation to yeast nitrogen nutrition during alcoholic fer-

mentation. Riv Vitic Enol 1983; 36:431.

43. Ferreira da Silva L, Kamiya NF, Oliveira MS, Alterthum F. Comparison of preservation methods applied to yeasts used for ethanol production in Brazil. Rev Microbiol Sao Paulo 1992; 23:177-182.

44. Campbell I. Culture, storage, isolation and identification of yeasts. In: Campbell I, Duffus JH, eds. Yeast: A Practical Approach, 2nd ed. Oxford: IRLPress, 1991:1-2.

45. Berny JF, Hennebert GL. Viability and stability of yeast cells and filamentous fungus spores during freeze-drying. Effects of protectants and cooling rates. Mycologia 1991; 83:805-815.

46. Hough JS. The Biotechnology of Malting and Brewing, 1st paperback ed. Cambridge: Cambridge University Press, 1991:102.

47. Loureiro V. Portuguese contribution on immobilized yeast cell, for sparkling wine production. In: Colagrande O, ed. Sviluppi della Biotecnologia nella Produzione dello Spumante Classico. Atti del 4° Simposio Internationale sul Vino, Pavia. Pinerolo Italy: Chiriotti Publishers, 1990:74-77.

48. Smith D. Tolerance to freezing and thawing. In: Jennings DH, ed. Stress Tolerance of Fungi, 1st ed. New York: Marcel Dekker, Inc, 1993:146, 151.

49. Argiriou T, Kanellaki M, Voliotis S, Koutinas AA. Kissiris-supported yeast cells: High biocatalytic stability and productivity improvement by successive preservations at 0°C. J Agric Food Chem 1996; 44:4028-4031.

50. Atlas RM, Bartha R. Effects of abiotic factors and environmental extremes on microorganisms. In: Brady EB, senior ed. Microbial Ecology. Fundamentals and Applications, 3rd ed. USA: The Benjamin/Cummings Publishing Company. 1993:215.

51. Russell PD. Fermenter and bio-reactor design. In: King RD, Cheetham PSJ, eds. Food Biotechnology 1. London New York: Elsevier Applied Science, 1987:6-8.

52. Mauricio JC, Moreno J, Medina M, Ortega J. Fermentation of Pedro Ximenez musts at various temperatures and different degrees of ripeness. Belg J Food Chem Biotechnol 1986; 41:71-76.

53. Kusewicz D. Characteristics of fermentation abilities in some varieties of wine yeast of cryophilic types. Acta Aliment Pol 1975; 1:235-246.

54. Kishimoto M, Shinohara T, Soma E, Goto S. Selection and fermentation properties of cryophilic wine yeasts. J Ferment Bioeng 1993; 75:451-453.

55. Bakoyianis V, Kanellaki M, Kaliafas A, Koutinas AA. Low temperature wine making by immobilized cells on mineral kissiris. J Agric Food Chem 1992; 40:1293-1296.

56. Bardi EP, Koutinas AA. Immobilization of yeast on delignified cellulosic material for room temperature and low temperature wine making. J Agric Food Chem 1994; 42:221-226.

57. Bardi EP, Bakoyianis V, Koutinas AA, Kanellaki M. Room temperature and low temperature wine making using yeast immobilized on gluten pellets. Process Biochem 1996; 31:425-430.

58. Jackson RS. Wine Science. Principles and Applications, 1st ed. London: Academic Press, 1994:245.

59. Kishimoto M. Fermentation characteristics of hybrids between the cryophilic wine yeast *Saccharomyces bayanus* and the mesophilic wine yeast *Saccharomyces cerevisiae*. J Ferment Bioeng 1994; 77:432-435.

60. Hara S, Iimura Y, Oyama H, Kozeki T, Kitano K, Otsuka KI. The breeding of cryophilic killer wine yeasts. Agric Biol Chem 1981; 45:1327-1334.

61. Eustace R, Thornton RJ. Selective hybridization of wine yeasts for higher yields of glycerol. Can J Microbiol 1987; 33:112-117.

62. Thornton RJ. Selective hybridization of pure culture wine yeasts. II. Improvement of fermentation efficiency and inheritance of SO$_2$ tolerance. Eur J Appl Microbiol Biotechnol. 1982; 14:159-164.

63. Watari J, Takata Y, Ogawa M, Nishikawa N, Kaminura M. Molecular cloning of a flocculation gene in *Saccharomyces cerevisiae*. Plasmid DNA purification from *E. coli* culture. Agr Biol Chem 1989; 53:901-903.

64. Guerzoni ME, Marchetti R, Giudici P. Modifications of aroma components of wines obtained

by fermentation with *Saccharomyces cerevisiae* mutants. Bull OIV 1985; 58:228-234.

65. Lee S, Villa K, Patino H. Yeast strain development for enhanced production of desirable alcohols/esters in beer. J Amer Soc Brew Chem 1995; 53:153-156.

66. Navarro JM, Durant G. Modification of yeast metabolism by immobilization onto porous glass. Eur J Appl Microbiol 1977; 4: 243-254.

67. Doran PM, Bailey JE. Effects of immobilization on growth, fermentation properties and macromolecular composition of *Saccharomyces cerevisiae* attached to gelatin. Biotechnol Bioeng 1986; 28:73-87.

68. Gallazo JL, Bailey JE. Fermentation pathway kinetics and metabolic flux control in suspended and immobilized *Saccharomyces cerevisiae*. Enzyme Microbiol Technol 1990; 12:162-172.

69. Hilge-Rotmann B, Rehm HJ. Comparison of fermentation properties and specific enzyme activities of free and calcium-alginate-entrapped *Saccharomyces cerevisiae*. Appl Microbiol Biotechnol 1990; 33:54-58.

70. Shimobayashi Y, Tominaga K. Application of biotechnology in the food industry. I. Brewing of white wine by a bioreactor. Hokaidoritsu Kogyo Shikenjo Hokoku 1986; 285:199-204.

71. Nakanishi K, Yokotsuka K. Fermentation of white wine from Koshu grape using immobilized yeast. Nippon Shokuhin Kogyo Gakkaishi 1987; 34:362-369.

72. Mori S. Fruit wine or sake manufacture by bioreactor. Jpn Kokai Tokkyo Koho JP 6261, 577; 18 March.1987.

73. Fumi M, Trioli G, Colagrande O. Preliminary assessment on the use of immobilized yeast cells in sodium alginate for sparkling wine processes. Biotechnol Lett 1987; 9:339-342.

74. Hamdy MK. Method for rapidly fermenting alcoholic beverages. Patent Cooperation Treaty Int Appl WO 9005, 189; 17 May 1990.

75. Ageeva NM, Merzhanian AA, Sobolev EM. Effect of yeast adsorption on the functional activity of the yeast cells and composition of wine. Mikrobiologiya 1985; 54:830-834.

76. Otsuka K. Wine making. Jpn Kokai Tokkyo Koho 80, 159, 789; 12 December 1980.

77. Lommi H, Advenainen J. Method using immobilized yeast to produce ethanol and alcoholic beverages. Eur Pat Appl EP 361, 165; 4 April 1990.

78. Kolpakchi AP, Isaeva VS, Zhvirblyanskaya A Yu, Kazantsev EN, Serova EN, Rattel NN. Fixation of brewer's yeast to polymer materials. Prikl. Biokhim. Microbiol. 1976; 12:866-870.

79. Kolpakchi AP, Isaeva VS, Kazantsev EN, Fertman GI. Improvement in the fermentation of brewing wort with yeast fixed on a support. Ferment Spirt Prom - st 1980; 2:9-14.

80. Pardonova B, Polednikova M, Sedova H, Kahler M, Ludvik J. Biocatalyst for beer production. Brauwissenschaft 1982; 35:254-258.

81. Shindo S, Kamimura M. Immobilization of yeast with hollow PVA gel beads. J Ferment Bioeng 1990; 70:232-234.

82. Moll M. Continuous brewing of beer. US Patent No 4,009,286, 1977.

83. Moll M, Duteutre B. Continuous beer preparation using adsorbed yeasts. Cell Immobilisees Colloque, Paris, France, 1979:173-185.

84. Godtfredsen SE, Ottesen M, Svensson B. Application of immobilized yeast and yeast comobilized with amyloglucosidase in the brewing process. Proc Congr Eur Brew Conv, Copenhagen Denmark 1981; 18:505-511.

85. Linko YY, Linko P. Continuous ethanol production by immobilized yeast reactor. Biotechnol Lett 1981; 3:21-26.

86. Onaka T, Nakanishi K, Inoue T, Kubo S. Beer brewing with immobilized yeast. Bio/Technology 1985; 3:467-470.

87. Nakanishi K, Murayama H, Sato H, Nagara A, Yasui T, Mitsui S. Continuous beer brewing with yeast immobilized on granular ceramic. Hakko Kogaku Kaishi 1989; 67:509-514.

88. Bardi EP, Koutinas AA, Soupioni MJ, Kanellaki ME. Immobilization of yeast on delignified cellulosic material for low temperature brewing. J Agric Food Chem 1996; 44:463-467.

89. Bardi E, Koutinas AA, Kanellaki M. Room and low temperature brewing with yeast immobilized on gluten pellets. Process Biochem 1997; 32:691-696.

90. Kana K, Kanellaki M, Psarianos C, Koutinas AA. Ethanol production by *Saccharomyces cere-*

visiae immobilized on mineral kissiris. J Ferment Bioeng 1989; 68:144-147.

91. Koutinas AA, Gourdoupis C, Psarianos C, Kaliafas A, Kanellaki M. Continuous potable alcohol production by immobilized *Saccharomyces cerevisiae* on mineral kissiris. Appl Biochem Biotechnol 1991; 30:203-216.

92. Koutinas AA, Kanellaki M. Continuous potable alcohol production by *Zymomonas mobilis* on γ-alumina pellets. J Food Sci 1990; 55:525-527,531.

93. Iconomou L, Kanellaki M, Voliotis S, Agelopoulos K, Koutinas AA. Continuous wine making by delignified cellulosic materials supported biocatalyst. Appl Biochem Biotechnol 1996; 60:303-313.

94. Bardi EP. Wine making and brewing by immobilized cells on delignified cellulosic material and gluten pellets. PhD Thesis, University of Patras, Greece, 1997.

95. Bakoyianis V, Koutinas AA, Agelopoulos K, Kanellaki M. Comparative study of kissiris, γ-alumina and Ca-alginate as supports of cells for batch and continuous wine making at low temperatures. J Agric Food Chem 1997; 45:4884-4888.

96. Bardi E, Koutinas AA, Psarianos C, Kanellaki M. Volatile by-products formed in low-temperature wine -making using immobilized yeast cells. Process Biochem 1997; 32:579-584.

97. Suomalanien H, Lehtonen M. The production of aroma compounds by yeast. J Inst Brew 1979; 85:149-156.

98. Nykanen L. Formation and occurrence of flavor compounds in wine and distilled alcoholic beverages. Amer J Enol Viticult 1986; 37:84-96.

99. Rapp A, Mandery H. Wine aroma. Experentia 1986; 42:873-884.

100. Stashenko H, Macku C, Takayuki S. Monitoring volatile chemicals formed from must during yeast fermentation. J Agric Food Chem 1992; 40:2257-2259.

101. Shinohara T, Saito K, Yanagida F, Goto S. Selection and hybridization of wine yeasts for improved winemaking properties: Fermentation rate and aroma productivity. J Ferment Bioeng 1994; 77:428-431.

102. Wagener WWD, Wagener GWW. The influence of esters and fusel alcohols content upon the quality of dry white wines. S Afr J Agric Sci 1968; 11:469-476.

103. Soles RM,Ough CS, Kunkee RE. Ester concentration differences in wine fermented by various species and strains of yeasts. Am J Enol Viticult 1982; 33:94-98.

104. Holloway P, Subden RE. Volatile metabolites produced in a Riesling must by wild yeast isolates. Can Inst Sci Technol J 1991; 24:57-59.

105. Cabezudo MD, Gorostiza EF, Herraiz M, Fernandez-Biarange J, Garcla-Dominguez JA, Molera MJ. Mixed columns made to order in gas chromatography. IV. Isothermal selective separation of alcoholic and acetic fermentation products. J Chromat Sci 1978; 16:61-67.

106. Lee CY, Acree TE, Butts RM. Determination of methanol in wine by gas chromatography. Anal Chem 1975; 47:747-748.

107. Gallazzo JL, Bailey JE. In vivo nuclear magnetic resonance analysis of immobilization effects on glucose metabolism of yeast *Saccharomyces cerevisiae*. Biotechnol Bioeng 1989; 33:1283-1289.

108. Hough JS, Briggs DE, Stevens R, Young TW. Malting and Brewing Science, vol. 2. New York: Chapman and Hall Ltd Eds. 1982, reprinted 1987:692.

109. Pajunen E, Gronqvist A, Lommi H. Continuous secondary fermentation and maturation of beer in an immobilized yeast reactor. MBAA Technical Quart 1989; 26:147-151.

110. Yamauchi Y, Okamoto T, Murayama H, Kajino K, Nagara A, Noguchi K. Rapid maturation of beer using an immobilized yeast bioreactor. 2. Balance of total diacetyl reduction and regeneration. J Biotechnol 1995; 38:109-116.

111. Russell I, Stewart GG. Contribution of yeast and immobilization technology to flavor development in fermented beverages. Food Technol 1992; 46:146-150.

112. Norton S, D'Amore T. Physiological effects of yeast cell immobilization: Application for brewing. Enzyme Microbiol Technol 1994; 16:365-375.

113. Bardi EP, Soupioni M, Koutinas AA, Kanellaki M. Effect of temperature on the formation of volatile by-products in brewing by immobilized cells. Food Biotechnol 1996; 10:203-217.

Cold-resistant plant development by genetic manipulation of membrane lipids

O. Ishizaki-Nishizawa

Central Laboratories for Key Technology, Kirin Brewery Co., 1-13-5 Fukuura, Kanazawa, Yokohama, Kanagawa 236, Japan

1
Introduction

Plants are unable to escape from environmental stress by moving as animals do. Therefore, they have developed great capabilities of surviving under abiotic stress, such as low temperatures. Plants from temperate regions are able to tolerate chilling temperature (low nonfreezing temperature) and also freezing temperature after they are kept under low nonfreezing temperature, which is known as cold acclimation.[1,2] These species are classified as chilling-resistant plants. On the contrary, many species from tropical and subtropical regions have difficulty surviving under temperatures below approximately 15°C.[1,2] These species are classified as chilling-sensitive plants and include many economically important crops such as rice, maize and soybean. This diversity indicates that the level of capability of higher plants to survive under low-temperature stress is genetically determined.

Low temperatures induce many changes in plant cellular components. These changes are thought to be involved in the acclimation or protection of plants under low-temperature stress. Plants from temperate regions increase the content of sugar, soluble protein, proline and other organic acids in cells during the cold-acclimation process.[3–5] It was reported that various sugars, certain soluble proteins and proline are able to act as cryoprotectants.[6–10] These increased compounds are thought to directly contribute to low-temperature tolerance of acclimated cells. However, the precise role of the low-temperature-induced compounds *in vivo* is still uncertain.

Low temperatures activate a number of cold-regulated genes, such as those encoding late-embryogenesis-abundant proteins,[11] osmotin,[12] dehydrin,[13] alcohol dehydrogenase[14] and translation elongation factors.[15] Some of these gene products are thought to be effective for the direct protection of the cells from freezing and dehydration, and the other gene products seem to be involved in the metabolic changes that result in the increase of the low-temperature-induced compounds mentioned above. Therefore, overexpression of any one of these genes was applied to improve low-temperature resistance. However, no significant improvement on low-temperature resistance was observed in those genetically engineered plants. The *in vivo* activities and functions of most cold-regulated genes are still unknown.

Lyons[16] and Raison[17] suggested that the primary site of low-temperature damage was the cell membrane. They proposed the membrane hypothesis that the thermotropic phase transition (liquid-crystalline-to-gel phase transition) of membrane lipids results in dysfunction of biological membrane activities, such as selective permeability and sustaining of membrane-bound enzyme activities. Low temperatures induce a number of alterations in cell membranes, including the extent of fatty acid unsaturation,[18] the composition of glycerolipids[19] and the positional redistribution of fatty acids within lipid molecules.[20] All of these changes are thought to be effective to decrease the thermotropic phase transition temperature of membrane lipids. However, bulk membrane lipids even in chilling-sensitive plants would not be expected to initiate phase transition above 0°C.[21] For these reasons, the thermotropic phase separation of membrane lipids did not seem to be a crucial factor in determining the low-temperature sensitivity of higher plants.

2
Low-temperature damage and membrane lipid unsaturation

It has been well established in most organisms that the unsaturation level of membrane lipids alters with changes in growth temperature.[18,22] Low temperatures generally increase the degree of membrane lipid unsaturation, which follows the activation of enzymes such as desaturases of fatty acids. The increase in the level of membrane lipid unsaturation is correlated to the sustained activity of membrane-bound enzymes at low temperatures.[17,23] Several *Arabidopsis thaliana* mutants with a reduced level of fatty acid unsaturation, *fad2* (defective in cytoplasm Δ12 desaturase),[24] *fad5* (defective in plastid *sn*-2-palmitoyl monogalactosyl diacylglycerol (MGDG) desaturase),[25] *fad6* (defective in plastid Δ12 desaturase)[25] and *fab1* (defective in plastid palmitoyl-acyl carrier protein (16:0-ACP) elongase),[26] showed sensitivity to chilling temperatures more than wild-type plant. For these reasons, the unsaturation level of membrane lipids has been considered to be critical in determining the low-temperature sensitivity of higher plants.

3
Genetic manipulation of membrane lipids to confer low-temperature resistance to higher plants

3.1
High decrease of saturation level of total membrane lipids by expressing a broad-specificity Δ9 desaturase

3.1.1
Background

Nobody has succeeded in demonstrating the phase separation above 0°C in bulk membrane lipids even in chilling-sensitive plants,[21] nor has the correlation between the saturation level of total membrane lipids and low-temperature sensitivity in higher plants been established.[27] However, membrane lipids having only saturated fatty acids show phase transition at temperatures above approximately 30°C.[28] By contrast, the presence of only one centrally positioned *cis*-double bond in the fatty acids in membrane lipids effectively decreases the phase transition temperature to approximately 0°C.[28] Thus, there has been sustained interest in the possible role of the saturation level of membrane lipids as a determinant factor of low-temperature sensitivity in higher plants.

Three enzymatic activities are known to be involved in desaturation of saturated fatty acids in higher plants.[29] The first desaturation reaction from stearoyl-ACP (18:0-ACP) to oleoyl-ACP (18:1*c*9-ACP) is accomplished by stearoyl-ACP desaturase which is a soluble protein localized in the stroma of plastid. The other two enzymes are membrane-bound proteins localized in plastid. One of them desaturates palmitic acid (16:0) at *sn*-2 in phosphatidylglycerol (PG) to *trans*-3-hexadecenoic acid (16:1*t*3). Because physical properties of 16:1*t*3 are very similar to those of a saturated fatty acid, 16:1*t*3 is often considered equivalent to a saturated fatty acid. The other enzyme desaturates 16:0 at *sn*-2 in MGDG to *cis*-7-hexadecenoic acid (16:1*c*7). Thus, saturated fatty acids esterified to other lipids or other positions on the glycerol backbone are not desaturated at all in higher plants.

By contrast, in cyanobacteria, desaturation of saturated fatty acids is only accomplished by a Δ9 desaturase which introduces a *cis*-double bond at the Δ9 position of saturated fatty acids linked to any position of all membrane lipids.[30] The *cis*-9-monounsaturated fatty acids esterified to lipids are preferentially utilized as substrate by desaturases specific for the position at Δ6 and Δ12.[31] Therefore, we predicted that the saturation level of total membrane lipids in higher plants might be highly reduced by expressing the Δ9 desaturase from cyanobacteria, and then the transgenic plants might be low-temperature resistant.

3.1.2
Cloning and identification of Δ9 desaturase gene from Anacystis nidulans R-2

Higher plants contain a high amount of 16:0 in membrane lipids.[32] Fatty acid composition analyses showed that the Δ9 desaturases in a cyanobacterium, *Anacystis nidulans* R-2 (*Synechococcus* PCC 7942), have a high substrate specificity for lipid-linked 16:0.[33] For these reasons, we isolated a Δ9 desaturase gene from *A. nidulans* R-2. The Δ9 desaturase gene product was expressed in *E. coli* and the total fatty acid composition of *E. coli* was analyzed. The results signified that the Δ9 desaturase introduces a *cis*-double bond at the Δ9 position of saturated fatty acids esterified to lipids; the enzyme shows higher specificity for 16:0 but also utilizes stearic acid (18:0) as substrate.[34]

3.1.3
Expression of Δ9 desaturase in tobacco plants

The desaturases of cyanobacteria utilize ferredoxin as an intermediate electron donor which is indispensable for expression of the enzymatic activity.[35] In higher plants, lipid biosynthetic activities are mainly confined to the plastid and endoplasmic reticulum (ER). Within plastids, lipids are synthesized by the "procaryotic pathway" and ferredoxin is utilized as the electron donor.[36] Lipids are also synthesized in the ER by the "eucaryotic pathway" and cytochrome b_5 is utilized as the electron donor, as in microsomes of animal cells and yeast.[37] For this reason, expression of the functional Δ9 desaturase in higher plants requires that the protein must be transported to plastids. To accomplish this, the transit peptide of the pea RuBisCO small subunit[38] was placed at the 5' end of the Δ9 desaturase coding region. The chimeric gene was stably introduced into tobacco plant, *Nicotiana tabacum* cv. Xanthi, under transcriptional control of the cauliflower mosaic virus 35S promoter, and a transgenic tobacco plant, which constitutively expressed the Δ9 desaturase, was obtained.[34]

3.1.4
Changes in saturation level of membrane lipids

Figure 1 shows the fatty acid composition of total lipids from leaves and roots of wild-type and the transgenic tobacco plant.[34] The level of saturated fatty acids (16:0 + 16:1*t*3 + 18:0) in leaves and roots of the transgenic plant was reduced to 72 and 61% of that of wild-type plant, respectively. By contrast, the level of monounsaturated fatty acids (16:1*c*9 plus 18:1*c*9) in leaves and roots of the transgenic plant was 16.6- and 8.6-fold increased relative to that of wild-type plant, respectively. This result indicates that the Δ9 desaturase of *A. nidulans* R-2 localized in plastids significantly reduces the saturation level of total membrane lipids in photosynthetic and nonphotosynthetic tissues in higher plants. The level of polyunsaturated fatty acids (16:2 + 18:2 + 16:3 + 18:3) in leaves and roots of the transgenic plant was reduced to 91 and 95% of that of wild-type plant, respectively.

Figure 2 shows the fatty acid composition of seven kinds of major individual membrane lipids, monogalactosyl diacylglycerol (MGDG), digalactosyl diacylglycerol (DGDG), sulfoquinovosyl diacylglycerol (SQDG), phosphatidyl glycerol (PG), phosphatidylcholine (PC), phosphatidylethanolamine (PE), and phosphatidylinositol (PI), from leaves of wild-type and the transgenic plant.[34] The level of saturated fatty acids in individual membrane lipids in the transgenic plant was reduced to 48–86% of that

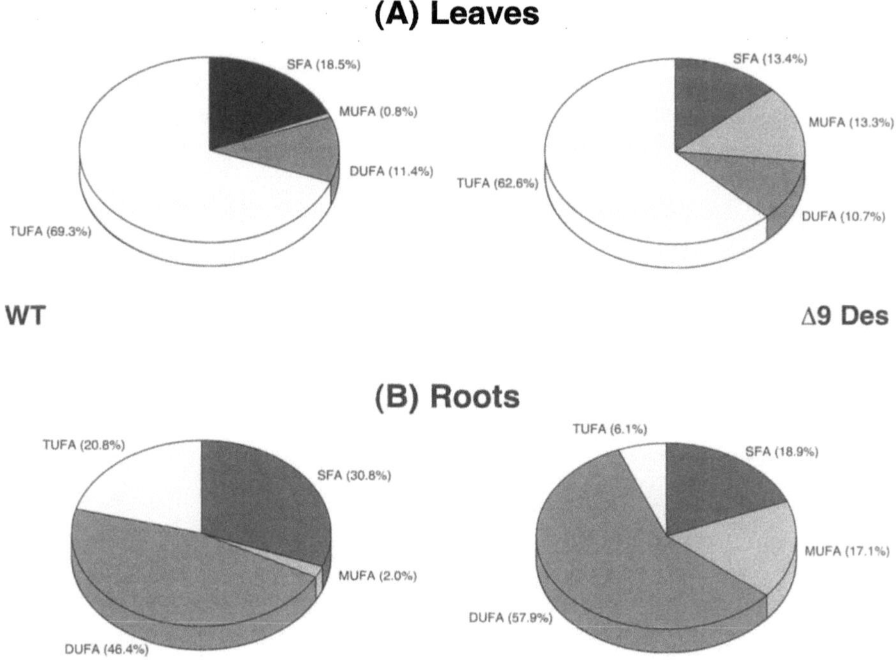

Fig. 1. Fatty acid composition of total lipids from leaves (**A**) and roots (**B**) of wild-type and the transgenic plant (reprinted with permission from ref. 34)
WT wild-type tobacco plant, *Δ9 Des* transgenic tobacco plant expressing Δ9 desaturase of *A. nidulans* R-2, *SFA* saturated fatty acids, *MUFA* monounsaturated fatty acids, *DUFA* diunsaturated fatty acids, *TUFA* triunsaturated fatty acids

of wild-type plant. By contrast, the level of monounsaturated fatty acids in individual membrane lipids in the transgenic plant was 11–25% higher than that of wild-type plant. This result indicates that the saturation level of membrane lipids is significantly reduced in all biomembranes in leaf cells by expressing the Δ9 desaturase of *A. nidulans* R-2 in plastids in higher plants. Although the Δ9 desaturase of *A. nidulans* R-2 is thought to be localized in plastids in the transgenic plant, a high amount of monounsaturated fatty acids is also detected in membrane lipids, PC, PE and PI, which are synthesized only in the endoplasmic reticulum.

3.1.5
Low-temperature resistance

When wild-type and the transgenic plant grown on agar media in a plastic box were exposed to 1°C under constant light for 11 d, wild-type plant exhibited chlorosis (decomposition of chloroplast), while the transgenic plant showed no injury.[34] This

(A) Plastidial lipids:PG

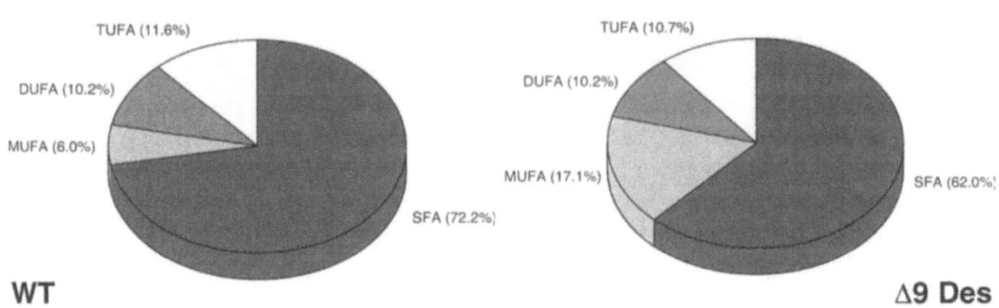

SQDG

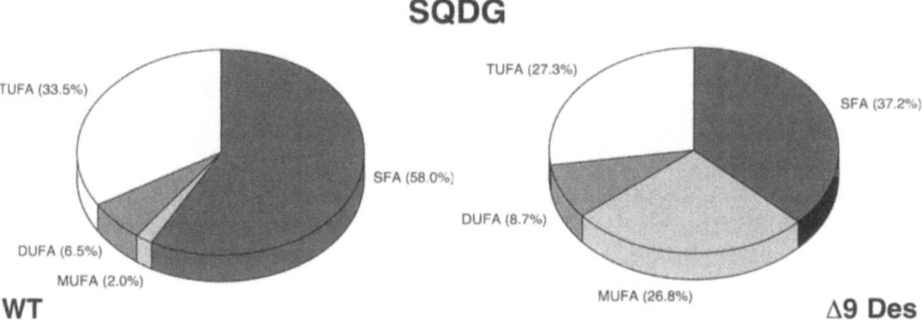

MGDG

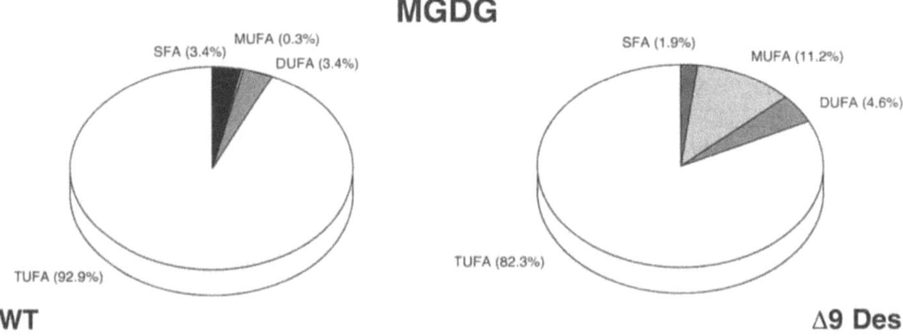

DGDG

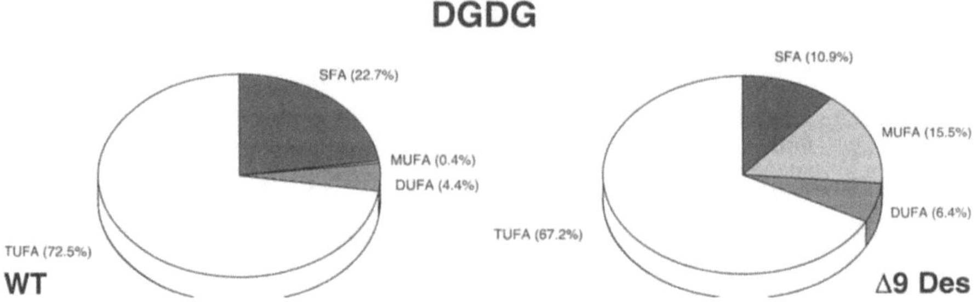

(B) Cytoplasmic lipids: PC

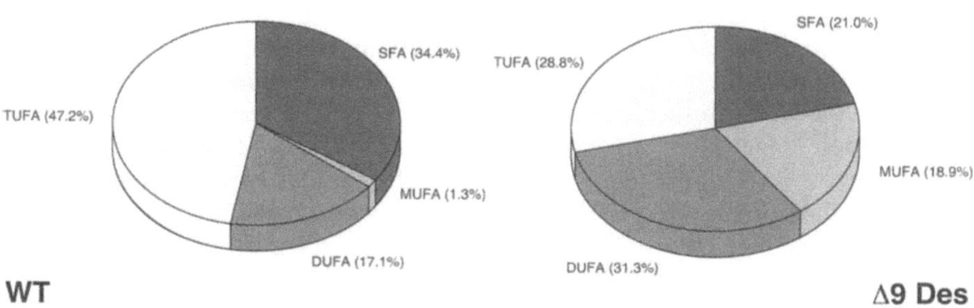

PE

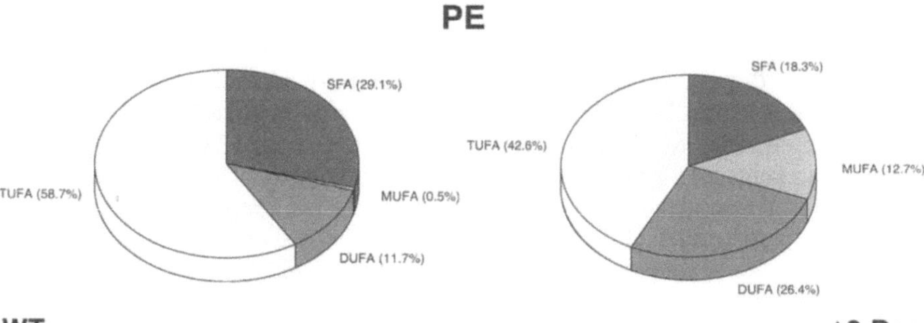

PI

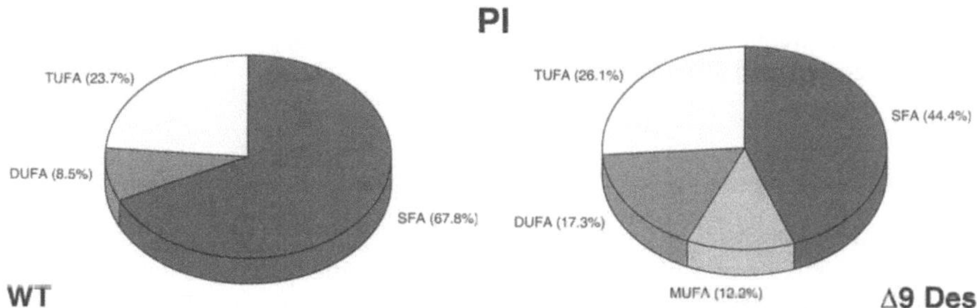

Fig. 2. Fatty acid composition of individual membrane lipids from leaves of wild-type and the transgenic plant (reprinted with permission from ref. 34)
(A) Plastidial lipids, membrane lipids synthesized in plastids; *PG* phosphatidyl glycerol, *SQDG* sulfoquinovosyl diacylglycerol, *MGDG* monogalactosyl diacylglycerol, *DGDG* digalactosyl diacylglycerol; (B) Cytoplasmic lipids, membrane lipids synthesized in the endoplasmic reticulum; *WT* wild-type tobacco plant, *Δ9 Des* transgenic tobacco plant expressing Δ9 desaturase of *A. nidulans* R-2, *SFA* saturated fatty acids, *MUFA* monounsaturated fatty acids, *DUFA* diunsaturated fatty acids, *TUFA* triunsaturated fatty acids, *PC* phosphatidylcholine, *PE* phosphatidylethanolamine, *PI* phosphatidylinositol

indicates that the transgenic plant is resistant to a short period of exposure to low chilling temperature. Seeds of wild-type and the transgenic plant were germinated on agar media in a petri dish at 10°C under constant light for 52 d. Although wild-type plant seedlings had very pale green cotyledons, the transgenic plant seedlings were the same as the seedlings germinated at normal temperature (25°C).[34] This indicates that the transgenic plant is able to develop chloroplasts normally at the germination stage under moderate chilling temperature. Seeds were germinated and grown on agar media in a petri dish at 10°C under constant light for 70 d. Although the growth rate of wild-type plant was not significantly less than that of the transgenic plant, chlorosis was developed on almost the whole region of the primary and secondary leaves of wild-type plant. By contrast, the primary and secondary leaves of the transgenic plant mostly showed no chlorosis.[34] These results indicate that the transgenic plant is so chilling-resistant as to be able to grow normally under moderate chilling temperature. In these low-temperature sensitivity analyses, plant materials germinated or grown on agar media in a plastic box or a petri dish were used to make the humidity condition constant, because the low-temperature injury of higher plants exhibits humidity dependency.

Seeds of wild-type and the transgenic plant were germinated and grown on soil at 15°C under a photo-period of 16 h light and 8 h dark for 12 weeks. Height of the transgenic plant was approximately 15% taller than that of wild-type plant (unpubl. data). This indicates that the growth rate of the transgenic plant is higher than wild-type plant under chilling temperature. All of these results signify that the transgenic plant is able to grow like naturally chilling-resistant plant species under chilling temperatures in the range of 0–15°C. The significant decrease of the saturated fatty acids level in most membrane lipids is thought to be important in significantly enhancing chilling resistance of higher plants.

3.2
Decrease of saturation level of phosphatidylglycerol by expressing glycerol-3-phosphate acyltransferase from chilling-resistant plants

3.2.1
Background

Fatty acid composition of seven kinds of major membrane lipids (MGDG, DGDG, SQDG, PG, PC, PE and PI) was compared among some higher plant species whose level of chilling sensitivity is diverse from sensitive to resistant.[39] This analysis suggested that chilling sensitivity of higher plants is correlated with the saturation level of fatty acids in a specific phospholipid in plastid membrane, PG. Fatty acid composition of PG was compared among many plant species to evaluate the correlation.[40] The comparison indicated that the saturated molecular species of PG, 1-16:0-2-16:0-PG and 1-16:0-2-16:1t3-PG, account for less than 20% in the chilling-resistant plant species and more than 25% in the chilling-sensitive plant species. The phase transition temperatures of saturated molecular species of PG were above approximately 30°C. On the contrary, unsaturated molecular species of PG exhibit phase separation at temperatures below 0°C. For these reasons, PG was thought to be one of the determinant factors of chilling sensitivity of higher plants.[39] However, it was also unlikely that a very minor component of saturated molecular species of PG in plastid membrane lipids could initiate phase separation even in the small area of plastid membrane, which results in low-temperature injury, because the proportion of PG in plastid membrane lipids is less than 10%.[41]

The pathway of PG synthesis suggested that glycerol-3-phosphate acyltransferase (glycerol-P acyltransferase) in plastid was a key enzyme to determine the level of saturated molecular species in PG.[42,43] Glycerol-P acyltransferase esterifies the acyl group of acyl-ACP to the *sn*-1 position of glycerol-3-phosphate. If 18:1*c*9 of 18:1*c*9-ACP is esterified by glycerol-P acyltransferase, the PG is an unsaturated molecular species. By contrast, if 16:0 of 16:0-ACP is esterified, the PG is fated to be a saturated molecular species. The substrate selectivity of glycerol-P acyltransferase was compared between that from chilling-resistant plants, spinach and pea,[44] and the chilling-sensitive plant, squash.[45] The enzyme in the chilling-resistant plants had a rather strict selectivity for 18:1*c*9-ACP over 16:0-ACP. By contrast, the enzyme in the chilling-sensitive plant was nonspecific, transferring either of 18:1*c*9 or 16:0 at comparable rates. Therefore, we predicted that the level of saturated molecular species in PG in plastid membranes of chilling-sensitive plants might be reduced by expressing glycerol-P acyltransferase from chilling-resistant plants, and then the transgenic plants might be low-temperature resistant.

3.2.2
Expression of glycerol-P acyltransferase from A. thaliana in tobacco plants

Glycerol-P acyltransferase from *A. thaliana* (chilling-resistant plant) was revealed to have a substrate selectivity for 18:1*c*9-ACP over 16:0-ACP. We obtained a transgenic tobacco plant by expressing a glycerol-P acyltransferase cDNA from *A. thaliana* in *Nicotiana tabacum* var. Samsun.[46] The glycerol-P acyltransferase gene was under transcriptional control of the cauliflower mosaic virus 35S promoter.

3.2.3
Changes in saturation level of membrane lipids in leaves

There are no significant differences in the absolute amounts of seven kinds of major membrane lipids between wild-type and the transgenic plant. There are no significant differences between wild-type and the transgenic plant in the fatty acid composition of six kinds of major membrane lipids, except PG. While the content of saturated fatty acids (16:0 + 16:1*t*3 + 18:0) in PG in leaves was 68% in wild-type plant, it was 64% in the transgenic plant.[46] This result demonstrates that the saturation level of fatty acids in PG is able to be reduced by introduction and expression of an appropriate glycerol-P acyltransferase. Considering that the *sn*-2 position of PG is occupied exclusively by 16:0 or 16:1*t*3, the content of saturated molecular species in PG calculated from fatty acid composition was 36% for wild-type plant and 26% for the transgenic plant. The content of saturated molecular species in PG in leaves of *A. thaliana* was 20%. This indicates that the level of saturated molecular species in PG in leaves of the transgenic tobacco plant is shifted toward that in chilling-resistant plants.[46]

3.2.4
Low-temperature resistance

The sensitivity to low temperature of the transgenic plant was analyzed by photosynthetic activity and visible damage of the whole plant. In the former, leaves were exposed to 1°C for 4 h in the light before returning them to 27°C and monitoring the effects of this treatment on photosynthetic oxygen evolution. The photosynthetic activity of wild-type plant was reduced by 25±11% by chilling treatment at 1°C for 4 h, but in the transgenic plant only by 7±3%.[46] This result indicates that the photosynthetic activity of the transgenic plant is more resistant to low chilling temperature than that of wild-type plant. Recovery of photosynthesis after the temperature was raised again to 27°C was larger in the transgenic plant than in wild-type plant. Although photoinhibition of photosynthesis might be explained by phase separation of membrane lipids or loss of physiological function of PG in the membrane lipids localized around the photosynthetic reaction center (photosystem II protein complex), the mechanism of how the proportion of saturated fatty acids in PG affects the photosynthetic activity in higher plant is still unclear.

For the second chilling-sensitivity analysis, the plant grown on agar media in a plastic box was exposed to 1°C for 10 days to evaluate visible damage to the whole plant. Damage such as chlorosis and total deterioration of leaves was obvious after the plants were kept at 25°C for 2 days after chilling treatment. Leaves of wild-type plant suffered partial chlorosis. By contrast, the transgenic plant showed no injury.[46] This indicates that the transgenic plant is more resistant to low chilling temperature than wild-type plant. The resistant phenotype of the transgenic plant is probably based on a recovery of photosynthesis at normal temperature, which is better than that of wild-type plant.

3.3
Slight increase of triunsaturated fatty acid content by expressing plastid ω-3 desaturase

Another group reported that the content of triunsaturated fatty acids (16:3 + 18:3) in leaves was approximately 1.1-fold increased relative to that of wild-type plant by overexpressing a plastid ω-3 desaturase from *A. thaliana* in plastids in the tobacco plant *Nicotiana tabacum* cv. SR1.[47] The transgenic plant exhibited an improved growth rate compared to that of wild-type plant at normal temperature (25°C) after a short period (7 d) of exposure to low chilling temperature (1°C). However, it is very unlikely that the alteration of chilling sensitivity in this case is caused by a mechanism related to membrane lipids, because the alteration of triunsaturated fatty acids content is too small.

4

Conclusions and discussion

Three strategies for genetic manipulation of membrane lipids have been reported to confer low-temperature resistance to higher plants. In Δ9 desaturase strategy, the level of saturated fatty acids in total membrane lipids in leaves and roots was reduced to 72 and 61% of that of wild-type, respectively; the level of saturated fatty acids in individual membrane lipids in leaves was reduced to 48–86% of that of wild-type; the level of saturated fatty acids in PG in leaves was reduced to 86% of that of wild-type, and the level of saturated molecular species in PG was calculated to be 54% of that of wild-type; the level of monounsaturated fatty acids in individual membrane lipids in leaves was 11–25% higher than that of wild-type; the level of triunsaturated fatty acids in total membrane lipids in leaves and roots was reduced to 90 and 29% of that of wild-type, respectively.[34]

By contrast, in glycerol-P acyltransferase strategy, the level of saturated fatty acids in PG in leaves was slightly reduced to 94% of that of wild-type, which resulted in the reduction of saturated molecular species in PG to 78% of that of wild-type.[46] The reduction level of saturated molecular species in PG in Δ9 desaturase strategy was 2.1-fold higher than that in glycerol-P acyltransferase strategy. In ω-3 desaturase strategy, the level of triunsaturated fatty acids in total membrane lipids in leaves was slightly increased to 1.1-fold of that of wild-type.[47] These results indicate that Δ9 desaturase strategy is much more effective in altering the saturation or unsaturation level of fatty acids in membrane lipids in higher plants than glycerol-P acyltransferase or ω-3 desaturase strategy.

The transgenic plant expressing the Δ9 desaturase is resistant to a short-period of exposure to low chilling temperature (1°C). Seeds of the transgenic plant are also able to germinate and grow normally under chilling temperatures in the range of 10–15°C.[34] These results indicate that the transgenic plant expressing the Δ9 desaturase is like naturally chilling-resistant plant species. By contrast, the transgenic plant expressing glycerol-P acyltransferase or ω-3 desaturase shows improved recovery of photosynthetic activity or improved growth rate at normal temperature (25°C) after a short period of exposure to low chilling temperature (1°C).[46,47] These results indicate that these two kinds of transgenic plants require normal temperature to recover from the damage due to a short period of exposure to low chilling temperature. This signifies that the level of chilling resistance of these two kinds of transgenic plants is quite different from that of naturally chilling-resistant plants and the transgenic plant expressing the Δ9 desaturase. This suggests that Δ9 desaturase strategy is the most effective to confer low-temperature resistance to higher plants in the three strategies.

The content of polyunsaturated fatty acids in leaf membrane lipids in *A. thaliana* mutants, *fad5* and *fab1*, was 90 and 88% of that of wild-type plant, respectively.[25,26] The content of triunsaturated fatty acids in leaf membrane lipids in *A. thaliana* mutants, *fad2* and *fab1*, was 94 and 90% of that of wild-type plant, respectively.[24,26] These mutants were reported to be low-temperature sensitive compared to wild-type plant. By contrast, the content of polyunsaturated and triunsaturated fatty acids in leaf membrane lipids in the transgenic tobacco plant expressing the Δ9 desaturase was 91 and 90% of that of wild-type plant, respectively.[34] However, the transgenic tobacco plant significantly exhibits a low-temperature-resistant phenotype. If the polyunsaturated or triunsaturated fatty acid contents were a crucial factor to determine the low-temperature sensitivity in higher plants, the transgenic tobacco plant would be sensitive to low temperature. This suggests that the level of polyunsaturated or triunsaturated fatty acids is not a crucial factor determining low-temperature sensitivity in higher plants. For this reason, another mechanism different from that related to membrane lipids is probably required for explaining the improvement of low-temperature sensitivity in the transgenic tobacco plant expressing ω-3 desaturase. The level of saturated fatty acids in membrane lipids is thought to be more crucial for determining the function of membrane rather than the level of triunsaturated fatty acids, because the physical properties of membrane lipids are heavily dependent on the saturation level of fatty acids in lipids.

Nobody has succeeded in answering the question how the saturation level of fatty acids in membrane lipids determines the low-temperature sensitivity of higher plants. Moon et al.[48] compared the level of photoinhibition between wild-type and the transgenic tobacco plant which is sensitive to chilling temperature more than wild-type plant, because the transgenic plant expresses squash (a chilling-sensitive plant) glycerol-P acyltransferase. They suggested that the recovery of photoinhibition in wild-type leaves was better than that in the transgenic tobacco leaves. However, they failed to show the difference in recovery of photoinhibition in thylakoid membranes between wild-type and the transgenic tobacco plant. These results suggest that the difference in recovery of photoinhibition in leaves between wild-type and the transgenic tobacco plant is not based on the difference in fatty acid saturation level in PG. It is very likely that a large amount of squash glycerol-P acyltransferase protein depresses the activity of endogenous components of photosynthesis in leaves, because this soluble protein is overexpressed and localized in plastids in the transgenic tobacco plant. For this reason, the comparison between wild-type and the transgenic plant which is resistant to low temperature more than wild-type plant is required to eliminate the negative effect of the overexpressed foreign protein from the evaluation of low-temperature resistance in transgenic plants. The transgenic tobacco plant expressing the Δ9 desaturase seems to be very suitable for analyzing the mechanism of saturation level of fatty acids in membrane lipids for determining the low-temperature sensitivity, because the phenotype of low-temperature resistance of the transgenic tobacco plant is much more distinct than the other two strategically different transgenic tobacco plants expressing glycerol-P acyltransferase from *A. thaliana* or ω-3 desaturase.

The Δ9 desaturase strategy is also the most promising for obtaining transgenic low-temperature resistant plants from many economically important crops, and quite a few trials to introduce and express the Δ9 desaturase in many kinds of plant species will reveal the capability of the strategy to confer the low-temperature resistance to higher plants.

5
References

1. Wang CY, ed. Chilling Injury of Horticultural Crops. Boca Raton: CRC Press, 1990.
2. Thomashow MF. *Arabidopsis thaliana* as a model for studying mechanisms of plant cold tolerance. In: Meyerowitz EM, Somerville CR, eds. Arabidopsis. New York: Cold Spring Harbor Laboratory Press, 1994:807-834.
3. Levitt J. Responses of Plants to Environmental Stress. Chilling, Freezing and High Temper-

ature Stresses. New York: Academic Press, 1980.

4. Sakai A, Larcher W. Frost Survival of Plants: Responses and Adaptations to Freezing Stress. New York: Springer-Verlag, 1987.

5. Guy CL. Cold acclimation and freezing stress tolerance: Role of protein metabolism. Annu Rev Plant Physiol Plant Mol Biol 1990; 41:187-223.

6. Volger HG, Heber U. Cryoprotective leaf proteins. Biochim Biophys Acta 1975; 412:335-349.

7. Rudolph AS, Crowe JH. Membrane stabilization during freezing: The role of two natural cryoprotectants, trehalose and proline. Cryobiology 1985; 22:367-377.

8. Carpenter JF, Crowe JH. The mechanism of cryoprotection of proteins by solutes. Cryobiology 1988; 25:244-255.

9. Hincha DK, Heber U, Schmitt JM. Freezing ruptures thylakoid membranes in leaves, and rupture can be prevented in vitro by cryoprotective proteins. Plant Physiol Biochem 1989; 27:795-801.

10. Hincha DK, Heber U, Schmitt JM. Proteins from frost-hardy leaves protect thylakoids against mechanical freeze-thaw damage in vitro. Planta 1990; 180:416-419.

11. Gilmour SJ, Artus NN, Thomashow MF. cDNA sequence analysis and expression of two cold-regulated genes of *Arabidopsis thaliana*. Plant Mol Biol 1992; 18:13-21.

12. Zhu B, Chen THH, Li PH. Expression of an ABA-responsive osmotin-like gene during the induction of freezing tolerance in *Solanum commersonii*. Plant Mol Biol 1993; 21:729-735.

13. Bartels D, Schneider K, Terstappen G, Piatkowski D, Salamini F. Molecular cloning of abscisic acid-modulated genes which are induced during desiccation of the resurrection plant *Craterostigma plantagineum*. Planta 1990; 182:27-34.

14. Jarillo JA, Leyva A, Salinas J, Martinez-Zapater JM. Low temperature induces the accumulation of alcohol dehydrogenase mRNA in *Arabidopsis thaliana*, a chilling-tolerant plant. Plant Physiol 1993; 101:833-837.

15. Dunn MA, Morris A, Jack PL, Hughes MA. A low-temperature-responsive translation elongation factor 1a from barley (*Hordeum vulgare* L.). Plant Mol Biol 1993; 23:221-225.

16. Lyons JM. Chilling injury in plants. Annu Rev Plant Physiol 1973; 24:445-466.

17. Raison JK. The influence of temperature-induced phase changes on kinetics of respiratory and other membrane-associated enzymes. J Bioenerg 1973; 4:258-309.

18. Cossins AR. Homeoviscous adaptation of biological membranes and its functional significance. In: Cossins AR, ed. Temperature Adaptation of Biological Membranes, London: Portland, 1994:63-76.

19. Lynch DV, Thompson GA Jr. Low temperature-induced alterations in the chloroplast and microsomal membranes of *Dunaliella salina*. Plant Physiol 1982; 69:1369-1375.

20. Thompson GA Jr, Nozawa Y. The regulation of membrane fluidity in *Tetrahymena*. Biomembranes 1984; 12:397-432.

21. Lynch DV. Chilling injury in plants: The relevance of membrane lipids. In: Katterman F, ed. Environmental Injury to Plants. San Diego: Academic Press, 1990:17-34.

22. Cossins AR, ed. Temperature Adaptation of Biological Membranes. London: Portland, 1994.

23. Raison JK. Membrane lipids: structure and function. In: Stumpf PK, ed. The Biochemistry of Plants, vol. 4. New York: Academic Press, 1980.

24. Miquel M, James D Jr, Dooner H, Browse J. *Arabidopsis* requires polyunsaturated lipids for low-temperature survival. Proc Natl Acad Sci USA 1993; 90:6208-6212.

25. Hugly S, Somerville C. A role for membrane lipid polyunsaturation in chloroplast biogenesis at low temperature. Plant Physiol 1992; 99:197-202.

26. Wu J, Lightner J, Warwick N, Browse J. Low-temperature damage and subsequent recovery of fab1 mutant *Arabidopsis* exposed to 2°C. Plant Physiol 1997; 113:347-356.

27. Somerville C. Direct tests of the role of membrane lipid composition in low-temperature-induced photoinhibition and chilling sensitivity in plants and cyanobacteria. Proc Natl Acad Sci USA 1995; 92:6215-6218.

28. Silvius JR. Thermotropic phase transitions of pure lipids in model membranes and their modification by membrane proteins. In: Jost PC, Griffith OH, eds. Lipid-Protein Interac-

tions, vol. 2. New York: Wiley, 1982:239-281.

29. Ohlrogge J, Browse J. Lipid biosynthesis. Plant Cell 1995; 7:957-970.

30. Lem NW, Stumpf PK. In vitro fatty acid synthesis and complex lipid metabolism in the cyanobacterium *Anabaena variabilis*. Plant Physiol 1984; 74:134-138.

31. Murata N, Wada H. Acyl-lipid desaturases and their importance in the tolerance and acclimatization to cold of cyanobacteria. Biochem J 1995; 308:1-8.

32. Browse J, Warwick N, Somerville CR, Slack CR. Fluxes through the prokaryotic and eukaryotic pathways of lipid synthesis in the '16:3' plant *Arabidopsis thaliana*. Biochem J 1986; 235:25-31.

33. Bishop DG, Kenrick JR, Kondo T, Murata N. Thermal properties of membrane lipids from two cyanobacteria, *Anacystis nidulans* and *Synechococcus* sp. Plant Cell Physiol 1986; 27:1593-1598.

34. Ishizaki-Nishizawa O, Fujii T, Azuma M, Sekiguchi K, Murata N, Ohtani T, Toguri T. Low-temperature resistance of higher plants is significantly enhanced by a nonspecific cyanobacterial desaturase. Nature Biotechnol 1996; 14:1003-1006.

35. Wada H, Avelange-Macherel MH, Murata N. The desA gene of the cyanobacterium *Synechocystis* sp. strain PCC6803 is the structural gene for $\Delta12$ desaturase. J Bacteriol 1993; 175:6056-6058.

36. McKeon TA, Stumpf PK. Purification and characterization of the stearoyl-acyl carrier protein desaturase and the acyl-acyl carrier protein thioesterase from maturing seeds of safflower. J Biol Chem 1982; 257:12141-12147.

37. Smith MA, Cross AR, Jones OTG, Griffiths WT, Stymne S, Stobart K. Electron-transport components of the 1-acyl-2-oleoyl-sn-glycero-3-phosphocholine $\Delta12$-desaturase in microsomal preparations from developing safflower (*Carthamus tinctorius* L.) cotyledons. Biochem J 1990; 272:23-29.

38. Schreier PH, Seftor EA, Schell J, Bohnert HJ. The use of nuclear-encoded sequences to direct the light-regulated synthesis and transport of a foreign protein into plant chloroplasts. EMBO J 1985; 4:25-32.

39. Murata N, Sato N, Takahashi N, Hamazaki Y. Compositions and positional distributions of fatty acids in phospholipids from leaves of chilling-sensitive and chilling-resistant plants. Plant Cell Physiol 1982; 23:1071-1079.

40. Roughan PG. Phosphatidylglycerol and chilling sensitivity in plants. Plant Physiol 1985; 77:740-746.

41. Murata N, Nishida I. Lipids in relation to chilling sensitivity of plants. In: Wang CY, ed. Chilling Injury of Horticultural Crops. Boca Raton: CRC Press, 1990:181-199.

42. Sparace SA, Mudd JB. Phosphatidylglycerol synthesis in spinach chloroplasts: Characterization of the newly synthesized molecule. Plant Physiol 1982; 70:1260-1264.

43. Murata N. Molecular species composition of phosphatidylglycerols from chilling-sensitive and chilling-resistant plants. Plant Cell Physiol 1983; 24:81-86.

44. Frentzen M, Heinz E, McKeon TA, Stumpf PK. Specificities and selectivities of glycerol-3-phosphate acyltransferase and monoacylglycerol-3-phosphate acyltransferase from pea and spinach chloroplasts. Eur J Biochem 1983; 129:629-636.

45. Frentzen M, Nishida I, Murata N. Properties of the plastidial acyl-(acyl-carrier-protein):glycerol-3-phosphate acyltransferase from the chilling-sensitive plant, squash (*Cucurbita moschata*). Plant Cell Physiol 1987; 28:1195-1201.

46. Murata N, Ishizaki-Nishizawa O, Higashi S, Hayashi H, Tasaka Y, Nishida I. Genetically engineered alteration in the chilling sensitivity of plants. Nature 1992; 356:710-713.

47. Kodama H, Hamada T, Horiguchi G, Nishimura M, Iba K. Genetic enhancement of cold tolerance by expression of a gene for chloroplast ω-3 fatty acid desaturase in transgenic tobacco. Plant Physiol 1994; 105:601-605.

48. Moon BY, Higashi S, Gombos Z, Murata N. Unsaturation of the membrane lipids of chloroplasts stabilizes the photosynthetic machinery against low-temperature photoinhibition in transgenic tobacco plants. Proc Natl Acad Sci USA 1995; 92:6219-6223.

The potential use of cold-adapted rhizobia to improve symbiotic nitrogen fixation in legumes cultivated in temperate regions

D. Prévost[1]*, P. Drouin[1] and H. Antoun[2]

[1] Soils and Crops Research and Development Centre, Agriculture and Agri-Food Canada, 2560 Hochelaga Blvd, Sainte-Foy, Quebec, Canada G1V 2J3
[2] Department of Soil Science and Agri-Food Engineering, Faculty of Agriculture and Food Science, Pavillion Charles-Eugène Marchand, Laval University, Sainte-Foy, Quebec, Canada G1K 7P4

1
Introduction

Nitrogen is one of the most limiting factors for plant growth, and nitrogen fertilizer is one of the major cost for crop production. Legumes can meet most of their nitrogen requirement through the process of biological nitrogen fixation in which atmospheric nitrogen (N_2) is reduced to ammonium (NH_4) by the enzyme nitrogenase present in the bacteria rhizobia. The site of nitrogen fixation is inside the nodule, a root structure where bacteroids of rhizobia take energy to reduce nitrogen from carbohydrates derived from plant photosynthesis. Legumes species are found under all ecosystems from tropical to arctic climates, and usually, nodulated species are associated with one or few rhizobial species. Rhizobia show high levels of phylogenetic diversity and they are actually grouped in five genera: *Azorhizobium, Rhizobium, Sinorhizobium, Mesorhizobium* and *Bradyrhizobium.*[1] For instance, strains of *Sinorhizobium meliloti* (former *Rhizobium meliloti*) are specific to alfalfa and cannot nodulate soybean, and soybean is usually nodulated by *Bradyrhizobium japonicum* or *Sinorhizobium fredii* which do not nodulate alfalfa. Cultivating legumes is beneficial for soil N-fertility and overcomes the high costs of N-fertilizers which cause groundwater pollution. Legumes have higher protein content than cereals, and constitute high quality protein or oil source for human and animal nutrition. They are also used for soil remediation or reforestation of devasted areas.

Legume productivity is thus influenced through the interactions of legume host, rhizobia and environmental factors. Improvement of legume productivity can be achieved through the management of cropping systems to minimize stresses and maximize yield, and by the development of legume cultivars and the selection of rhizobia with high efficiency to fix nitrogen.[2] Both macro- and microsymbionts should be tolerant to soil and environmental stress factors. Inoculation of legumes

---

* Corresponding author

with rhizobia has been practiced for a long time, and the first commercial legume inoculants were produced at the beginning of the 1930s. However, the competition for nodulation between indigenous soil rhizobia and commercial inoculant strains may reduce the economic benefits (increased crop yield) of inoculation.[3] The need to inoculate is not universal and some legumes may be effectively nodulated by indigenous soil rhizobia. Commercial inoculants consist of preparations of living rhizobia in a solid or liquid support that is applied to the seeds or in soil at sowing. The use of pre-inoculated legumes seeds is an approach that eliminates the inoculation procedure by the farmers. A review of different methods of inoculation and types of inoculants is given by Smith.[4]

Under cool temperate climates, the early growth of legumes occurs when soil temperatures are far below the optimum range for the growth of rhizobia and for nitrogen fixation by the symbiotic association.[5] For instance, Sprent[6] reported that the establishment of an effective N_2-fixing symbiosis one week earlier in the growing season could double the amount of N_2 fixed and thus increase legume productivity. In Canada, slow establishment and regrowth of alfalfa (a perennial legume) and poor early vegetative growth of soybean (an annual legume) in spring may be attributed to the inhibitory effect of cool soil conditions.[7,8] With the aim to improve nitrogen fixation of temperate legumes under low temperatures, selection of cold-adapted rhizobia has been considered as a valuable tool in many studies. Screening techniques often involve comparison of symbiotic activities between rhizobia isolated from a temperate legume species cultivated in northern regions to those originating from the same legume species cultivated in southern regions.[9] Psychrotrophic rhizobia from legumes indigenous to arctic ecosystems constitute also a good genetic reservoir to transfer cold-adaptation traits to rhizobia that are specific to legumes of agricultural importance.[10] The aim of this chapter is to present the actual knowledge of cold-adapted rhizobia and the benefits of using them in agriculture.

2

Effect of low temperature on the establishment of the symbiosis between legumes and rhizobia

In the plant-soil ecosystem, temperature is a very important environmental factor, that influences the different interactions occurring between plant, soil and microorganisms. That temperature effect can be indirect, through its interaction with other environmental parameters like moisture or oxygen, or direct by affecting the rate of biological reactions. The behavior of the free-living rhizobia in soil, and every step involved in the establishment of the symbiosis between rhizobia and legumes are therefore affected by cold temperature.

The optimum growth temperature for most rhizobia occurs between 25 and 30°C, with 35°C preferred by *S. meliloti* strains.[11] However, strains of clover rhizobia grow well at 10°C,[12] and some arctic rhizobia from *Astragalus* and *Oxytropis*[13] and strains of *Rhizobium leguminosarum* bv. viciae isolated from *Lathyrus* spp. found in arctic regions[14] are able to grow at 5°C. The survival of rhizobia in soil is more affected by elevated temperature, particularly under moist conditions, than by low temperature. In fact, as rhizobia are tolerant to temperatures below 4°C,[11] manufacturers recommend the storage of rhizobial inocula under refrigeration to insure a higher number of rhizobia per seed at sowing.

Nodulation of soybean (*Glycine max* (L.) Merrill) is drastically reduced at a root temperature of 15°C as compared to 25°C.[15] For soybean, nodulation occurs predominantly within the topmost 0–10 cm of root, Montanez et al.,[15] however, observed that, in general, about 17% of nodules were located below 10 cm of roots at 15°C compared to 26% at 25°C. These observations suggest that cold temperature might also affect the ability of the different strains to migrate in the rhizosphere.

In a study on the effect of temperature on the early symbiotic establishment events between soybean and *B. japonicum*, Zhang and Smith[16] observed that at root zone temperatures between 25 and 17.5°C, the infection processes were progressively delayed as temperature declined. At root zone temperatures lower than 17°C, the infection steps were strongly inhibited. In legumes, the formation of nodules is the result of a series of complex signals exchange between plant and bacteria. Flavonoids are secreted by plant roots and they cause the induction of nodulation genes *(nod)* in compatible rhizobia.[17] As a result, rhizobia produce lipo-chito-oligosaccharidic signal molecules known as Nod-factors, that induce various plant responses including root-hair deformation, cortical cell-division and nodule formation.[18,19] Low root zone temperature inhibits the biosynthesis of genistein, an isoflavone identified as one of the major compounds in soybean root extract responsible for induction of *nod* genes, in soybean plants.[20] In *B. japonicum*, higher concentrations of genistein were required for maximum expression of *nod* genes at 10 and 15°C than at 25°C.[21] McKay and Djordjevic[22] examined Nod metabolite production and excretion in *R. leguminosarum* bv. trifolii, under a range of environmental conditions reported as having an adverse effect on nodulation (pH 5, cold temperature, and low levels of phosphorus). They observed that by lowering the incubation temperature from 28 to 18°C the relative concentration and number, and in particular the amount of Nod metabolite excreted, were significantly reduced in strains showing a temperature sensitive nodulation phenotype with *Trifolium subterraneum*. The ability of different strains of rhizobia to produce and release Nod factors is probably a major determinant of nodule formation and occupancy at low temperature, and it can explain in part the inhibitory effect of low temperature on nodule formation and nitrogen fixation in legumes.

Cold root temperatures can significantly influence the competition between strains of rhizobia.[23] For example, when two strains of *R. leguminosarum* bv. trifolii where used as inoculum with clover plants, the majority of nodules were formed by one strain at 12°C, but the other strain was more competitive at 25°C.[24]

Most legumes cultivated in temperate area have an optimum growth temperature ranging from 15 to 25°C, and they are exposed to a wide range of temperature during their life cycle. Annual and perennial legumes are exposed to cold temperature early in spring and perennial legumes are also exposed to extreme cold during overwintering. Extreme temperatures are known to reduce nitrogenase activity in nodules of annual and perennial legumes. From 5 to 25°C the nitrogenase activity of alfalfa-attached nodules increased linearly with a Q_{10} value of 1.7.[25] Layzell et al.[26] also recorded Q_{10} value ranging from 1.3 to 2.4 for nitrogenase activity (C_2H_2 reduction) in intact nodulated soybean plants at root temperatures between 15 and 25°C. Lindström[27] performed a study with three different legumes, to analyze the different factors affecting nitrogenase activity measured under field conditions in Finland. Generally, there was a good correlation between nitrogenase activity and plant growth rate, indicating that all factors influencing growth rate in the field also affected nitrogenase activity. With red clover *(Trifolium pratense)* and alfalfa *(Medicago sativa)* nitrogenase activity was still detected in November when soil temperature was 1.5°C and air temperature 0.5°C. In a large study performed with 226 isolates of *S. meliloti*, Rice et al.[7] found that alfalfa nodulation occurred at 9°C, but there was no significant N_2-dependent plant growth at this temperature. However, several isolates were able to produce N_2-dependent plant growth at root temperatures between 10 and 12°C. Lynch and Smith[29] did not observe any difference between the soybean cv. Mapple Arrow and a cold tolerant Evans isoline at a root zone temperature of 19°C. Both cultivars were equally limited by the low root zone temperature and at 19°C N_2-fixation was substantially reduced (30–40%), 44 days after inoculation. However, a cold-tolerant line of *Phaseolus vulgaris* had a superior growth at low temperature and this was attributed to its ability to nodulate well, and to form larger nodules which fixed more N under low temperature stress.[30] When *Vicia faba* L. was grown at 10°C as compared to 18°C, nodules developed much more slowly and were much larger, but the larger size did not compensate for the lower nitrogenase activity measured in the cold grown nodules.[31] Finally, low root temperatures delay nodule senescence and this provides a probable explanation for overwintering in nodules observed in cold regions.[32]

3

Improvement of symbiotic nitrogen fixation by rhizobia isolated from temperate legumes

Improvement of productivity of forage legumes, such as alfalfa, or grain legumes, such as beans, has been extensively performed through the selection of plant cultivar. However, some studies demonstrated the importance of rhizobial strain selection to improve growth of legumes under cool climates.

3.1
Clover

Earlier studies showing the differential response of rhizobia to temperature have been conducted with clover which is one of the most important legumes under European northern temperate conditions. The effect of temperature on competition for nodulation of white clover *(Trifolium repens)* amongst two strains similarly effective at 20°C clearly demonstrated that the host preference for rhizobial strains can be altered by temperature.[24] Surprisingly, the strain isolated from United Kingdom was relatively more competitive at the low temperature of 12°C than the Icelandic strain. Low root temperature also affected differently the structure of nodules of white clover formed by two fully effective rhizobial strains from white clover.[33] Strain TA1, originating from a cold environment in Tasmania, was the only one to form bacteroids at 7°C and differed from strain SU297 which was isolated from northern New South Wales. The persistence of nitrogenase activity at low temperature also depended on the *Rhizobium* strain, and strain TA1 was the most efficient and maintained greater enzymatic activity for a longer period than strain SU297.[34] In another study with white clover,[5] strains of rhizobia isolated from sub-arctic Scandinavia (68° to 70°N lat.) showed a faster growth, nodulated earlier and showed a better nitrogenase activity at 10°C than isolates from more southern areas, while no significant differences were observed at 20°C. With red clover *(Trifolium pratense)*, there was no correlation between geographical origin, growth rate at 10°C, and nodulation rate under simulated cold and warm climates of rhizobia isolated beween the latitudes 60° and 63°N in Finland.[9] However, under the cold climate, most northern strains grew faster at 10°C and had higher nitrogenase activity than the most southern strains. It was concluded that growth rate in pure culture at low temperature alone does not allow the prediction of the symbiotic ability under cold conditions. In the latter study, none of the strains was able to grow at 5°C.

3.2
Alfalfa

With the aim to improve symbiotic nitrogen fixation of alfalfa grown in north-western Canada where soil temperatures are often lower than 10°C at the 5-cm depth in May, Rice and Olsen[35] studied the effect of low root temperature on the competitive abilities of the two commercial strains of *S. meliloti* BALSAC and NRG-185. These two strains were equally competitive when alfalfa inoculated with both was grown under constant conditions (both shoot and root at 20°C, 16 h light, and 15°C, 8 h dark), but responded differently to various root-environment temperatures during nodulation. In fact, immunoassay of nodule strain occupancy showed that 63% of nodules were occupied by both strains at 8°C, and that strain NRG-185 alone occupied from 9 to 75% of nodules with increasing temperatures from 8 to 25°C, while proportions of nodules containing only strain BALSAC were relatively constant at 25% from 8 to 21°C. Growth rate of the strains under low temperatures was not determined in this study, but may have been a factor involved in the differential nodulation. These results showed that root-environment temperatures affected the competitive abilities of rhizobia and that selection of *S. meliloti* strains should take into account the effect of low temperatures at the time of nodule initiation. In a subsequent study by the same authors,[7] selection of *S. meliloti* strains capable of initiating efficient nodulation at low temperatures was carried among 226 isolates including NRG-185 and BALSAC. There was no apparent relationship between the doubling times at 10°C and the origin of the isolates (northwestern Canada, Alaska, strains from commercial inoculants), and some isolates produced nodules when whole plants of alfalfa were grown at 10°C. A further screening of 20 isolates showing variation in doubling times and abilities to nodulate at 10°C was done with alfalfa grown under controlled root temperatures. The isolates were divided in three groups based on the interaction between the number (effective and ineffective) and the weight of nodules and the plant dry matter yield. In a field experiment with three strains, one strain (NRG-34), previously classified as effective in its response to low temperature, produced more nodules and greater plant dry matter yield than the two others strains. Moreover, this strain was a strong competitor for nodulation under field conditions, having a higher nodule occupancy rate (80 and 60% nodules occupied 145 days and 412 days after planting respectively) than strain NRG-185 (50 and 25%), which is known to compete with indigenous rhizobia in the soils of northwestern Canada. These results suggest that evaluation of rhizobia by laboratory and growth chambers procedures is a valuable approach to select strains of *S. meliloti* showing a good performance for nodulation at low temperatures.

3.3
Soybean

Soybean is a very important crop that is now cultivated in eastern Canada because of the introduction of some varieties adapted to cool and short growing seasons. However, as stated for alfalfa, root temperature is still a dominant factor controlling competition for nodulation, symbiotic effectiveness and yield of soybean. In a study to evaluate the distribution of strains of *B. japonicum* in soybean nodules developed at different soil temperatures, Weber and Miller[36] found that strains belonging to serogroups which were infrequently recovered at low temperatures (10°C) became dominant at 30°C whereas serogroups forming the majority of nodules at 10 or 15°C formed fewer nodules at 30°C. Subsequent studies also showed marked differences between soybean strains for their competition and their rates of nitrogen fixation suggesting that specific *B. japonicum* should be selected for growing areas where soil temperatures are favorable for their establishment and corresponding to those encountered during plant growth.[15] For instance, Lynch and Smith[8] conducted field and laboratory studies to determine whether adaptability to cool soil conditions existed among *Bradyrhizobium* strains and among soybean cultivars under Canadian growing conditions. The relative growth rates of four rhizobial isolates obtained from Hokkaido, northern Japan, and two commercial strains used in Canada were fastest for two Japanese isolates (H5 and H30) and one commercial strain (532C) at 15°C and at 25°C. In field experiments, levels of nodule mass, shoot total nitrogen and nitrogen fixed were significantly comparable for soybean inoculated with each of the two Japan isolates H5 and H30, and with each of the commercial strains (532C and USDA 110). Under laboratory conditions for the determination of the effect of low root temperature (16–25°C), differences between *B. japonicum* strains (H5, H30 and 532C) were consistent across temperatures, and the commercial strain 532C always performed better than the two strains from Japan (H5 and H30). This better efficiency of strain 532C may be due to enhanced N_2-fixation per gram nodule and not due to increased nodulation.[29] These results suggest that strains of *B. japonicum* from cold environments are unlikely to enhance soybean N_2-fixation and growth under cool temperatures, and the symbiotic effectiveness at different temperatures appears to be unrelated to the growth in liquid cultures. Montanez et al.[15] also observed significant strain effects on the soybean symbiosis at the lowest (15°C) and the highest plant growth temperature (35°C). Two strains which were effective at 25 and 30°C were found to be ineffective at 15°C, and these were not originating from cold environments.

4
Selection of rhizobia from arctic legumes

4.1
Rhizobia from *Astragalus* and *Oxytropis* legume species

The adaptation of legumes to the extreme environment of the high arctic is of interest because of the likelihood that specific characteristics have evolved which may be useful in northern agricultural regions. The three species of arctic legumes (*Astragalus alpinus, Oxytropis maydelliana* and *Oxytropis arctobia*) indigenous to Sarcpa Lake (68°32'N, 83°19'W), Northwest Territories, Canada, have been reported to bring a significant contribution to the nitrogen budget of their environment.[37] This area is a tundra under continuous permafrost characterized by short growing season, low soil and air temperatures, long photoperiods and low soil nitrogen. Nitrogenase activity in nodules is clearly adapted to the prevailing soil temperatures, since activity was detectable down to 0°C and probably below -4°C.[38] Nodule development and basic structure arrangement of arctic nodules were found to be similar to that of the cylindrically-shaped nodules formed on temperate species of legumes. However, the host cytoplasm of the nodules contain unique lipid droplets that may be used to support nitrogen fixation under cool temperatures.[39,40]

Strains of rhizobia isolated from the legume species *Astragalus* and *Oxytropis* showed high genotypic diversity by the MRSP-analysis of 16S rRNA genes and were not classified according to their host plant or their geographic origin. Most of them had 16S rDNA genotypes similar or very close to species of the genus *Mesorhizobium (loti, ciceri)*.[41] The ability of rhizobia isolated from arctic species of *Astragalus* and *Oxytropis* to grow at 5°C in comparison to those isolated from temperate species was an indication of their adaptation to low temperatures.[13] Since their minimal and maximal growth temperatures were 0°C and 27–30°C respectively,[42] they could be classified as psychrotrophs, also defined as psychrotolerant according to Russell.[43,44] Arctic rhizobia could not nodulate legumes of agricultural importance in Canada such as soybean (unpubl. data) or clover,[13] but few strains could form empty nodules on alfalfa.[45] However, they could nodulate two temperate forage legumes: sainfoin *(Onobrychis viciifolia)* and cicer milkvetch *(Astragalus cicer)*. Further evaluation of the symbiotic effectiveness (shoot dry matter weight of plants nodulated by a specific strain) of 48 arctic strains on sainfoin showed that 5 strains were as effective as commercial strains.[46] The association arctic rhizobia and sainfoin was then used in subsequent studies to establish the advantages of using cold-adapted strains with a temperate legume.

4.2
Rhizobia from *Lathyrus* legume species

Another study was conducted with rhizobia isolated from two legume species of *Lathyrus* indigenous to northern Quebec, Canada. *Lathyrus japonicus* is a perennial legume indigenous to Kuujjuarapik (55°20'N, 77°50'W) in the Hudson's Bay arctic zone, and *L. pratensis* is an introduced European species that can grow in relatively cold regions of the forest boreal zone (Val d'Or, 48°07'N, 77°50'W). By the analysis of the 16S rRNA genes, these strains were identified as *R. leguminosarum* bv. viciae, a species that is also associated with the temperate legume genera *Pisum*, *Vicia* and *Lathyrus*.[14] However, most strains from *L. japonicus* (arctic) and a few from *L. pratensis* were in general able to grow at 5°C, while other *L. pratensis* strains and reference temperate strains showed no growth or very poor growth at this temperature. Like arctic strains from *Astragalus* and *Oxytropis*, some strains of *L. japonicus* and *L. pratensis* could be classified as psychrotrophs.[47] A few strains with different capacities to grow at low temperatures were selected for a symbiotic study with *Lathyrus sativus*, an annual temperate legumes species that showed a good potential to be used as green manure in western Canada.[48]

4.3
Potential use of arctic rhizobia with temperate legumes

As stated above, the association arctic rhizobia (from *Astragalus* and *Oxytropis*) and sainfoin was used as a model to determine the agronomic potential of cold-adapted rhizobia to improve nodulation and nitrogen fixation of temperate legumes under low temperatures. Sainfoin is a perennial forage legume that has potential in western Canada and in United States.[49] Its forage is similar in quality to that of alfalfa, but its major problem is a slow growth rate probably resulting in a reduction of the energy for nitrogen fixation.[50] However, for the purpose of the study, evaluation of arctic strains was always made in comparison to effective temperate strains (including one commercial strain).

The nodulating competitiveness of arctic (N31, N10) and temperate (116A15, SM2) rhizobia in association with sainfoin was investigated at root zone temperatures of 9, 12 and 15°C and shoots were kept at 20°C day and night.[51] Sainfoin was inoculated with arctic and temperate strains individually or in mixture. At 9°C, arctic strains occupied more than 65% of sainfoin nodules, while at 15°C the converse was apparent, except for one inoculum mixture (N10+116A15). Moreover, arctic strains elicited more nodules than temperate strains at 9°C, and less nodules at 15°C. At each temperature, values for symbiotic effectiveness (shoot dry weight, nitrogenase activity) and nodule number in plants inoculated by a mixture did not differ significantly than those from the most effective strain in the mixture, and were related to the proportion of this strain in the mixture. Since arctic rhizobia show shorter generation times and greater cell yields at 5 and 10°C than temperate strains, the overall advantage of arctic strains at 9°C may be related to a differential multiplication in the rhizosphere.

In another study, the effect of temperatures (5, 10, 15 and 20°C) was evaluated on nitrogenase activity of detached nodules and whole plants of sainfoin grown at optimal temperatures.[52] Evidence of an adaptation of nitrogen fixing-activity of arctic strains was demonstrated by the fact that they expressed at 5 and 10°C an average of 12 and 33% of their activity at 20°C, while temperate strains showed an average of only 3.7% and 22%. Also, the Q_{10} value (5–15°C) of 3.3 for the arctic strain N31 suggests that the nitrogenase system is affected by temperature like simple enzymatic reaction ($Q_{10}=2$), while the high Q_{10} value of 23.2 for temperate strains may indicate either that other factors influencing nitrogenase were affected or that nitrogenase itself was more sensitive.

Further studies under controlled and field conditions were conducted to ascertain whether this characteristic could improve the growth yield of sainfoin under low temperature.[53] Under controlled conditions simulating spring (3 weeks) and summer (9 weeks) temperatures, two arctic (N31, N10) and two temperate strains (SM2, 116A15) showed the same symbiotic effectiveness as measured by dry matter yield on sainfoin. Sainfoin is slow to develop an efficient N_2-fixing system, and maybe the simulated spring period of three weeks was too short to reflect the cold-adaptation of the arctic strains. Regrowth (second harvest) of sainfoin under optimal summer temperatures was similar for both types of strains. The effect of low temperature was evaluated on regrowth of well-established plants (after the second harvest), and the better efficiency of arctic strains under the low temperature regime was strongly evident, representing up to three times the shoot dry matter yield obtained with temperate strains. Nitrogenase activity measured during regrowth followed the same trends as yield.

Field evaluation is the most important test to demonstrate the agronomic advantages of cold-adapted rhizobia for the growth of legumes because it encompasses interactions between soil and environmental factors. Very few field studies have been reported in literature, except the evaluation of Japan isolates on soybean by Lynch and Smith,[8] just stated above. Field tests comparing arctic (N10, N31) and temperate (commercial inoculant) rhizobia in symbiosis with sainfoin were conducted in two sites located in eastern and two sites located in western Canada.[53] Variations in symbiotic efficiencies were observed among the sites, probably due to differences in agronomic practices for seeding and weeding and in edaphic conditions, the difficulty to determine if there was a competition problem with indigenous rhizobia in soils, and other factors such as temperature, moisture regimes and length of the growing period. In both sites, growth of sainfoin during the seeding year was too poor to evaluate yield. In western Canada, one site did not show any advantage of inoculation, whatever the strain used, and the other site showed a better efficiency of the arctic strain N31 over a commercial inoculant for the two years after seeding. In eastern Canada, the higher efficiency of the arctic strain was shown 1 or 2 years after seeding depending of the site. The effect was more significant at a coldest site (northern), where sainfoin nodulated by the arctic strain N31 produced 2 and 3 times more dry matter yield than the commercial inoculant at the first and second harvest respectively.

Symbiotic efficiency of cold-adapted rhizobia from *Lathyrus* were also evaluated under controlled cool conditions with *Lathyrus sativus*, an annual legume species used as green manure in Canadian prairies.[47] Results showed that a commercial strain, 175P1 (from *Vicia dasycarpa*) not able to grow at 0°C, was more efficient than strain LP0610 (from *L. pratensis*), cold-adapted (growth at 0°C) under an optimal temperature regime of 20/15°C (day/night), showing 49% more nitrogenase activity and 38% more shoot dry weight. However, at the lower temperature regime of 15/7°C, the cold-adapted strain LP0610 was slightly more efficient than the commercial strain 175P1, producing 7% more shoot dry weight and 8% more nitrogenase activity. Since some other cold-adapted or non-adapted rhizobia from *L. japonicus* or *L. pratensis* were not effective on *L. sativus*, it may be worthy to find a legume host showing more compatibility with *Lathyrus* rhizobia. *Lens culinaris*, *Vicia sativa* and *Vicia faba* could be alternative legume species to be used since all *Lathyrus* strains could nodulate them.[14] Nevertheless, results suggest a putative advantage of cold-adapted rhizobia to improve nitrogen fixation of *Lathyrus* at low temperatures, but emphasize the advantages to select cold-adapted strains that are as efficient (under optimal growth conditions) as commercial or reference strains with the legume species under study.

5
Relation between cold-adaptation for growth and symbiotic cold-adaptation

The examples reported above show that it is possible to improve competition, nodulation and nitrogen fixation at low temperatures by selecting adapted rhizobia from temperate and arctic ecosystems. This improvement in symbiotic properties is reflected on the growth of legumes. Plant dry matter yield of field-grown alfalfa was increased by 23% when inoculated with temperate strain NRG-34 selected on the basis of the interaction between growth rate, competitivity, nodulation and efficiency at a root temperature of 12°C in comparison to a commercial strain.[28] From the studies with arctic rhizobia from *Astragalus* and *Oxytropis*, it was possible to improve growth of a temperate legume by 30% under controlled and field conditions, and these strains also showed better competitivity and nitrogen fixation at low temperatures. Its seems that cold-adaptation for growth in pure culture at 10°C and symbiotic effectiveness (nodulation and nitrogenase activity) are closely related.[5,9] However, there was not always a correlation between the geographic origin of rhizobia and their growth rate at low temperatures, probably because northern and southern areas of isolation of the two studies were not situated between the same range of latitudes. Growth capacity of strains isolated from temperate legumes was determined at temperatures not lower than 10°C[7] or 15°C,[8] except in one study where rhizobia tested at 5°C[9] did not grow, as stated above.[42] Rhizobia isolated from indigenous arctic legumes where clearly classified as psychrotrophs on the basis of their growth temperature range (min. 0°C, max. ~30°C) while temperate rhizobia used for comparison could be classified as mesophiles (min. 4–7°C, max. 30–38°C).

In a programme to select rhizobia to improve nitrogen fixation at low temperatures, it seems that the capacity of rhizobia to grow at low temperature is not the only factor to predict a better competitivity, nodulation and nitrogen fixation at low temperature. For instance, the 226 isolates of *S. meliloti* could be classified in three groups showing variability in the interaction between growth rate and nodule formation.[7] Arctic rhizobia from *Astragalus* and *Oxytropis* showed in general very good growth at 5°C, but, in symbiosis with sainfoin grown under optimal conditions, only five strains (out of 48 tested) were as effective as the two effective temperate strains used for comparison.[46] However, in nodules of sainfoin plants developed at optimal temperature with two arctic and two temperate strains of similar efficiency (very effective) and with one less effective arctic strain, the relative nitrogenase activity at 5°C (percentage of that at 25°C) of all three arctic strains was greater than that of the two temperate strains, suggesting an adaptation of the arctic strains to low temperature for nitrogenase activity.[52]

The mechanisms that may be involved in cold-adaptation of nitrogenase in rhizobia have not been deeply investigated either in pure culture or in symbiosis. The bacteria *Azotobacter vinelandii*, *Clostridium pasteurianum* and other soil diazotrophs showed a similar response of nitrogenase activity to temperature changes.[54] However, like arctic rhizobia, diazotrophic *Pseudomonas* species originating from Canadian high Arctic were unique in their capacity to fix nitrogen at temperatures as low as 9 or even 4°C.[55] For rhizobial studies, intact nodules of alfalfa maintained high nitrogenase activity to much lower temperatures than the isolated enzymes,[56] suggesting that legumes possess a compensating mechanism for maintenance of nitrogenase activity problably due to an increase in adenylate energy charge in nodules. In arctic rhizobia in pure cultures, the total amount of nucleotides increased with the lowering of temperatures from 25 to 15°C.[10] However, arctic rhizobial strain N31 did not show higher affinity than temperate strain SM2 for the transport of succinate, a source of energy for nitrogenase at low temperatures.[57] So, it is unlikely that arctic rhizobia will use more efficiently sainfoin photosynthates than temperate rhizobia to support nitrogen fixation at low temperatures. However, the different structure of sainfoin nodules formed by both strains may be a factor that influences the substrate uptake: bacteroids of strain N31 are spherical and included in low numbers in the membrane envelope whereas those of the temperate strain SM2 are elongated and included in large numbers.[40]

Changes in fatty acid composition at low temperatures could be part of a mechanism to maintain membrane fluidity, and hence the ability to transport nutrients at low temperatures.[43] However, there was no relationship between the proportion of unsaturated fatty acids of bacteria, bacteroids or nodules, and the efficiency of two strains of *R. leguminosarum* in the fababean symbiosis at two temperature regimes; but, these strains did not differ in their growth rates at low remperatures.[58] Finally, the relative efficiency of the nitrogen fixing system may differ between cold-adapted and non-adapted rhizobia. In fact, in a study with different rhizobial strains with *Lotus pedunculatus*,[59] temperatures affected differently the ratio between H_2 involved in the enzymatic system of nitrogenase and the electron allocation to this enzyme.

6
Prospective research avenues

Under temperate agricultural conditions, the cold periods encountered during the growing season can significantly limit the establishment of the symbiosis between legumes and rhizobia. In this chapter we have presented many evidences indicating that it is possible to improve the performance of the symbiosis under cold stress. To realize such improvement, one has to select among each symbiotic partners individuals that perform well under cold conditions. As the first step of the symbiosis is the critical step in nodule formation, it will be worthy to check if legume lines that synthesize more flavonoids under cold temperature have better nodulations than their parents. As a direct correlation between plant growth and nitrogenase activity under cold conditions was observed,[27] legumes exhibiting better growth under stress should also in theory have a higher nitrogenase activity. Castonguay et al.[60] also reviewed the different biotechnological strategies for the improvement of cold temperature tolerance in forage legumes.

Rhizobia isolated from Canadian arctic legumes, are adapted to cold and supplied a very nice model illustrating strains that perform well under low temperature with the temperate legume sainfoin. In order to be able to use some of the interesting traits found in these arctic rhizobia with other agronomically important crop like alfalfa, Cloutier et al.[45,61,62] have described 11 nodulation genes in the arctic strain N33. This strain has a content in nodulation genes similar to that of *S. meliloti*, but it can form only a few, white empty nodules on alfalfa.[45] These results suggest that N33 and *S. meliloti* probably excrete similar Nod factors, however the nodulation genes of N33 are induced by flavonoid compounds (formononetin and p-coumaic acid) different from those inducing *S. meliloti*. Will it be possible to efficiently nodulate alfalfa with strain N33? To answer such question it is necessary to pursue the studies of nodulation genes in this strain in order to determine if any specificity genes are present, and to determine the structure of the Nod factor(s) produced by this strain.

Survival of rhizobia to winter freezing would be advantageous for early nodulation of perennial legumes at spring. However, in a study on heat-shock proteins (HSPs) and cold-shock proteins (CSPs) comparing three arctic strains (psychrotrophs) from *Astragalus* and *Oxytropis* and three temperate strains (mesophiles) from alfalfa and sainfoin, survival of arctic strains to freezing at -5 and -10°C was slightly lower (64 and 58%) than that of temperate strains (75 and 71%) although arctic strains produced more CSPs under freezing conditions (-10°C).[42] In these experiments, survival was estimated in pure cultures after a short period of 3 h (corresponding to the shock treatment), but the real potential of survival should be investigated from long-term frozen soils containing indigenous bacteria (with and without temperate rhizobia).

Finally, other rhizobia isolated from alpine or arctic legumes also have many potential genetic traits that can be used to improve the symbiosis under cold stress. However many studies need to be performed to appraise the behavior of these strains at each step of the symbiosis under cold conditions.

Future research needs with cold adapted rhizobia should target the following:

1. determination of the molecular basis of competition, early root colonization and infection of legumes by cold-adapted rhizobia;
2. isolation, characterization and comparison of the functions of the cold-shock and cold-adaptation proteins produced by cold-adapted rhizobia and temperate rhizobia;
3. elucidation of the mechanisms of action involved in the cold adaptation of the nitrogenase activity (C_2H_2 reduction) in nodules formed by cold-adapted rhizobia;
4. determination of the possibility to use plant growth promoting rhizobacteria (PGPR) in co-inoculation with rhizobia to improve nodulation at suboptimal root zone temperature as observed in soybean symbiosis.[63]

7
References

1. Young JPW, Haukka KE. Diversity and phylogeny of rhizobia. New Phytol 1996; 133:87-94.
2. Herridge DF, Rupela OP, Serraj R, Beck DP. Screening techniques and improved biological nitrogen fixation in cool season food legumes. Euphytica 1994; 73:95-108.
3. Barran LR, Bromfield ESP. Competition among rhizobia for nodulation of legumes. In: Mckersie BD, Brown DCW, eds. Biotechnology and Improvement of Forage Legumes. New York: CAB International, 1997:343-374.
4. Smith RS. Legume inoculant formulation and application. Can J Microbiol 1992; 38:485-492.
5. Ek-Jandér J, Fåhraeus G. Adaptation of *Rhizobium* to subarctic environment in Scandinavia. Plant and Soil 1971; special volume:129-137.
6. Sprent JI. The Biology of Nitrogen-Fixing Organisms. New York: McGraw-Hill Book Company Ltd, 1979.
7. Rice W, Olsen PE, Collins MM. Symbiotic effectiveness of *Rhizobium meliloti* at low root temperature. Plant and Soil 1995; 170:351-358.
8. Lynch DH, Smith DL. Early seedling and seasonal N_2-fixing symbiotic activity of two soybean [*Glycine max* (L.) Merr.] cultivars inoculated with *Bradyrhizobium* strains of diverse origin. Plant and Soil 1993; 157:289-303.
9. Lipsanen P, Lindstrom K. Adaptation of red clover rhizobia to low temperatures. Plant and Soil 1986; 92:55-62.
10. Bordeleau LM, Prévost D. Nodulation and nitrogen fixation in extreme environments. In: Graham PH, Sadowsky MJ, Vance CP, eds. Symbiotic Nitrogen Fixation. The Netherlands: Kluwer Academic Publishers, 1994:115-125.
11. Trinick MJ. Biology. In: Broughton WJ, ed. Nitrogen Fixation, vol. 2: *Rhizobium*. Oxford: Clarendon Press 1982:76-146.
12. Graham PH. Stress tolerance in *Rhizobium* and *Bradyrhizobium*, and nodulation under adverse soil conditions. Can J Microbiol 1992; 34:475-485.
13. Prévost D, Bordeleau LM, Caudry-Reznick S, Schulman HM, Antoun H. Characteristics of rhizobia isolated from three legumes indigenous to the Canadian high arctic: A*stragalus alpinus, Oxytropis maydelliana,* and *Oxytropis arctobia.* Plant and Soil 1987; 98:313-324.
14. Drouin P, Prévost D, Antoun H. Classification of bacteria nodulating *Lathyrus japonicus* and *Lathyrus pratensis* in northern Quebec as strains of *Rhizobium leguminosarum* biovar viciae. Int J Syst Bacteriol 1996; 46:1016-1024.
15. Montanez A, Danso SKA, Hardarson G. The effect of temperature on nodulation and nitrogen fixation by five *Bradyrhizobium japonicum* strains. Appl Soil Ecol 1995; 2:165-174.
16. Zhang F, Smith DL. Effects of low root zone temperatures on the early stages of symbiosis establishment between soybean [*Glycine max* (L.) Merr.] and *Bradyrhizobium japonicum.* J Exp Bot 1994; 45:1467-1473.

17. Redmond JW, Batley M, Djordjevic MA, Innes RW, Keumpel PL, Rolfe BG. Flavones induce expression of nodulation genes in *Rhizobium*. Nature 1986; 323:632-635.

18. Lerouge P, Roche P, Faucher C, Maillet F, Truchet G, Promé J-C, Dénarié J. Symbiotic host-specificity of *Rhizobium meliloti* is determined by a sulphated and acylated glucosamine oligosaccharide signal. Nature 1990; 344:781-784.

19. Relic B, Talmont F, Kopcinska J, Golinowski W, Promé J-C, Broughton WJ. Biological activity of *Rhizobium* sp. NGR234 Nod-factors on *Macroptilium atropurpureum*. Mol Plant-Microbe Interact 1993; 6:764-774.

20. Zhang F, Smith DL. Genistein accumulation in soybean (*Glycine max* [L.] Merr.) root systems under suboptimal root zone temperatures. J Exp Bot 1996; 47:785-792.

21. Zhang F, Charles TC, Pan B, Smith DL. Inhibition of the expression of *Bradyrhizobium japonicum nod* genes at low temperatures. Soil Biol Biochem 1996; 28:1579-1583.

22. McKay IA, Djordjevic MA. Production of nod metabolites by *Rhizobium leguminosarum* bv. trifolii are disrupted by the same environmental factors that reduce nodulation in the field. Appl Environ Microbiol 1993; 59:3385-3392.

23. Dowling DN, Broughton WJ. Competition for nodulation of legumes. Ann Rev Microbiol 1986; 40:131-157.

24. Hardarson G, Jones DG. Effect of temperature on competition amongst strains of *Rhizobium trifolii* for nodulation of two white clover varieties. Ann Appl Biol 1979; 92:229-236.

25. Heichel GH, Vance CP. Physiology and morphology of perennial legumes. In: Broughton WJ, ed. Nitrogen Fixation, vol. 3: Legumes. Oxford: Clarendon Press, 1983:99-143.

26. Layzell DB, Rochman P, Canvin DT. Low root temperatures and nitrogenase activity in soybean. Can J Bot 1984; 62:965-971.

27. Lindström K. Analysis of factors affecting in situ nitrogenase (C_2H_2) activity of *Galega orientalis*, *Trifolium pratense* and *Medicago sativa* in temperate conditions. Plant and Soil 1984; 79:329-341.

28. Rice WA, Olsen PE, Collins MM. Symbiotic effectiveness of *Rhizobium meliloti* at low root temperature. Plant and Soil 1995; 170:351-358.

29. Lynch DH, Smith DL. The effects of low root-zone temperature stress on two soybean (*Glycine max*) genotypes when combined with *Bradyrhizobium* strains of varying geographic origin. Physiol Plantarum 1994; 90:105-113.

30. Thomas RJ, Sprent JI. The effects of temperature on vegetative and early reproductive growth of a cold -tolerant and cold sensitive line of *Phaseolus vulgaris* L. 1-Nodulation, growth and partitioning of dry matter, carbon and nitrogen. Ann Bot 1984; 53:579-588.

31. Fyson A, Sprent JI. The development of primary root nodules on *Vicia faba* grown at two temperatures. Ann Bot 1982; 50:681-692.

32. Sutton Wd. Nodule development and senescense. In: Broughton WJ, ed. Nitrogen Fixation, vol. 3: Legumes. Oxford: Clarendon Press, 1983:144-212.

33. Roughley RJ. The influence of root temperature, *Rhizobium* strain and host selection on the structure and nitrogen-fixing efficiency of the root nodules of *Trifolium subterraneum*. Ann Bot 1970; 34:631-646.

34. Roughley RJ, Dart PJ. Reduction of acetylene by nodules of *Trifolium subterraneum* as affected by root temperature, *Rhizobium* strain and host cultivar. Arch Microbiol 1969; 69:171-179.

35. Rice WA, Olsen PE. Root-temperature effects on competition for nodule occupancy between two *Rhizobium meliloti* strains. Biol Fertil Soils 1988; 6:137-140.

36. Weber DF, Miller VL. Effect of soil temperature on *Rhizobium japonicum* serogroup distribution in soybean nodules. Agron J 1972; 64:796-798.

37. Karagatzides JD, Lewis MC, Schulman HM. Nitrogen fixation in the high arctic at Sarcpa Lake, Northwest territories. Can J Bot 1985; 63:974-979.

38. Schulman HM, Lewis MC, Tipping EM, Bordeleau LM. Nitrogen fixation by three species of Leguminosae in the Canadian High Arctic tundra. Plant Cell Environ 1988; 11:721-728.

39. Newcomb W, Wood SM. Fine structure of nitrogen-fixing leguminous root nodules from Canadian Arctic. Nord J Bot 1986; 6:609-626.

40. Prévost D, Bordeleau LM, Antoun H. Effet des souches arctiques de *Rhizobium* sur la structure des nodules du sainfoin *(Onobrychis viciifolia)* et de légumineuses arctiques *(Astragalus* et *Oxytropis* spp). Can J Bot 1989; 67:3164-3168.

41. Laguerre G, van Berkum P, Amarger N, Prévost D. Genetic diversity of rhizobial symbionts from legume species within the genera *Astragalus, Oxytropis* and *Onobrychis.* Appl Environ Microbiol 1997; 63:4748-4758.

42. Cloutier J, Prévost D, Nadeau P, Antoun H. Heat and cold shock protein synthesis in arctic and temperate strains of rhizobia. Appl Environ Microbiol 1992; 58:2846-2853.

43. Russell NJ. Cold adaptation of microorganisms. Phil Trans R Soc Lond 1990; 326:595-611.

44. Russell NJ. Biochemical differences between psychrophilic and psychrotolerant microorganisms. In: Guerrero R, Pedrod-Alio C, eds. Trends in Microbial Ecology. Madrid: Spanish Society for Microbiology, 1993:29-32.

45. Cloutier J, Laberge S, Castonguay Y, Antoun H. Characterization and mutational analysis of *nod*HPQ genes of *Rhizobium* sp. strain N33. Mol Plant-Microbe Interact 1996; 9:720-728.

46. Prévost D, Bordeleau LM, Antoun H. Symbiotic effectiveness of indigenous arctic rhizobia on a temperate forage legume: Sainfoin *(Onobrychis viciifolia).* 1987; 104:63-69.

47. Drouin P. Caractérisation phénotypique et génotypique et étude des mécanismes d'adaptation au basses températures de souches de *Rhizobium* isolées de *Lathyrus japonicus* et *Lathyrus pratensis.* PhD thesis, Laval University, Canada, 1997.

48. Biederbeck VO, Bouman OT, Looman AE, Slinkard AE, Bailey LD, Rice WA, Jansen HH. Productivity of four annual legumes green manure in dryland cropping systems. Agron J 1993; 85:1035-1043.

49. Hanna MR, Smoliak S, Wilson DB, Cooke DA, Goplen BP. Le sainfoin dans l'Ouest du Canada. Publication 1470, Agriculture Canada, 1980.

50. Hume LJ, Withers NJ, Rhoades DA. Nitrogen fixation in sainfoin *(Onobrychis viciifolia).* 2. Effectiveness of the nitrogen-fixing system. NZJ Agric Res 1985; 28:337-348.

51. Prévost D, Bromfield ESP. Effect of low root temperature on symbiotic nitrogen fixation and competitive nodulation of *Onobrychis viciifolia* (sainfoin) by strains of arctic and temperate rhizobia. Biol Fert Soils 199; 12:161-164.

52. Prévost D, Antoun H, Bordeleau LM. Effects of low temperatures on nitrogenase activity in sainfoin *(Onobrychis viciifolia)* nodulated by arctic rhizobia. FEMS Microbiol Ecol 1987; 45:205-210.

53. Prévost D, Bordeleau LM, Michaud R, Lafrenière C, Waddington J, Biederbeck VO. Nitrogen fixation efficiency of cold-adapted rhizobia in sainfoin *(Onobrychis viciifolia)*: Laboratory and field evaluation. In: Graham PH, Sadowsky MJ, Vance CP, eds. Symbiotic Nitrogen Fixation. The Netherlands: Kluwer Academic Publishers, 1994:171-176.

54. Waughman GJ. The effect of temperature on nitrogenase activity. J Exp Bot 1977; 28:949-960.

55. Lifshitz R, Kloepper JW, Scher FM, Tipping EM, Laliberté M. Nitrogen fixing Pseudomonads isolated from roots of plants grown in the Canadian high Arctic. Appl Environ Microbiol 1986; 51: 251-255.

56. Miller RW, Al-Jobore A, Bernt WB. Properties of the alfalfa-*Rhizobium meliloti* symbiotic nitrogenase system in vivo and in vitro. Biochem Cell Biol 1986; 64:556-564.

57. Bigwaneza PC, Prévost D, Bordeleau LM, Antoun H. Effect of temperature on succinate transport by an arctic and a temperate strain of rhizobia. Can J Microbiol 1993; 39: 907-911

58. Théberge, MC, Prévost D, Chalifour FP. The effect of different temperatures on the fatty acid composition of *Rhizobium leguminosarum* bv. viciae in the faba bean symbiosis. New Phytol 1996; 134:657-664.

59. Pankurst CE, Layzell DB. The effect of bacterial strain and temperature changes on the nitrogenase activity of *Lotus pedunculatus* root nodules. Physiol Plantarum 1984; 62:404-409.

60. Castonguay Y, Laberge S, Nadeau P, Vezina L-P. Temperature and drought stress. In: McKersie BD, Brown DCW, eds. Biotechnology and Improvement of Forage Legumes. New York: CAB International, 1997:175-202.

61. Cloutier J, Laberge S, Prévost D, Antoun H. Sequence and mutational analysis of the common *nod* BCIJ region of *Rhizobium* sp. *(Oxytropis arctobia)* strain N33, a nitrogen-fixing microsymbiont of both arctic and temperate legumes. Mol Plant-Microbe Interact 1996; 9:523-531.
62. Cloutier J, Laberge S, Antoun H. Sequence and mutational analysis of the 6.7-kb region containing nodAFEG genes of *Rhizobium* sp. strain N33: Evidence of DNA rearrangements. Mol Plant-Microbe Interact 1997; 10:401-406.
63. Zhang F, Dashti N, Hynes RK, Smith DL. Plant growth promoting rhizobacteria and soybean {*Glycine max* (L.) Merr.} nodulation and nitrogen fixation at suboptimal root zone temperatures. Ann Bot (London) 1996; 77:453-459.

Plant protection by cold-adapted fungi

P. T. W. Wong[1]* and J. H. McBeath[2]*

[1] Agricultural Research Institute, Private Bag, Wagga Wagga, New South Wales, Australia 2650
[2] Plant Pathology and Biotechnology Laboratory, University of Alaska, Fairbanks, POB 757200, Fairbanks, Alaska 99775-7200, USA

1
Introduction

Biocontrol of plant diseases and pests is generally regarded as a safer alternative to chemical pesticides but despite intensive research in the last two decades, there are few commercial products. The early optimism arising from numerous promising laboratory and greenhouse experiments has given way to a sober realization of the difficulties of commercializing biocontrol agents (BCAs). With few exceptions, there appears to be little likelihood of broad-spectrum BCAs and each pest or disease may have to be countered by a specific antagonistic microrganism, which proliferates and acts in the infection court or ecological niche of the target species. Therefore, a thorough understanding of the ecology of the BCA and the plant pathogen or pest provides the best chance of delivering a commercial product.

In recent years, the screening process for identifying potential BCAs has been refined to take into account more fully the biotic and abiotic factors that impact on the pathogen-antagonist interaction. Antagonists have been selected from habitats which have shown a natural suppression of a pathogen or pest, e.g. take-all or fusarium suppressive soils. There is now less reliance on mass *in vitro* selection involving dual cultures on agar plates incubated at normal laboratory temperatures (20–25°C) and much effort has also been directed to elucidating mechanisms of suppression in the hope that some basic principles of biological control may be developed.

In this chapter, the effect of cold-adaptation by BCAs and of low temperatures on the pathogen–BCA interaction will be considered. This is especially relevant to areas of high latitude and altitude, pests and diseases of winter crops and post-harvest diseases in cold storage (1–4°C). Some of these BCAs are obligate psychrophiles and grow optimally at or near the freezing point of water.[1] However, many are mesophiles that have adapted to low temperatures (0–5°C) but grow optimally at 15–20°C. Because many of the economically destructive plant pathogens such as *Botrytis cinerea, Pythium ultimum* Trow, *Phytophthora infestans* (Mont.) de

---

* Corresponding authors

Bary, *Rhizoctonia solani* Kuhn, and *Sclerotinia sclerotiorum* (Lib.) De Bary, etc. are mesophiles with strong cold adaptation, a similar adaptation to cold by BCAs allows them to antagonize or outcompete the pathogens or pests in their natural environments. The selection of BCAs capable of growth and activity at realistic field temperatures has led to the commercialization or near-commercialization of a few cold-adapted fungi (Table 1).

2
Biocontrol of snow molds

Snow molds are obligate or facultative psychrophilic fungi which can attack many economic plants under a cover of snow. They comprise over a dozen species and occur commonly in high latitude regions. These pathogens infect host plants in late fall or early winter when the soil is not yet frozen. During the long winters, under a thick snow layer, snow mold fungi proliferate and spread in host tissues under the dark, humid conditions. Death of the plant is due to maceration of plant tissues[2,3] and depletion of carbohydrate reserves.[4,5] Snow molds are difficult to control. The application of fungicides is costly and often ineffective. Because of the fragility of the environments in which they occur, chemical use also elicits great concern. As such, much research has focused on the biocontrol of snow molds using saprophytic fungi, which antagonize the pathogens.[6–10]

Table 1. Control of plant pathogens and an insect pest by cold-adapted fungi

Biocontrol fungus	Pathogen or pest	Mechanism	Commercial status
Psychrophiles			
Typhula phacorrhiza	*T. ishikariensis* *T. incarnata*	Competition	Experimental
Acremonium boreale	*Microdochium nivale* *Myriosclerotinia borealis*	Antibiosis	Experimental
Humicola grisea var. *grisea*	*Pythium paddicum* *P. iwayamai*	Antibiosis	Experimental
Cold-adapted mesophiles			
Trichoderma atroviride	*Rhizoctonia solani* *C. psychromobidus* *S. sclerotiorum* *Phytophthora* spp. *Pythium* spp. etc.	Antibiosis, Mycoparasitism	Commercial "Plant-Helper"
Phialophora sp. (lobed hyphopodia)	*Gaeumannomyces* *graminis* var. *tritici*	Competition, Induced resistance	Near commercial
Metarhizium anisopliae	*Adoryphorus couloni*	Mycoparasitism	Commercial "Bio-Green"

Smith and Davidson[4] described a cold-tolerant fungus, *Acremonium boreale* Smith and Davidson, which suppressed *Typhula ishikariensis* Imai, *Microdochium nivale* (Ces. ex Berl. and Vogl.) Samuels and Hallet and a sclerotial Low Temperature Basidiomycete (sLTB), subsequently identified as *Coprinus psychromobidus* Traquair. It was an invasive primary saprophyte of a wide range of grasses and dicotyledonous plants but only caused minimal damage to cold-hardened grass seedlings in pot experiments. On agar, it could grow at -6°C and showed good growth at 10°C. It was antagonistic to the snow molds in a temperature range of -3 to 10°C but the antagonism varied considerably with the pathogenic species. In Alaska (latitude 54°N to 71.5°N), agricultural activities are located mostly in the sub-arctic region where snow-bound days range from 170 to 210 per annum with a maximum soil temperature of 12 to 15°C. Snow molds are the most serious diseases of winter wheat (*Triticum aestivum* L.) and turf grasses. The most prevalent snow mold fungi are: *Myriosclerotinia borealis* (Bubak and Vleugel) Kohn, *M. nivale* and *C. psychromobidus*. Unlike other regions in the far north, *Typhula* spp. have not been found in Alaska. The activities of hyperparasites on snow molds were noticed in the early 80's in studies on winter wheat. McBeath (unpubl. data) found that disease severity on winter wheat caused by snow molds was far worse than the blank control when the fungicide benomyl was used at sub-lethal dosages and it was suspected that biological factors may be involved. Later, *Trichoderma atroviride* Karsten (see next section) was found to be an effective antagonist of snow molds. In studies to test the efficacy of *T. atroviride* in controlling *C. psychromobidus*, conducted under conditions of cold (about 4°C), dark and 100% relative humidity for 8 weeks, no survivors (100% mortality) were found when winter wheat seedlings germinated from untreated seeds were challenged with *C. psychromobidus*.[10] A significant improvement in viability of winter wheat seedlings was observed when the seedlings were previously protected with *T. atroviride* biotype 603 or CHS 861. Plant mortality in infected plants treated with biotype 603 (70% mortality) was slightly more than that in infected plants treated with CHS 861 (50% mortality) but no difference was observed between uninfected plants treated with CHS 861(39% mortality) and the blank control (39% mortality). Examination of the sheath of *T. atroviride*-treated winter wheat grown in snow mold contaminated soils revealed a dramatic reduction in the number of sclerotia formed by *C. psychromobidus*. *T. atroviride* seemed to forge a symbiotic relationship with the roots of winter wheat plants; *T. atroviride* mycelia, chlamydospores and chlamydospore primordia-like structures were observed in the root cells of winter wheat with no adverse effects on the plants.

In Canada, Burpee et al.[7] successfully controlled Typhula blight in creeping bentgrass (*Agrostis palustris* Huds.) with non-pathogenic *T. phacorrhiza* Reichard ex Fr. isolates. When grain inocula of the BCAs were applied at the rate of 7×10^3 colony-forming units (CFU) m^{-2} to the diseased turf, control of the disease was comparable to that achieved by chemical control using pentachloronitrobenzene (PCNB) applied at the rate of 30 kg ha^{-1}.[7,8] The BCA survived and reproduced in the turfgrass thatch and suppressed the disease for up to 16 months following two annual applications of the grain inocula. This residual suppression of the disease was most promising and a distinct advantage over chemical fungicides, which required more frequent applications. However, the application rates of the BCA in these experiments equate to 500–2,000 kg ha[-1], which are obviously too high to be economic even in the turfgrass industry.[8] To increase the efficacy of the biocontrol, 33 new isolates of *T. phacorrhiza* were compared with the two earlier isolates. Twenty-two isolates provided significantly greater (P≤0.05) control than the standard isolate T011 while six gave more than 80% disease suppression.[11] There was, therefore, considerable variability in the efficacy of the isolates but it was not established whether the greater disease suppression in these new isolates was related to their greater growth rates at 0°C. Wu et al.,[12] studying 382 other isolates of *T. phacorrhiza* from corn soils in Ontario, Canada, confirmed the great variability in suppressive activity of these isolates but could not correlate it with their growth rates at 10°C. It may have been better to correlate their efficacy with the growth rates at or near 0°C since these are the critical temperatures at which the BCA and pathogens interact.

The mechanism of disease suppression has not been fully elucidated but laboratory studies indicate that antibiotics are not produced by *T. phacorrhiza*, nor is it parasitic on other *Typhula* species.[13] Competition for substrates and nutrients is a probable mechanism since the growth of *T. ishikariensis* and *T. incarnata* was reduced on agar media that had been previously exposed to the BCA. When nutrients were subsequently supplied to these previously colonized plates, the growth and production of sclerotia of the pathogens were significantly increased. It appears that during the saprophytic and/or parasitic phases of growth, some isolates of the BCA may compete more successfully against the pathogens for substrate possession and essential nutrients. Burpee[13] also observed that *T. phacorrhiza* colonized the surfaces of turfgrass leaves and penetrated leaf tissues via the stomata. He postulated that competition for infection sites may be involved.

Matsumoto and Tajima[9] found that isolates of *T. phacorrhiza* from grasses after snow melt near Sapporo, Japan, suppressed *T. ishikariensis* in pot and field experiments. In general, they were more effective against *T. ishikariensis* biotype B than biotype A. However, isolates from the Tenpoku district of northern Hokkaido, Japan, which has more than 140 d of snow cover per annum suppressed both biotypes. It was suggested that they exerted natural biocontrol of the snow molds and may have contributed to the longevity of sown perennial ryegrass (*Lolium perenne* L.) pastures in the district. In contrast, isolates from the Sapporo district, which has a shorter period of snow cover of 120 d per annum, were only effective against biotype B and perennial ryegrass pastures did not persist for more than 3–4 years. It was inferred that the greater adaptation to the longer period of snow cover in northern Hokkaido made the isolates more effective as BCAs but there was no direct evidence for this.

Matsumoto[15] cited the work of Y. Sudate and T. Fujiwara (unpubl. data) which showed that a cold-tolerant isolate of *Humicola grisea* Traaen var. *grisea* suppressed the growth of *T. ishikariensis, T. incarnata, M. nivale, Pythium paddicum* Hirane and *P. iwayamai* Ito in culture and reduced significantly the amount of turf disease in pot experiments. The application of the BCA at the rate of 50 g m[-2] (7.5×10^6 CFU g[-1]) to turf field plots, naturally infested with *T. ishikariensis* biotype B, gave disease control equivalent to that of organic copper sprays. Antibiotics appeared to be involved in the suppression but they have not been identified.

3
Control of plant diseases by *Trichoderma atroviridae*

Trichoderma atroviride Karsten CHS 861 was selected from a large number of BCAs isolated from the sub-arctic region of Alaska. The fungus is a mesophile but is well adapted to cold temperatures. It has a temperature range of 4°C (or less) to 33°C, which is comparable to that of many economically destructive plant pathogens. For instance, the optimal temperature of *B. cinerea* is 16–21°C, but it can inflict serious damage on leaves, fruits and other plant tissues at temperatures of 16°C or less, under 100% relative humidity. The temperature range of *Pythium* spp. is 7–30°C, but the pathogens are most destructive on seeds and seedlings at temperatures of 12°C or less. *P. infestans* (5–30°C), *R. solani* (1–32°C) and *S. sclerotiorum* (4–26°C) can cause damage on tubers, roots and stems under cool temperatures when plant tissues are particularly vulnerable. The ability of *T. atroviride* to function under similar environmental conditions and especially at very low temperatures (4–10°C) makes it a highly effective plant disease control agent. The wild type CHS 861 is also naturally resistant to the fungicides metalaxyl and pentachloronitrobenzene (PCNB) but is very sensitive to benomyl. Several *T. atroviride* biotypes have been developed that possess characteristics similar to the wild type but are highly resistant to benomyl and can, therefore, be used in conjunction with this fungicide, if desired.[10]

Trichoderma atroviride CHS 861 has been found to suppress the growth and development of a broad spectrum of pathogenic fungi with various degrees of efficacy. In the studies of the interaction of *T. atroviride* and snow mold fungi, the clear zone of inhibition created by *T. atroviride* was very pronounced in *C. psychromobidus* and in *M. borealis* but not in the case of *M. nivale*. However, a peculiar "bundled" mycelial formation was observed in *M. nivale* but not with the other pathogens. *T. atroviride* was found to be able to inhibit the growth of *Armillaria mellea* (Vahl. ex Fr.) Kummer but did not grow over the top of *A. mellea*; nor did it seem to be able to cause lysis of the mycelia. To the contrary, suppression of growth and lysis of mycelia was observed readily on *C. psychromobidus, B. cinerea, F. solani, M. nivale, M. borealis, Phytophthora* spp. (*P. cactorum* (Lebert and Cohn) J. Schrot., *P. capsici* Leonian, *P. infestans, P. parasitica* Dastur), *Pythium* spp. (*P. aphanidermatum* (Edson) Fitzpatrick, *P. dissotocum, P. ultimum*), *R. solani, S. sclerotiorum, T. incarnata* Lasch ex Fr., *T. idahoensis,* and *T. ishikariensis* and *Verticillium dahliae* Kleb.

Very little or no differences in their reactions to *T. atroviride* were found among strains of susceptible fungi. For instance, no differences were observed in their reactions to the hyperparasitism of A1 and A2 strains of *P. infestans*. All of the strains of *R. solani* tested were found very susceptible to *T. atroviride*. Hyphae of *T. atroviride* coiled around and penetrated hyphae of *R. solani*. Only slight variations in the degrees of their responses to *T. atroviride* suppression were observed in the isolates of *R. solani* tested, even from different anastomosis groups.

Trichoderma atroviride has been found capable of utilizing many plant pathogenic fungi as food sources. In the studies of hyperparasitism, clumps of *T. atroviride* conidia were dusted onto vigorously growing aerial mycelia of *C. psychromo-*

bidus, B. cinerea and *V. dahliae*. No fungistatic effect to *T. atroviride* was ever observed from the plant pathogenic fungi tested; conidia of *T. atroviride* germinated readily on the surface of hypha of host fungi and their germ tubes penetrated into or through the hypha of plant pathogens (Fig. 1A). No specialized penetration structures were observed. Proliferation of *T. atroviride* occurred through branching and re-branching of hyphae both inside and outside the cells of the host fungi. The presence of *T. atroviride* in the cells of host fungi may be as single strands of hyphae (Fig. 1B) or as a bundle of hyphae in either direction (Fig. 1C). Hyphae of *T. atroviride* penetrated freely through the cell walls and became intertwined with the hyphae of the host fungi. Hyphae of host fungi may also be penetrated by several side branches originating from hyphae of *T. atroviride* growing alongside of it (Fig. 1D). In addition to direct parasitization, proteinaceous, antimycotic substances also seemed to play a significant role in causing the deterioration of cells of host fungi. Cytoplasm coagulation and leaching, cell desiccation, blistering and degradation of cell walls and lysis occurred to hyphal cells of host fungi with or without the immediate presence of *T. atroviride*. Cell deterioration and lysis were most noticeable in *V. dahliae*. The leaching of cytoplasm and pigmentation from sclerotial cells resulted in the bleaching of color from dark brown to hyaline, and the degradation of cell walls resulted in the lost of structural integrity of microsclerotia. In the cases of *C. psychromobidus* and *B. cinerea*, a deep crater created by the lysis of hyphae cells occurred at the site of inoculation. The crater was surrounded by vigorous growing hyphae of *T. atroviride*, which eventually overtook the entire colonies of these host fungi.

In *M. borealis*, the sclerotia not only provides a means of survival for the snow mold fungus but also serves as its primary source of inoculum. In light and scanning electron microscopic studies, *T. atroviride* has been found to be capable of parasitizing the sclerotia of *M. borealis* and using it as a food source. The hyphae of *T. atroviride* penetrated the sclerotia of *M. borealis* mainly through the nicks and cracks occurring in the rind. A mycelial "lysis" similar to that observed in *C. psychromobidus* and *B. cinerea* has been observed on the medulla cells resulting in the rapid loss of the integrity of the medulla. The sclerotia then became cavernous and hollow-centered. Sporulation of the *T. atroviride* has been observed both on the exterior and interior of the diseased sclerotia.[10]

T. atroviride showed no phytotoxicity to plants. In some cases, plants treated with *T. atroviride* demonstrated marked response in growth. Growth promotion characteristics have been observed in plants germinated from *T. atroviride*-treated seeds in the absence of plant pathogens. Significant increases of weight, height, as well as numbers of flowers and pods were found on *T. atroviride*-treated pea plants. Significant increases in the germination of pea seeds, grown in *Pythium*-infested soils, were observed in those treated with CHS 861 (66% germination) or biotye 603 (44% germination) when compared to the seeds without *Trichoderma* treatment (11% germination). A slight decline in germination was found in biotype 603-treated seeds (78% germination) grown in untreated soils compared to those of CHS 861 (100% germination) and blank controls (100% germination). No statistical differences were found among the fresh weights of plants germinated from CHS 861- or biotype 603-treated seeds when compared to an untreated control. No

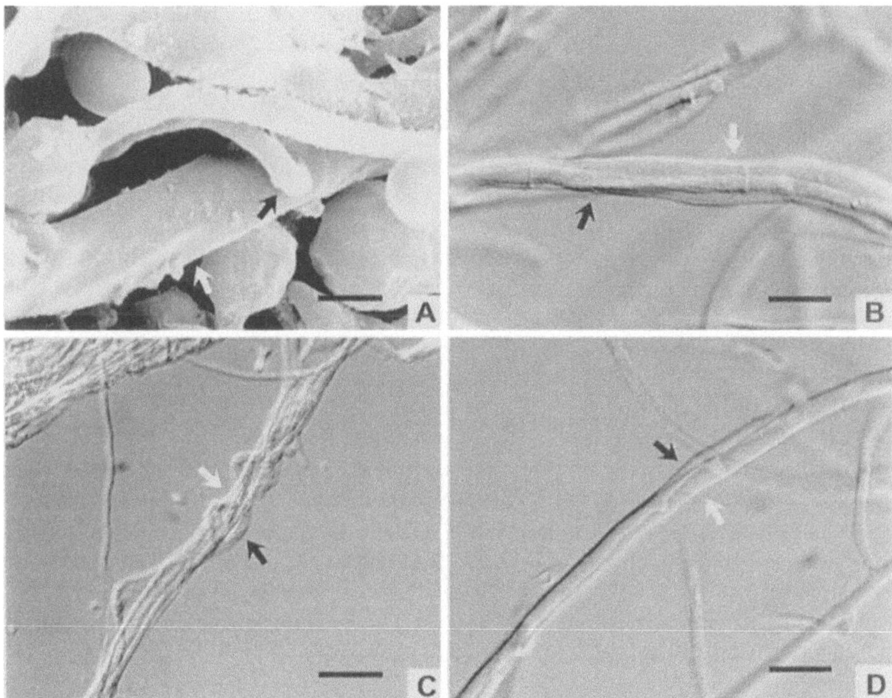

Fig. 1. Interactions of *Trichoderma atroviride* (black arrow) and plant pathogenic fungi (white arrow)

(**A**) Narrow hyphae of *T. atroviride* penetrating into the larger hyphae of *Coprinus psychromobidus*, bar = 20 µm; (**B**) Two slightly overlapping *T. atroviride* hypha in the hyphal cells of *Microdochium nivale*, bar = 40 µm; (**C**) Strands of *T. atroviride* hypha in the *Rhizoctonia solani* hyphal cells and intertwining with the hyphae of *Rhizoctonia solani*, bar = 40 µm; (**D**) Hyphae of *T. atroviride* in and out of the hyphae of *Phytophthora infestans*, bar = 40 µm

statistically significant differences were found in the fresh weight of pea plants germinated from CHS 861- or biotype 603-treated seeds grown in soils with or without *Pythium* infestation.[10]

Field trials to test the efficacy of *T. atroviride* in controlling *R. solani* on potatoes were conducted in Alaska and Montana. The field trial sites were selected from fields known to have heavy infestations of *R. solani*. Results of the field trials in Alaska using a potato cultivar very susceptible to *R. solani* indicated that *T. atroviride* treatment improved the quality of potatoes considerably. Significantly more black scurf were found on potato tubers produced from the blank control (51%) than *T. atroviride* CHS 861 (29%), biotype 453 (7%), biotype 603 (28%) and PCNB (16%). In the years when disease incidence was low, no significant differences in yields of potatoes were found between treatments of *T. atroviride* and the blank control, whereas yields of potatoes in PCNB treatment were significantly lower. In the field efficacy trials conducted in Montana, significantly more black scurf were found on potato tubers produced from the treatments of the blank control (73%) and carrier control (50%). Treatments of *T. atroviride* CHS 861 (39%), 453 (39%), 603 (39%) and 901 (27%) were comparable in their effectiveness in controlling black scurf as compared to Top SMZ (35%) and a bi-nucleate *Rhizoctonia*. (Mc Beath, unpubl. data).

Sclerotinia stem rot, caused by *S. sclerotiorum*, is one of the most serious diseases on garden flowers in high latitude region. The pathogen produces many large sclerotia on the surface and in the pith of the stem. The sclerotia survive the harsh winter in Alaska well and remain viable in the soil for many years. Each spring, these sclerotia serve as the primary inoculum of the disease. In many cases, this disease is so severe that it makes the cultivation of flowers, especially petunia, impossible. Control of this disease is extremely difficult. The systemic fungicide Topsin M (thiophanate methyl) is effective but it is very costly. Furthermore, it has no effect on the sclerotia of the fungus.

S. sclerotiorum is susceptible to mycoparasitism by *T. atroviride*. The hyphae of *T. atroviride* penetrated the hyphae of *S. sclerotiorum* and caused cytoplasmic leaching and coagulation, cell wall degradation and cell lysis. *T. atroviride* can parasitize the sclerotia of *S. sclerotiorum* and use them as a food source. In efficacy studies, treatment variables included *T. atroviride* in potting mix, *T. atroviride* in both potting mix and in field soils, Topsin M (applied as a drench) and a blank control. Petunia plants treated with Topsin M and *T. atroviride* stayed in bloom longer. These plants were also significantly healthier than the blank control as measured by fresh weight and dry weight. Application of Topsin M and *T. atroviride* in potting mix as well as in field soils yielded significantly fewer sclerotia per plant.[15] The BCA has been commercialized recently under the trade name of "Plant-Helper".

4
Control of take-all by *Phialophora* sp. (lobed hyphopodia)

Take-all disease of cereals, caused by *Gaeumannomyces graminis* (Sacc.) Arx and Olivier var. *tritici* Walker (*Ggt*), is a devastating root-rotting disease worldwide. Control options are limited to crop rotation and, in some countries, the induction of the "take-all decline" or suppressive soil phenomenon, where the prolonged monoculture of cereals results in a reduction of the disease to an economically acceptable level. However, severe and unacceptable levels of disease occur in the first 2–3 years before decline sets in and other control measures are required. Biocontrol of take-all has been actively researched in the last two decades and, although many bacterial and fungal antagonists of *Ggt* have shown promise, none has been commercialized.[16]

Several non-pathogenic fungi, closely related to the pathogen, such as *G. graminis* var. *graminis* (*Ggg*), *Phialophora* sp. (lobed hyphopodia) and *P. graminicola* (Deacon) Walker have protected wheat against take-all in greenhouse and field experiments in Australia, Europe and the USA.[17–22] In New South Wales, Australia, several isolates of *Ggg* and *Phialophora* sp. (lobed hyphopodia) increased wheat yields by 10–45% when 45 kg ha⁻¹ of oat grain inocula of these fungi were mixed with wheat seed and sown into take-all infested soil.[19] However, the yields of the protected wheat only approached those of healthy wheat when the levels of take-all were low to moderate (yield losses of 10–25%). With severe disease (yield losses of >30%), the yields obtained by this form of biological control, although statistically significant, were not acceptable to farmers.

Wong[23] considered cold-tolerance in the BCAs to be an important attribute for the biocontrol of take-all because the soil temperatures in the root zone of wheat in much of the Australian wheat belt are between 5 and 15°C during the early growing period, when the roots have to be pre-colonized by the BCA for protection. Earlier isolates of *Ggg* grew poorly at 10°C and not at all at 5°C, and this could be a reason for the lower level of take-all control obtained in those field experiments.[19] In greenhouse experiments, Wong[24] demonstrated that cold-tolerant isolates of *Ggg* and *Phialophora* sp. (lobed hyphopodia) were more effective than high-temperature isolates at a temperature regime (7–13°C) resembling winter soil temperatures in Australia. Lower soil temperatures (0–10°C) would prevail in wheat areas in the northern hemisphere, especially during the early growth period of winter cereals.

Recently, Wong et al.[25] confirmed in field experiments that highly cold-adapted isolates of *Ggg* and *Phialophora* sp. (lobed hyphopodia) provided consistent and high levels of protection against take-all. These strains had growth rates of 1.0–1.4 mm d^{-1} on agar plates at 5°C, which were equal to or greater than those of pathogenic *Ggt* (0.5–1.3 mm d^{-1}). The growth rates on agar had earlier been shown to correlate well with growth rates on wheat roots.[24] The competitive advantage of these cold-adapted isolates allowed them to colonize the root systems more quickly and extensively. Microscopic examination revealed the prevalence of the characteristic lobed hyphopodia on seminal roots, the subcrown internode, crown roots and even leaf sheaths of the tillers. Wheat plants, heavily colonized by these root-inhabiting fungi, were not adversely affected and grain yields were not significantly different to those of healthy uninoculated wheat. The fastest-growing isolate (KY) of *Phialophora* sp. (lobed hyphopodia) is in the process of being commercialized by an Australian biotechnology company, Bio-Care Technology Pty Ltd.

The suppression of the take-all fungus by closely-related non-pathogenic fungi is a cross-protection phenomenon.[23,26] There was no evidence of antibiosis, parasitism or the production of phytoalexins or fungitoxic compounds.[27] The presence of *Ggg* and *Phialophora* spp. in the wheat root cortex induced greater lignification and suberization of the endodermis and xylem vessels directly below the colonized cortex (Speakman and Lewis[28]; Weir and Wong, unpubl. data). However, as this is only a localized response, a large proportion of the the root system especially in the crown region would have to be colonized in order to provide adequate protection. Therefore, competition for root cortical tissues is critical in this form of biocontrol. This has been confirmed by the application of increasing rates of the BCA inoculum in field experiments when a dose response was obtained.[25] This host-induced resistance, being structural and permanent, should provide greater protection than mechanisms mediated by the production of antibiotics, which may be labile in soil.

5
Pest control by *Metarhizium anisopliae*

Many isolates of an entomopathogenic fungus, *Metarhizium anisopliae* (Metschn.) Sorokin, have been shown to infect a large number of insect pests but only a few applications have been commercialized.[29] These include the control of spittle bugs (*Mahanarva posticata* Stål) in Brazil, the German cockroach (*Blatella germanica* L.) in the United States of America and the Tasmanian redheaded cockchafer (*Adoryphorus couloni* Burmeister) in Australia.[30] The restricted growth and activity of entomopathogenic fungi at low temperatures was cited as a possible reason for the lack of their commercial development in cool temperate regions.[30] In southern Australia, where the redheaded cockchafer is a pest, field soils have a mean annual temperature range of 5–18°C. The larvae of this pest cause serious damage to pastures and field crops, especially when numbers exceed 300 larvae m^{-2}. There are no chemical pesticides registered for their control and Rath et al.[31,32] have shown good suppression of the insect pest by a specially selected cold-active isolate of *M. anisopliae* (DAT F-001). Field experiments have confirmed that the incorporation of 6.3x10^4 spores g^{-1} soil into Tasmanian pastures during the fall reduced the survival of larvae by 81.5% in the following late spring. However, the fall applications of the BCA did not reduce larval populations quickly enough to avoid pasture damage. The populations of the BCA remained at or near their original inoculation levels during the duration of the field experiments (12–30 months) and may provide longer-term suppression of subsequent pest generations.

Although the pathogenicity of the isolates is an obvious pre-requisite for successful biocontrol, abiotic factors such as temperature, moisture and pH must also be considered for the development of *M. anisopliae* as a myco-insecticide. In a survey of 204 isolates of the fungus from Tasmanian soils, only 36 (18%) could germinate at 5°C and less than half of these (15/36) were pathogenic to *A. couloni*.[33] They differentiated 16 strains on germination temperature and these fell into three broad groups: those able to germinate at 5°C (cold-active), those able to germinate at 37°C (heat-active) and those unable to germinate at either of these temperatures (meso-thermoactive). McCammon and Rath[34] found that the cold-active strains generally germinated faster than other strains at mid-range temperatures (10–15°C). The germination rates of these strains were also generally higher than most of the other strains although not always significantly so. The ability of *M. anisopliae* spores to germinate and grow at various temperatures has been correlated with the virulence of some isolates[35] but, in the case of DAT F-001, it was pathogenic to larvae of *A. couloni* at 5, 10 and 15°C[36]). However, to be expected, the death rate of the larvae was fastest at 15°C and slowest at 5°C. From the temperature data and a study of the carbohydrate utilization of a large number of isolates of *M. anisopliae*, Rath et al[37] showed that the cold-active isolates were distinct from *M. anisopliae* var. *anisopliae* and named the new variety, *M. anisopliae* var. *frigidum*.

In a later study, Roddam and Rath[38] surveyed the fungal flora of subantarctic Macquarie Island (54°S, 159°E) for the occurrence of *M. anisopliae* and its possible greater adaptation to the constantly low temperatures that prevailed on the island (4.8±2.2°C). They found that *M. anisopliae* occurred at lower numbers in Macquarie Island soils than in Tasmanian soils[31] or in subarctic or arctic soils in Finland.[39] When the Macquarie Island isolates were compared with the Tasmanian ones, none of the former isolates grew, germinated or sporulated faster than the cold-active Tasmanian isolates. They were, therefore, similar cold-adapted mesophiles or facultative psychrophiles with their highest growth rates at 15–20°C. Roddam and Rath[38] concluded that these subantarctic isolates were unlikely to be more effective than the cold-active isolate DAT F-001, which has recently been commercialized by an Australian biotechnology company, Bio-Care Technology Pty Ltd, and is marketed in Australia as "Bio-green".

6
Conclusions

The recognition of the role of temperature in the antagonist-pathogen interaction has led to the deliberate selection of cold-adapted BCAs for use in temperate soils, where temperatures in the range of 0–5°C are common. This strategy has yielded commercial products such as "Bio-Green" *(M. anisopliae)* for the control of a pasture pest in southern Australia and "Plant-Helper"*(T. atroviride)* for the control of many plant diseases (Table 1). A cold-tolerant isolate of *Phialophora* sp. (lobed hyphopodia) is also close to commercialization. In all these situations, the fast germination rate of spores and/or enhanced growth rate of the mycelia in the soil, the infection court or the target organism conferred a competitive advantage on the BCAs. The ability to continue to produce antibiotics, cell wall degrading enzymes and toxins or induce host resistance at these low temperatures (0–5°C) make them superior to those isolates that are incapable of growth or activity during the colder months.

For snow molds, it is the cessation of plant growth and metabolism under snow cover that makes the hosts vulnerable to these pathogens. Therefore, the rapid colonization and prior possession of the limited substrates (decaying tissues) ensure that the BCAs can exclude the pathogens when they arrive at the substrates, especially if the former also produce antibiotics as do *A. boreale* and *H. grisea* var. *grisea*. These BCAs, being obligate or facultative psychrophiles, are eminently suited to the ecological niche, which is relatively free of general antagonism because of the zero or sub-zero temperatures.

The great variability in the suppressive activity of *T. phacorrhiza* isolates suggests that selection for superior isolates may be possible. It would be interesting to see if selection on the basis of mycelial growth rates at 0°C can be correlated with the suppression of pathogenic *Typhula* spp. The field results on the use of *T. phacorrhiza* for the control of pathogenic *Typhula* spp. in turfgrasses and pasture grasses are promising but the occurence of isolates of *T. phacorrhiza* pathogenic to winter wheat in eastern Ontario, Canada (Schneider and Seaman[40]) may make its registration as a commercial product problematical in wheat-growing areas.

There is now a trickle of BCAs reaching the marketplace. With the exception of *T. atroviride*, the BCAs are generally specific to a disease or a pest. The selection for cold-tolerance has been critical in the development of BCAs that have to be active in winter or under snow. In other situations, e.g. in greenhouses or in the tropics, high-temperature isolates would obviously be preferred. An eminent biocontrol protagonist, the late Professor Ralph (Tex) Baker, used to describe his high-temperature isolate of *Trichoderma harzianum* Rifai from Colombia, which was effective against a number of soilborne diseases in greenhouse crops, as his "Spanish-speaking *Trichoderma*". He prophesized that what was needed for general use in temperate field soils, however, was "an Eskimo-speaking *Trichoderma*". The discovery in Alaska of superior cold-adapted isolates of *Trichoderma atroviride* confirms his vision.

Acknowledgements. We thank Drs N. Matsumoto and A. C. Rath for valuable discussions and Professor B. J. Deverall for reviewing the manuscript.

7
References

1. Deverall BJ. Psychrophiles. In: Ainsworth GC, Sussman AS, eds. The Fungi, vol. 3. New York: Academic Press, 1968:129-135.
2. McBeath JH, Wenko L. A simple, versatile method to determine extracellular enzymes in snow molds. Phytopathology 1986; 76:1143.
3. McBeath JH, Mehdizadegan F, Lockwood H. Anther culture selection to enhance snow mold disease resistance in winter wheat. In: Nijkamp HJJ, Van Der Plas LHW, Van Aartrijk J, eds. The Progress in Plant Celluar and Molecular Biology. Current Plant Science and Biotechnology in Agriculture, The Netherlands: Kluwer Academic Publishers, 1990:258-263.
4. Kiyomoto RK, Bruehl GW. Carbohydrate accumulation and depletion by winter cereals differing in resistance to *Typhula idahoensis*. Phytopathology, 1977; 67:206-211.
5. Amano Y, Osanai SI. Winter wheat breeding for resistance to snow mold and cold hardiness. III. Varietal differences of ecological characteristics on cold acclimtion and relationships of them to resistance. Bull Hokkaido Prefect Agric Expt Sta 1983; 50:83-97.
6. Smith JD, Davidson JGN. *Acremonium boreale* n. sp., a sclerotial, low-temperature-tolerant, snow mold antagonist. Can J Bot 1979; 57:2122-2139.
7. Burpee LL, Kaye LM, Goulty LG, Lawton MB. Suppression of gray snow mold on creeping bentgrass by an isolate of *Typhula phacorrhiza*. Plant Disease 1987; 71:97-100.
8. Lawton MB, Burpee LL. Effect of rate and frequency of application of *Typhula phacorrhiza* on biological control of Typhula blight of creeping bentgrass. Phytopathology 1990; 80:70-73.
9. Matsumoto N, Tajimi T. Biological control of *Typhula ishikariensis* on perennial ryegrass. Ann Phytopath Soc Japan 1992; 58:741-751.
10. McBeath JH. Cold Tolerant Trichoderma. International Application Published under the Patent Cooperative Treaty (PCT), International Publication Number WO 92.03056, 1992.
11. Nelson EB, Burpee, LL, Lawton MB. Biological control of turfgrass diseases In: Leslie AR, ed. Handbook of Integrated Pest Management for Turf and Ornamentals. Ann Arbor: Lewis Publishers, 1994:409-427.
12. Wu C, Hsiang T, Yang L, Liu LX. Evaluation of *Typhula phacorrhiza* for the biocontrol of grey snow mold in turfgrass. In: Tang W, Cook R J, Rovira A D, eds. Advances in Biological Control of Plant Diseases. Beijing: China Agricultural University Press, 1996:227-233.
13. Burpee LL. Interactions among low-temperature-tolerant fungi: prelude to biological control. Can J Plant Pathol 1994; 16:247-250.
14. Matsumoto N. Biological control of snow mold. In: Li PH, Chen TH, eds. Plant Cold Hardiness. New York: Plenum, 1998:343-350.

15. McBeath JH, Matheke G, Wagner P. Control of petunia Sclerotiinia stem rot with *Trichoderma atroviride*. Phytopathology 1996; 86:S37.

16. Wong PTW. Biocontrol of wheat take-all in the field using soil bacteria and fungi. In: Ryder MH, Stephen PM, Bowen GD, eds. Improving Plant Productivity with Rhizobacteria. Adelaide: CSIRO Division of Soils, 1994:24-28.

17. Deacon JW. Control of the take-all fungus by grass leys in intensive cereal cropping. Plant Pathol 1973; 22:88-94.

18. Wong PTW. Cross-protection against the wheat and oat take-all fungi by *Gaeumannomyces graminis* var. *graminis*. Soil Biol Biochem 1975; 7:189-194.

19. Wong PTW, Southwell RJ. Field control of take-all of wheat by avirulent fungi. Ann Appl Biol 1980; 94:41-49.

20. Speakman JB. Control of *Gaeumannomyces graminis* var. *tritici* in wheat by isolates of the *G. graminis* var. *graminis/Phialophora* sp. (lobed hyphopodia) complex under field conditions Phytopathol Z 1984; 109:188-191.

21. Rothrock CS. Effect of chemical and biological treatments on take-all of winter wheat. Crop Prot 1988; 7:20-24.

22. Duffy BK, Weller DM. Use of *Gaeumannomyces graminis* var. *graminis* alone and in combination with fluorescent *Pseudomonas* spp. to supress take-all of wheat. Plant Disease 1995; 79:907-911.

23. Wong PTW. Biological control by cross-protection. In: Asher MJC, Shipton PJ, eds. Biology and Control of Take-all. London: Academic Press, 1981:417-431.

24. Wong PTW. Effect of temperature on growth of some avirulent fungi and cross-protection against the wheat take-all fungus. Ann Appl Biol 1980; 95:291-299.

25. Wong PTW, Mead JA, Holley, MP. Enhanced field control of wheat take-all using cold tolerant isolates of *Gaeumannomyces graminis* var. *graminis* and *Phialophora* sp. (lobed hyphopodia). Plant Pathol 1996; 45:285-293.

26. Tivoli B, Lemaire JM, Jouan, B. Premunition du Ble contre *Ophiobolus graminis* Sacc. par des souches peu agressives du meme parasite. Ann Phytopathol 1974; 6:395-406.

27. Deverall BJ, Wong PTW, McLeod, S. Failure to implicate antifungal substances in cross-protection of wheat against take-all. Trans Br Mycol Soc 1979; 72:233-236.

28. Speakman JB, Lewis BG. Limitation of *Gaeumannomyces graminis* by wheat root responses to *Phialophora radicicola*. New Phytol 1978; 80:373-380.

29. Gillespie AT. The use of fungi to control pests of agricultural importance. In: Burge MN, ed. Fungi in Biological Control Systems. Manchester: Manchester University Press, 1988:37-80.

30. Rath AC. *Metarhizium anisopliae* for control of the Tasmanian pasture scarab *Adoryphorus couloni*. In: Jackson TA, Clare TR, eds. The Use of Pathogens in Scarab Pest Management. Andover: Intercept, 1992:217-227.

31. Rath AC, Koen TB, Yip HY. The influence of abiotic factors on the distribution and abundance of *Metarhizium anisopliae* in Tasmanian pasture soils. Mycol Res 1992; 96:378-384.

32. Rath AC, Koen TB, Anderson GC, Worledge D. Field evaluation of the entomogenous fungus *Metarhizium anisopliae* (DAT F-001) as a biocontrol agent for the redheaded pasture cockchafer, *Adoryphorus couloni* (Coleoptera: Scarabaeidae) Aust J Agric Res 1995; 46:429-440.

33. Yip HY, Rath AC, Koen TB. Characterization of *Metarhizium anisopliae* isolates from Tasmanian pasture soils and their pathogenicity to redheaded cockchafer (Coleoptera; Scarabaeidae: *Adoryphorus couloni*). Mycol Res 1992; 96:92-96.

34. McCammon SA, Rath AC. Separation of *Metarhizium anisopliae* strains by temperature dependent germination rates. Mycol Res 1994; 98:1253-1257.

35. Samuels KDZ, Heale JB, Llewellyn M. Characteristics relating to the pathogenicity of *Metarhizium anisopliae* toward *Nilaparvata lugens*. J Invertebr Pathol 1989; 53:25-31.

36. Rath AC, Anderson GC, Worledge D, Koen TB. The effect of low temperature on the virulence of *Metarhizium anisopliae* (DAT F-001) to the subterranean scarab *Adoryphorus couloni*. J Invertebr Pathol 1995; 65:186-192.

37. Rath AC, Carr CJ, Graham BR. Characterization of *Metarhizium anisopliae* strains by carbohydrate utilization (AP150CH). J Invertebr Pathol 1995; 65:152-161.
38. Roddam LF, Rath AC. Isolation and characterization of *Metarhizium anisopliae* and *Beauveria bassiana* from subantarctic Macquarie Island. J. Invertebr Pathol 1997; 69:285-288.
39. Vanninen I. Distribution and occurrence of four entomopathogenic fungi in Finland: effect of geographical location, habitat type and soil type. Mycol Res 1996;100:93-101.
40. Schneider EF, Seaman WL. *Typhula phacorrhiza* on winter wheat. Can J Plant Pathol 1986; 8:269-276.

Snow mold-crop-environment interactions

D. A. Gaudet* and A. Laroche

Research Centre, Agriculture and Agri-Food Canada, POB 3000, Lethbridge, Alberta, Canada T1J 4B1

1
Introduction

In temperate, boreal, and sub-arctic zones, plants such as winter cereals, grasses, and perennial forages must survive harsh winter conditions, frequently encountering exposure to lethal and sublethal low temperatures. However, in many regions, a deep, persistent snow cover insulates the root and crown zone, maintaining soil temperatures between 0° and -10°C despite very low ambient temperatures, thereby protecting plants against low temperature injury.[1,2] The protracted snow cover creates a dark, humid environment with constant temperatures that prevents photosynthesis and drastically reduces plant metabolism. These conditions favour development of psychrophilic fungi known as snow molds, which can cause extensive damage to agricultural, ornamental, and native plants. In this article, we outline the adaptations that plants and snow mold fungi have made to survive and thrive in a nival environment, and discuss the dynamic nature of the snow mold-plant interactions under snow, including the interaction between freezing tolerance and snow mold resistance.

2
Snow mold fungi

Snow mold damage in crops can appear as irregular white or brown patches in the field shortly after snow melt (Fig. 1 upper part). In severe cases, nearly 100% of the plants may be killed. Infected plants may appear white or gray and frequently have mycelial growth on the leaf surface (Fig. 1 upper and middle part). Fungal structures such as sclerotia may also be visible on the leaves (Fig. 1 lower part) enabling the fungus to survive the summer and serve as an inoculum source in subsequent infection cycles during the autumn and winter. The prevalence and severity of snow mold damage is generally governed by environmental conditions during the autumn and depth and duration of the snow cover during the winter.[3–5]

---

* Corresponding author

Fig. 1. Snow mold damage to winter cereals in the central and northern prairies in western Canada: (**upper part**) patchy occurrence of the Low Temperature Basidiomycete (LTB) fungus in winter wheat plots immediately after snow melting; (**middle part**) fall rye plants killed by the LTB fungus, and (**lower part**) sclerotia of *Sclerotinia borealis* on fall rye

The taiga and boreal zones in the northern hemisphere are characterized by very low ambient temperatures, and a persistent snow cover on frozen soil throughout the winter. In these regions where the snow cover can last as long as 200 d, snow scald (*Sclerotinia borealis* Bub. and Vleug.), speckled snow mold (*Typhula ishikariensis* Imai) and cottony snow mold (sterile Low Temperature Basidiomycete, LTB), are the most important snow molds.[5,6] Beneath the snow on frozen ground, the optimum temperature for development of LTB is -3°C[4,7] and of *Sclerotinia borealis* is -2.0°C.[1,8] In temperate or northern regions modified by a maritime climate, where milder winter temperatures and snow cover in excess of 70 d on unfrozen soil, pink snow mold (*Microdochium nivale* (Fr.) Samuels and Hallet), gray snow mold (*Typhula incarnata* Lasch ex Fr.), and snow rot (*Pythium iwayami* S. Ito. Hirane) are the most important snow molds.[1,3] Other fungi that have been reported causing damage to plants in nival environments are *Sclerotinia trifoliorum* Eriks. and *S. nivalis* I. Saito.[9]

Despite the requirement for prolonged exposure to zero or sub-zero temperatures for plant parasitism, most snow fungi grow optimally in sterile growth media from 10 to 15°C.[10,11] For example, the optimum temperature for infection of wheat, alfalfa, and grasses by the LTB fungus under non-sterile conditions is -3°C,[7,11] but the same fungus grows optimally on sterile growth media at 10–15°C and is capable of causing storage rot on apples and pears at temperatures between 5 and 15°C.[11] Conversely, the pink snow mold fungus, *M. nivale*, has an optimum in-vitro growth temperature of 25°C, and can develop on winter cereals at all stages of plant growth.[12] Matsumoto[13] considered those snow molds that infect the host only at low nival and sub-nival temperatures in nature as obligate snow molds. Those that infect over a wide range of temperatures from the nival to the higher temperatures during the growing season, were classified as facultative snow molds. This discrepancy between the optimum growth temperatures of snow mold fungi under sterile and non-sterile conditions has been attributed to the poor competitive ability of snow mold fungi with other soil microorganisms.[13] The adaptation of snow molds to growth at temperatures near 0°C has generally allowed them to escape microbial competition in the nival environment.[14] However, even in a nival environment, most snow molds are poor competitors and must depend on plant parasitism for their existence.[15] Furthermore, microbial antagonism of snow mold fungi by low temperature saprophytes has been observed. Development of *Typhula incarnata* and *Microdochium nivale* on turfgrasses are inhibited by some strains of the low temperature saprophyte *Typhula phacorrhiza* Reichard ex Fr.[16] A low temperature-tolerant bacterium belonging to the genus *Pseudomonas* is also a snow mold antagonist.[17]

Competition for limited resources under snow has also created intraspecific competitive interactions among biotypes within *T. incarnata*, *T. ishikariensis*, and LTB taxa.[13,18] This highly competitive interaction among snow mold isolates has created a complex ecological structure whereby different survival and reproductive strategies exist among biotypes (for a review, see ref. 13). Different vegetative incompatibility groups within biotypes of *T. ishikariensis* have adopted different survival strategies according to the prevailing cropping practices.[14] Biological diversity among incompatibility groups of biotype B of *T. ishikariensis* decreased with increasing intensity of agriculture; diversity was lowest in a continuous cropping regime, which represents the highest level of cropping intensity and was highest in pastures and meadows which represented the least intensive agricultural production.[14] Different vegetative incompatibility groups of snow molds tend to antagonize each other on the periphery of individual patches as they compete for available resources.[14]

3
Plant development

The ability of plants to withstand the numerous winter stresses is genetically determined by the plant, and varies among and within plant species.[19] A wide range of genetic diversity exists among winter wheat cultivars for freezing tolerance which is clearly multigenically controlled.[20] Conversely, limited genetic variation exists among winter cereal cultivars in levels of resistance to snow molds. Following extensive screening of the world germplasm collection of winter wheat, less than a dozen lines were identified as containing potentially useful levels of snow mold resistance.[3] Snow mold resistance in wheat appears to be conditioned by the additive effect of two or three loc.[21,22]

During the autumn and early winter, low average temperatures induce an acclimation process in plants which leads to increased levels of both freezing tolerance and snow mold resistance.[19,23–25] However, the relationship between freezing tolerance and snow mold resistance is not well understood. In the development of freezing tolerance in cereals, the first stage of hardening is gradual and occurs at temperatures above freezing. Six weeks at hardening temperatures between 0 and 8°C are required for winter wheat plants to attain peak levels of freezing resistance.[26] Additional freezing resistance can be achieved in a second stage of hardening if hardened plants are briefly subjected to mild sub-freezing temperatures such as -3°C.[27,28] By contrast, snow mold resistance is rapidly acquired following onset of hardening conditions.[29] Light is required to induce snow mold resistance[23,29] but is not necessarily required for induction of freezing resistance as wheat sprouted at low temperatures in the dark can develop high levels of freezing resistance.[30,31]

Maximum freezing resistance levels are maintained for 1 or 2 weeks in wheat after which levels begin to gradually decline as plants maintain low metabolic levels near freezing temperatures under snow;[32] this decline is more rapid in non-hardy than in cold hardy cultivars. Similarly, snow mold resistance gradually decreases throughout the winter under snow and the decline is more rapid in susceptible than in resistant cultivars.[33] Cold-acclimated plants subjected to growing conditions at warm temperatures rapidly deharden and lose their ability to withstand low temperatures[19,34] but retain some snow mold resistance.[24,35] Therefore, it may be possible to separate the effects of cold hardening and snow mold resistance by studying the expression of snow mold resistance under dehardening conditions.

Low temperature stress increases the susceptibility of plants to snow mold and, conversely snow mold stress reduces the plant's resistance to freezing temperatures. Tronsmo[36] observed that sublethal low temperature stress, between -6 and -8°C, increased the susceptibility in timothy to both *M. nivale* and *T. ishikariensis* and that snow mold susceptibility, induced by sublethal low temperature stress was less severe in a very cold-hardy cultivar than in a moderately hardy cultivar. Similarly, infection by the LTB fungus was more severe following exposure of winter wheat plants to sublethal low temperatures between -5 and -10°C.[37] In addition, sublethal levels of cottony snow mold damage decreased the freezing tolerance of partially and fully hardened winter wheat plants.[37] The impact of sublethal low temperature stress has not been determined under field conditions but it is likely that any stress that adversely affects the plant's metabolism also reduces a plant's ability to resist stresses from other sources.

Resistance to snow molds is clearly linked to developmental stage in winter cereals, legumes, and grasses. There appear to be at least two forms of resistance to snow molds linked to the stage of development in winter cereals:

1. resistance that increases with increasing plant age in both resistant and susceptible cultivars but is expressed to a greater extent in resistant cultivars, and,
2. resistance that is expressed in resistant cultivars in the pre-tillering stage.

For maximum expression of snow mold resistance, plants must be seeded early enough to attain an optimal size prior to the onset of winter.[3] The optimum date for planting winter wheat in the Pacific Northwest of the United States is mid- to late-August which permits development of large plants with numerous tillers.[3] When seeding is delayed until mid-September, cultivars that are normally resistant become susceptible to snow molds. However, further delay of seeding until October results in an increase in snow mold resistance. A similar relationship between early seeding and snow mold resistance has been observed in grasses[23] and alfalfa.[38] Under controlled environment conditions, older hardened grasses and winter cereals express high levels of resistance to snow molds.[29,39,40] Even susceptible cultivars can develop substantial levels of snow mold resistance if growing conditions permit the development of many tillers and large crowns.[3,29,39,40] Cavelier[41] and Gaudet and Kozub[40] observed that snow mold resistance is also expressed at the pre-tillering stage. The physiological basis for either form of resistance is unknown but the accumulation of physiologically active substances involved in snow mold resistance may develop earlier and reach higher levels in resistant cultivars than in susceptible ones.

4
Soluble carbohydrates

The hardening process in winter cereals is accompanied by a reduction in growth rate and plant height, a decrease in leaf surface, and an increase in plant dry weights in rye *(Secale cereale)* and wheat *(Triticum aestivum)*.[32,42] The increase in dry weight is due, in part, to a decrease in water content and increase in cytoplasmic content.[42,43] Concurrent with the first stage hardening in cereals is the accumulation in the leaves and crowns of large quantities of soluble mono-, di- and trisaccharides, and fructans, which are polymers of fructose possessing β-2-1 and β-2-6 linkages.[44–48] The large quantity of photosynthate in grasses exposed to hardening conditions is due to the continuation of active photosynthesis coupled with a reduced demand for both structural and non-structural carbohydrates at low temperatures.[49,50]

Levels of fructose, glucose, and sucrose are high in the crowns of wheat, rye, and oat seedlings during the initial stages of hardening but decrease during subsequent prolonged exposure to hardening conditions as shown here for wheat (Fig. 2). Simple carbohydrates are converted to fructans following induction of sucrose:sucrose fructosyltransferases (SST), sucrose phosphate synthase and sucrose:fructose fructosyltransferase (Fig. 3).[26,48,49,51–55] Fructans and sucrose can also be metabolized by different β-fructosidases (invertase, fructan endohydrolase and fructan exohydrolase (FEH)). Up to 45% of the total plant dry weight consists of fructans in snow mold resistant winter wheat cultivars and large winter wheat plants accumulate 1.5 times higher levels of fructans than small plants when exposed to hardening temperatures. Increased levels of soluble sugars have been associated with the development of both freezing resistance[27,28,51–59] and snow mold resistance.[33,41,60–62] Figure 2 illustrates the differences in accumulation of soluble simple sugars and fructans among three cultivars. The snow mold resistant cultivar CI14106 rapidly accumulates high fructan levels compared to the susceptible cultivars Norstar and Cappelle Desprez.

Differences in snow mold resistance among winter wheat cultivars have been attributed to the rate of carbohydrate accumulation in crowns during autumn.[61,63,64] Simple and complex carbohydrates may be involved in increasing the cryo-stability of the plasma membrane.[46,47,65–67] Winter wheat plants briefly exposed to -3°C metabolize fructan into fructose, glucose, and sucrose during the second stage of hardening in winter cereals to further increase the level of freezing resistance.[27]

A snow cover exceeding 50 cm depth is sufficient to prevent light transmission to the soil surface.[1] Winter cereals subjected to temperatures near 0°C in the dark undergo carbohydrate depletion through respiration during the winter and early spring.[25] Snow mold resistant cultivars accumulate higher levels of carbohydrate and metabolize them at a slower rate compared to susceptible cultivars and the amount of carbohydrate remaining in the spring after snow mold attack may be critical to winter survival.[21,33,61,62] The mechanism of enhanced carbohydrate accumulation and reduced depletion among snow mold resistant cultivars is not known but may be due to the differential activities of the enzymes SST and FEH, that are involved in the metabolism of fructan (Fig. 3).[33]

5
Mechanisms of resistance

There are no satisfactory hypotheses to explain the nature of snow mold resistance. The mechanism(s) of resistance against most snow mold fungi appears to be identical among winter wheat cultivars[21,63] and is independent of the freezing resistance potential of cultivars.[21,40,63] However, this is not always the case as a positive associ-

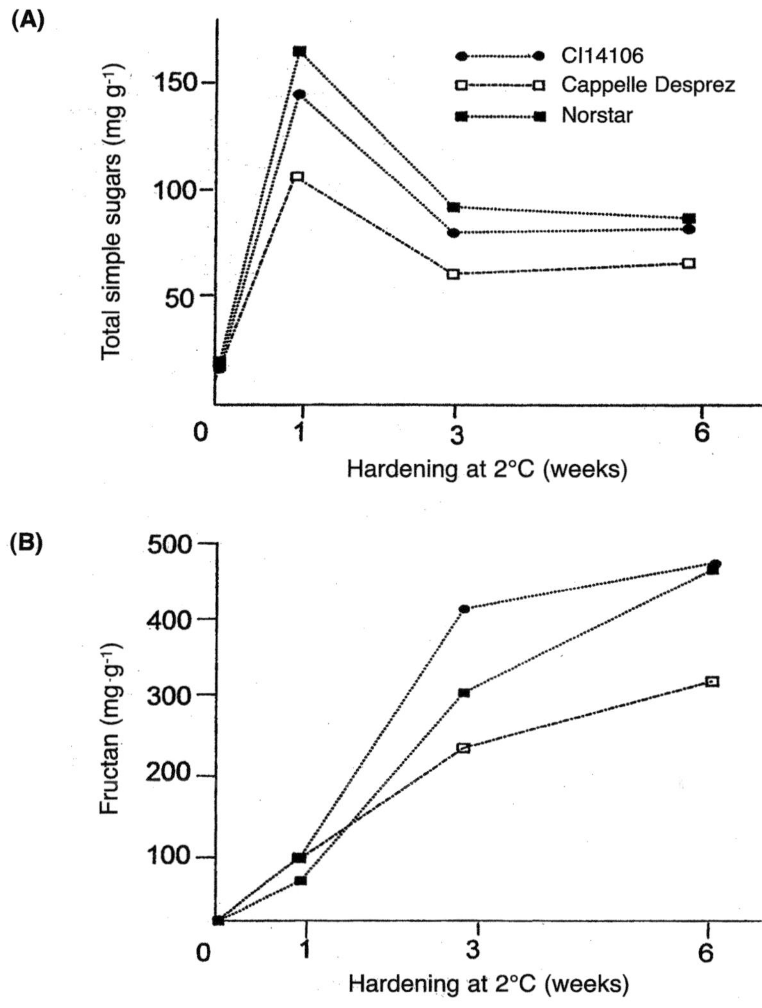

Fig. 2. Accumulation (mg carbohydrate g^{-1} dry weight) of total simple sugars (fructose + glucose + sucrose) (**A**), and fructan (**B**) in crown tissues, for three winter wheat cultivars following pre-hardening growth for 6 weeks at 20°C, and hardening for 0, 1, 3 or 6 weeks at 2°C. CI14106 is resistant, Norstar is intermediate, and Cappelle Desprez is susceptible to snow molds

ation between level of genotypic freezing resistance and resistance to *M. borealis*[21], and a strong correlation between resistance to *T. ishikariensis* and *M. nivale*, and freezing resistance in orchardgrass *(Dactylis glomerata)* and timothy *(Phleum pratense)*[36,59] have been recorded. Thus, the relationship between resistance to freezing temperatures and snow mold resistance may vary for individual host/pathogen combinations. Maintaining high fructan levels in the crowns throughout the winter appears to be important in overwintering ability of winter wheat in regions that

$$\text{fructose + glucose} \;\underset{E4}{\overset{E1}{\rightleftarrows}}\; \text{sucrose}$$

$$\text{sucrose + sucrose} \;\underset{E4}{\overset{E2}{\rightleftarrows}}\; \begin{array}{c}\text{1-kestose}\\\text{6-kestose}\end{array} \text{+ glucose}$$

$$\begin{array}{c}\text{1-kestose}\\\text{6-kestose}\end{array}\text{+ sucrose} \;\underset{E4}{\overset{E3}{\rightleftarrows}}\; \text{oligofructans + glucose}$$

$$\text{oligofructans + sucrose} \;\underset{E4}{\overset{E3}{\rightleftarrows}}\; \text{fructans + glucose}$$

$$\text{fructan} \;\rightarrow\; \text{fructan + frucose}$$

Fig. 3. Summary of biosynthesis and metabolism of fructans in cereals
E1 sucrose phosphate synthase, *E2* sucrose:sucrose 1-fructosyltransferase, sucrose:sucrose 6-fructosyltransferase, *E3* sucrose:sucrose 6-fructosyltransferase, *E4* β-fructosidase (invertase, fructan exohydrolase, fructan endohydrolase)

frequently encounter snow mold injury.[61,62,64] It is possible that snow mold fungi have a limited ability to degrade long chain fructan polymers that average in degree of polymerization between DP9 and DP-11 in winter wheat following prolonged exposure to hardening conditions.[68] Alternatively, a differential regulation of plant invertases may result in a lower concentration of simple soluble sugars which would reduce the levels of simple sugars available for pathogen development.

A different possible mechanism for snow mold resistance involves the role of decreasing plant water potentials in limiting the growth of the snow mold fungi.[63,69,70] Each fungus possesses a characteristic range of water potential values under which it can grow.[69] A decrease in plant water potential caused by carbohydrate accumulation has been proposed as a mechanism to reduce, if not prevent, the growth of some snow mold fungi. Growth of *M. nivale, T. ishikariensis* and *T. incarnata* is impaired at water potentials below -1.0 MPa while growth of *M. borealis* was optimal under water potentials between -1.0 and -2.0 MPa.[24,63,70] Water potential in grasses decreases during the hardening process, possibly explaining the less extensive growth of snow mold fungi on hardened plants.[70]

The production of antifungal substances induced during hardening as part of a generalized plant resistance response,[71] or as part of a specific response to invasion by snow mold fungi, may also be an important mechanism of resistance. Recently it has become apparent that sugars, particularly hexoses, can serve as physiological signals, repressing or activating plant genes involved in many essential metabolic processes including defence against plant pathogens.[72] The expression of systemic acquired resistance coincides with sucrose mediated inhibition of photosynthesis.[73]

The rapid increase in simple sugars following exposure of winter cereals to hardening temperatures, may serve as a stimulus for the accumulation of antifungal substances. A generalized increase in resistance to plant diseases including snow molds has been observed in hardened plants but not in unhardened plants.[23,24,39]

6
Conclusions

In summary, winter annual and perennial crop species grown in the northern boreal ecosystem must survive periods of protracted snow cover and low temperatures during the winter. In deep snow regions, plants are susceptible to damage caused by interaction of snow molds and low temperatures. Therefore, high levels of resistance to low temperatures and snow molds are requisite for crops adapted to these regions. Accumulation of soluble carbohydrates in plants during the fall is linked to both hardening and resistance to attack by snow molds. Snow mold resistant cultivars accumulate higher levels of carbohydrate and metabolize them at slower rates than susceptible cultivars. The quantity and quality of carbohydrates, particularly fructans, remaining in the spring after snow mold attack, may be critical for winter survival. However, the total accumulation of carbohydrates is largely dependent on the stage of development of the winter cereal plant at the beginning of the winter. In depth studies of the metabolism of soluble carbohydrates, particularly fructans, during plant growth, hardening and snow mold infection, may provide insight to survival mechanisms in plants subjected to stresses caused by snow molds and low temperatures.

7
References

1. Tomiyama K. Studies on the Snow Blight of Winter Cereals. Report No 47 Hokkaido Nat Agric Expt Sta, 1955.
2. Ylimäki A. The effect of snow cover on temperature conditions in the soil and overwintering of field crops. Ann Agricul Fenniae 1962; 1:192-216.
3. Bruehl GW. Developing wheat resistant to snow mold in Washington State. Plant Dis 1982; 66:1090-1095.
4. Gaudet DA , Bhalla MK, Clayton GW, Chen THH. Effect of cottony snow mold and low temperatures on winter wheat survival in central and northern Alberta. Can J Plant Pathol 1989; 11:291-296.
5. Nissinen O. Analysis of climatic factors affecting snow mould injury in first-year timothy (*Phleum pratence* L.) with special reference to *Sclerotinia borealis*. Acta Univ Oulu A 1996; 289:1-115.
6. Gaudet DA, Bhalla MK. Survey for snow mold diseases of winter cereals in central and northern Alberta 1983-87. Can Plant Dis Surv 1988; 68:15-20.
7. Gaudet DA. Effect of temperature on pathogenicity of sclerotial and non-sclerotial isolates of *Coprinus psychromorbidus* under controlled conditions. Can J Plant Pathol 1986; 8:394-399.
8. Arsvoll K. *Sclerotinia borealis*. Sporulation, spore germination and pathogenesis. Meld Norg LandbrHogsk 1976; 55:1-11.

9. Saito I. *Sclerotinia nivalis*, sp. Nov., the pathogen of snow mold of herbaceous dicots in northern Japan. Mycoscience 1997; 38:227-236.

10. Matsumoto N, Tajimi A. Life-history strategy in *Typhula incarnata* and *T. Ishikariensis* biotypes A, B, and C as determined by sclerotium production. Can J Bot 1988; 66:2485-2490.

11. Gaudet DA, Sholberg PL. Comparative pathogenicity of *Coprinus psychromorbidus* monokaryons and dikaryons on winter wheat, alfalfa, grasses, and pome fruit. Can J Plant Pathol 1990; 12:31-37.

12. Cook RJ. Fusarium diseases of wheat and other small grains in North America. In: Nelson PE, Toussoun TA, Cook RJ, eds. Fusarium Diseases, Biology, and Taxonomy. Penn State Univ Press, 1981:39-52.

13. Matsumoto N. Ecological adaptations of low temperature plant pathogenic fungi to diverse winter climates. Can J Plant Pathol 1994; 16:237-240.

14. Matsumoto N, Tajimi A. Effect of cropping history on the population structure of *Typhula incarnata* and *Typhula ishikariensis*. Can J Bot 1993; 71:1434-1440.

15. Jacobs DL, Bruehl GW. Saprophytic ability of *Typhula incarnata*, *T. Idahoensis*, and *T. Ishikariensis*. Phytopathology 1986; 76:695-698.

16. Burpee LL. Interactions among low-temperature tolerant fungi: Prelude to biological control. Can J Plant Pathol 1994; 16:247-250.

17. Matsumoto N, Tajimi A. Bacterial flora associated with the snow mold fungi, *Typhula incarnata* and *T. Ishikariensis*. Ann Phyopathol Soc Jap 1987; 53:250-253.

18. Lebeau JB. Antagonism between isolates of a snow mold pathogen. Phytopathology 1975; 65:877-880.

19. Levitt J. Responses of Plants to Environmental Stresses, vol 1. London: Academic Press, 1980.

20. Singh J. Laroche A. Freezing tolerance in plants. A biochemical overview. Biochem Cell Biol 1988; 66:650-656.

21. Amano Y. Studies on methods of breeding wheat for winter hardiness. Bull Hokkaido Pref Agric Exp Sta 1987; 64:1-79.

22. Iriki N, Kuwabara T. Half diallel analysis of field resistance of winter wheat to *Typhla ishikariensis* Biotype A in artificially infested plots. Japan J Breed 1993; 43:495-501.

23. Årsvoll K. Effects of hardening, plant age, and development in *Phleum pratense* and *Festuca pratensis* on resistance to snow mould fungi. Meld Norg LandbrHøgsk 1977; 56:1-14.

24. Tronsmo AM. Effects of dehardening on resistance to freezing and to infection by *Thyphula ishikariensis* in *Phleum pratense*. Acta Agric Scand 1985; 35:113-116.

25. Gaudet DA. Progress towards understanding interactions between cold hardiness and snow mold resistance and development of resistant cultivars. Can J Plant Pathol 1994; 16:241-246.

26. Roberts DWA. Duration of hardening and cold hardiness in winter wheat. Can J Bot 1970; 57:1511-1517.

27. Olien CR, Clark JL. Changes in soluble carbohydrate composition of barley, wheat, and rye during winter. Agron J 1993; 5:21-29.

28. Olien CR, Clark JL. Freeze-induced changes in carbohydrates associated with hardiness of barley and rye. Crop Sci 1995; 35:496-502.

29. Nakajima T, Abe J. Environmental factors affecting the expression of resistance in winter wheat to pink snow mold caused by *Microdochium nivale*. Can J Bot 1996; 73:1783-1788.

30. Roberts DWA. The effect of long exposure to low temperatures on the cold hardiness of sprouting wheat in the dark. Can J Plant Sci 1985; 65:893-900.

31. Thomas JB, Schallje GB, Roberts DWA. Prolonged freezing of dark-hardened seedlings for rating and selection of winter wheats for winter survival ability. Can J Plant Sci 1988; 68:47-55.

32. Roberts DWA. Identification of loci on chromosome 5A of wheat involved in control of cold hardiness, vernalization, leaf length, rosette growth habit, and height of hardened plants. Genome 1990; 33:247-259.

33. Yoshida M, Abe J, Moriyama M, Kuwabara T. Carbohydrate levels of winter wheat cultivars varying with freesing tolerance and snow mold resistance. Physiol Plantarum 1998;103:8-16.

34. Gusta LV, Fowler DB. Effects of temperature on dehardening and rehardening of winter cereals. Can J Plant Sci 1976; 56:673-678.
35. Nakajima T, Abe J. A method for assessing resistance to the snow molds *Typhula incarnata* and *Microdochium nivale* in winter wheat incubated at the optimum growth temperatures ranges of the fungi. Can J Bot 1990; 68:343-344.
36. Tronsmo AM. Predisposing effects of low temperature on resistance to winter stress factors in grasses. Acta Agric Scand 1984; 34:210-220.
37. Gaudet DA, Chen THH. Effect of freezing tolerance and low temperature stress on development of cottony snow mold *(Coprinus psychromorbidus)* in winter wheat. Can J Bot 1988; 66:1610-1615.
38. Hwang SF, Gaudet DA. Effects of plant age and late-season hardening on development of resistance to winter crown rot in first-year alfalfa. Can J Plant Sci 1995; 5:421-428.
39. Gaudet DA, Chen THH. Effects of hardening and plant age on development of resistance to cottony snow mold *(Coprinus psychromorbidus)* in winter wheat under controlled conditions. Can J Bot 1987; 65:1152-1156.
40. Gaudet D A, Kozub G C. Screening winter wheat for resistance to cottony snow mold under controlled conditions. Can J Plant Sci 1991; 71:957-965.
41. Cavelier M. La résistance de l'orge d'hiver aux attaques de *Typhula incarnata* Lash en fonction du stade de développement et de l'endurcissement des plantes au froid. Parasitica 1987; 43:131-153.
42. Krol M, Griffith M, Huner NPA. An appropriate physiological control for environmental temperature studies: comparative growth kinetics of winter rye. Can J Bot 1984; 62:1062-1068.
43. Huner NPA, Elfman B, Krol M, McIntosh A. Growth and development at cold-hardening temperatures. Chloroplast ultrastructure, pigment content and composition. Can J Bot 1984; 62:53-60.
44. Bancal P, Gaudillère JP. Rate of accumulation of fructan oligomers in wheat seedlings (*Triticum aestivum* L.) during the early stages of chilling treament. New Phytol 1989; 112:459-463.
45. Rybka Z. Changes in carbohydrate pool and osmolality in crowns and leaves of winter wheat seedlings during hardening to frost. Acta Physiol Plant 1993; 15:47-55.
46. Hurry, VM , Malmberg G, Gardeström P, Öquist G. Effects of a short-term shift to low temperature and of long-term cold hardening on photosynthesis and ribulose-1,5-bisphosphate carboxylase/oxygenase and sucrose phosphate synthase activity in leaves of winter rye (*Secale cereale* L.). Plant Physiol 1994; 106:983-990.
47. Hurry VM, Strand Å, Tobiæson M, Gardeström P, Öquist G, Cold hardening of spring and winter wheat and rape results in differential effects on growth, carbon metabolism, and carbohydrate content. Plant Physiol 1995; 109:697-706.
48. Santoiani CS, Tognetti JA, Pontis HG, Salerno GL, Sucrose and fructan metabolism in wheat roots at chilling temperatures. Physiol Plantarum 1993; 87:84-88.
49. Pollock CJ. Fructans and the metabolism of sucrose in vascular plants. New Phytol 1986; 104:1-24.
50. Pollock CJ, Cairns AJ. Fructan metabolism in grasses and cereals. Ann Rev Plant Physiol Plant Mol Biol 1991; 42:77-101.
51. Green DG. Soluble sugar changes occuring during cold hardening of spring wheat, fall rye and alfalfa. Can J Plant Sci 1983; 63:415-420.
52. Green DG, Ratzlaff CD. An apparent relationship of soluble sugars with hardiness in winter wheat varieties. Can J Bot 1975; 53:2198-2201.
53. Olien CR, Lester GE. Freeze-induced changes in soluble carbohydrates of rye. Crop Sci 1985; 25:288-290.
54. Livingston III D P. Nonstructural carbohydrate accumulation in winter oat crowns before and during cold hardening. Crop Sci 1991; 31:751-755.
55. Duchateau A, Bortlik K, Simmen U, Wiemken A, Bancal P. Sucrose:fructan 6-fructosyl-

transferase, a key enzyme for diverting carbon from sucrose to fructan in barley leaves. Plant Physiol 1995; 107:1249-1255.

56. Larsson S, Sugar content and membrane lipid modifications in winter wheat (*Triticum aestivum* L.) during cold hardening. Sveriges Utsadesforenings Tidskrift 1989; 99:143-149.

57. Livingston III DP, Olien CR, Freed RD. Sugar composition and freezing tolerance in barley crowns at varying carbohydrate levels. Crop Sci 1989; 29:1266-1270.

58. Koster KL, Lynch DV. Solute accumulation and compartmentation during the cold acclimation of Puma rye. Plant Physiol 1992; 98:108-113.

59. Tronsmo AM, Kobro G, Morgenlie S, Flengsrud R. Carbohydrate content and glycosidase activities following cold hardening in two grass species. Physiol Plantarum 1993; 88:689-695.

60. Bengtsson B. Soluble sugar changes during winter and resistance to snow mould in winter wheat. J Phytopathol 1989; 124:162-170.

61. Kiyomoto RK, Bruehl GW. Carbohydrate accumulation and depletion by winter cereals differing in resistance to *Typhula idahoensis*. Phytopathology 1977; 67:206-211.

62. Kiyomoto RK. Carbon dioxide exchange and total nonstructural carbohydrate in soft white winter wheat cultivars and snow mold resistant introductions. Crop Sci 1987; 27:746-752.

63. Bruehl GW, Cunfer B. Physiologic and environmental factors that affect the severity of snow mold of wheat. Phytopathology 1971; 61:792-799.

64. Yukawa T, Watanabe Y. Studies on fructan accumulation in wheat (*Triticum eastivum* L.) I. Relationship between fructan concentration and overwintering ability from aspect on the pedigree. Japan J Crop Sci 1991; 60:385-361.

65. Tognetti JA, Salerno GL, Crespi MD, Pontis HG. Sucrose and fructan metabolism of different wheat cultivars at chilling temperatures. Physiol Plantarum 1990; 78:554-559.

66. Guy CL, Huber JLA, Huber SC, Sucrose phosphate synthase and sucrose accumulation at low temperature. Plant Physiol 1992; 100:502-508.

67. Holaday AS, Martindale W, Alred R, Brooks A, Leegood RC, Changes in activity of enzymes of carbon metabolism in leaves during exposure of plants to low temperature. Plant Physiol 1992; 98:1105-1114.

68. Suzuki M, Nass HG, Fructan in winter wheat, triticale, and fall rye cultivars of varying cold hardiness. Can J Bot 1988; 66:1723-1728.

69. Cook RJ, Papendick RI, Role of water potential in microbial growth and development of plant disease, with special reference to postharvest pathology. HortScience 1978; 13:559-564.

70. Tronsmo AM, Host water potentials may restrict development of snow mold fungi in low temperature hardened grasses. Physiol Plantarum 1986; 68:175-179.

71. Hon W-C, Griffith M, Mlynarz A, Kwok YC, Yang DSC. Antifreeze proteins in winter rye are similar to pathogenesis-related proteins. Plant Physiol 1995; 109:879-889.

72. Jang JC, Sheen J. Sugar sensing in higher plants. Plant Cell 1994; 6:1665-1679.

73. Herbers K, Meuwley P, Frommer WB, Metraux JP, Sonnewald U. Systemic acquired resistance mediated by ectopic expression of invertase: possible hexose sensing in the secretory pathway. Plant Cell 1996; 8:793-803.

Simulation analysis of operating conditions for a municipal wastewater treatment plant at low temperatures

N. Funamizu* and T. Takakuwa

Department of Environmental Engineering, Hokkaido University, Kita 13, Nishi 8, Kita-ku, Sapporo 060, Japan

1
Introduction

The temperature of wastewater is an important parameter because of its effect on chemical and biological reaction rates in treatment processes. The wastewater temperature is usually higher than that of tapwater, because warm water is added from household use. The temperature of tapwater depends on geographic situation and type of water supply source: surface- or groundwater. In a combined sewer system, rain and/or snowmelts enter the sewer system, and this causes a change in the wastewater temperature. Depending on the geographic location, the mean annual temperature of wastewater varies from 10 to 20°C.[1] In a cold region, wastewater treatment plants (WWTPs) are operated in the range of far lower temperatures in winter than optimum temperatures for bacterial activity. The operational conditions of a WWTP should be specified according to the wastewater temperature. The purpose of this chapter is to demonstrate procedures to set operational conditions of a WWTP at low temperatures with the aid of a mathematical model. The simulation analysis work of the WWTP in Sapporo by the authors[2] is illustrated in this chapter.

Sapporo is a city that has combined sewer systems in a cold region. The Sewage system has 9 treatment plants, 17 pump stations with 7,400 km sewer pipes and covers 98.9% of about 1.8 million population. The so-called snow problem, which involves removing and dumping snow in the urban area, is one of the main issues of such a large city as Sapporo in a snowy-cold region. In response to the need to improve the combined sewer system and resolve the snow problem, the city of Sapporo set a new policy: combined sewer systems convey and treat snow as well as wastewater and stormwater. The plant operation related to this new policy is also discussed in this chapter.

---

* Corresponding author

2
Combined sewer system in a snowy-cold region

Combined sewer systems in Sapporo experience three types of storm events: rain, snow and snowmelt. Figure 1 shows monthly precipitation averaged for the last 10 years in Sapporo. Usually, rain occurs in May through November, the snow season is in January through March, and December, March and April are thaw periods. The ordinary precipitation in Sapporo is 1,240 mm per year, and snowfall is 480 cm per year. This amount of snowfall is equivalent to 40% of the annual precipitation. During wet weather, stormwater and wastewater are transported to the flow regulators through the combined sewer network, and then they are split into three paths: not treated, primary treated and fully treated discharges. In a snow season, there is no interaction between snowfall and combined sewer systems because snow is stored on site and some of it is transferred by trucking to stock yards. In one snow season in Sapporo $1,200 \times 10^6$ m^3 of snow was transported to stock yards by trucking. In a thaw period, snowmelt enters the combined sewer systems. Snowmelt decreases the sewage temperature and affects the biological reaction in a WWTP. Figure 2 depicts the monthly variation of water temperature at an aeration basin in the Sosei WWTP, Sapporo, showing that snowmelt causes low temperature.

The "Snow Problem" in Sapporo involves four issues. First, a public opinion survey shows that the citizens of Sapporo are not satisfied with the current service for removing snow from streets and residential areas. Thus, the request for these services stands in the foremost place. Second, it becomes difficult to provide enough sites to stock the transported snow. Third, snowmelt discharged from stock yards to receiving waters (rivers and groundwater) has resulted in pollution problems.

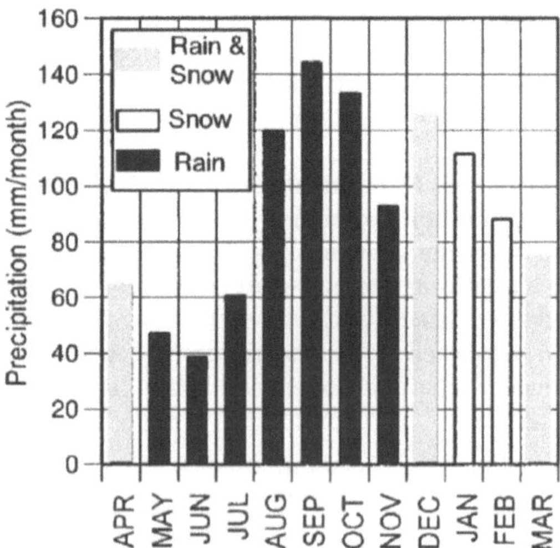

Fig. 1. Precipitation in Sapporo (avarage values for the last 10 years)

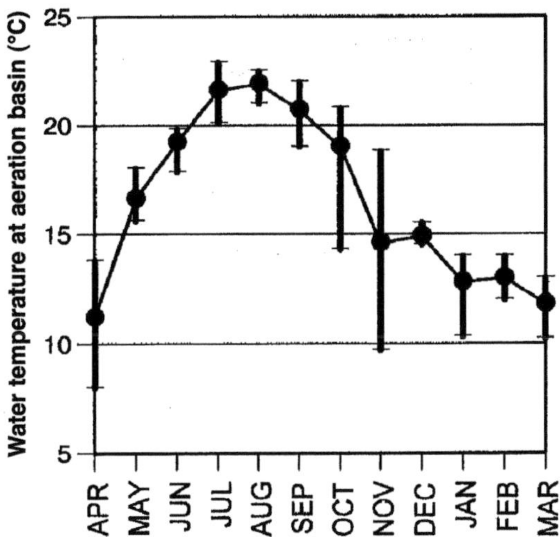

Fig. 2. Monthly variation of water temperature at the aeration basin in Sosei wastewater treatment plant in Sapporo

Fourth, snow-melting facilities for on-sites such as road heating systems are used in many places. This causes an increase in consumption of high-graded energy such as electricity and oil. Thus, there is a problem related to energy consumption.

These issues give rise to a better usage of the existing sewage system to transport and melt snow. Facilities for controlling combined sewer overflows can be converted into a snow disposal system in winter if they are originally designed for multipurpose systems. In a snowy-cold region, there are no rain events and no systems for rain are required in winter. However, in turn, we can utilize their large storage capacity for melting snow. In a snow-melting process we can also use heat energy of sewage. Figure 3 shows the heat energy flow through the water systems in Sapporo in February 1991. The temperature of tapwater was 3.8°C; water usage raised its temperature to 18.7°C. Mixing snowmelt caused a small drop in temperature and the effluent of WWTP had a temperature of 13°C. In evaluation of the sensitive heat energy of the effluent on the basis of the atmospheric temperature, the effluent had 1,308 TJ per month. Further, the total amount of sensitive heat energy in the effluent was about 9,000 TJ per year, and this value was comparable to about 25% of the annual heat energy consumption in Sapporo.[3] The quality of the heat energy in the effluent is, however, very low in a thermodynamic sense because of its lower temperature. Although it is difficult to use for power supply or heating, it is enough to melt snow.

Figure 4 shows the new strategy of Sapporo for combined sewer systems. One of the most remarkable plans is that the sewage system is used to transport and melt snow. The basin can be used for both retaining rainwater and melting snow. Melted snow goes through the sewer network. Treated water of a WWTP and sewage can be used as a heat source to melt snow and as a media to transport snow.

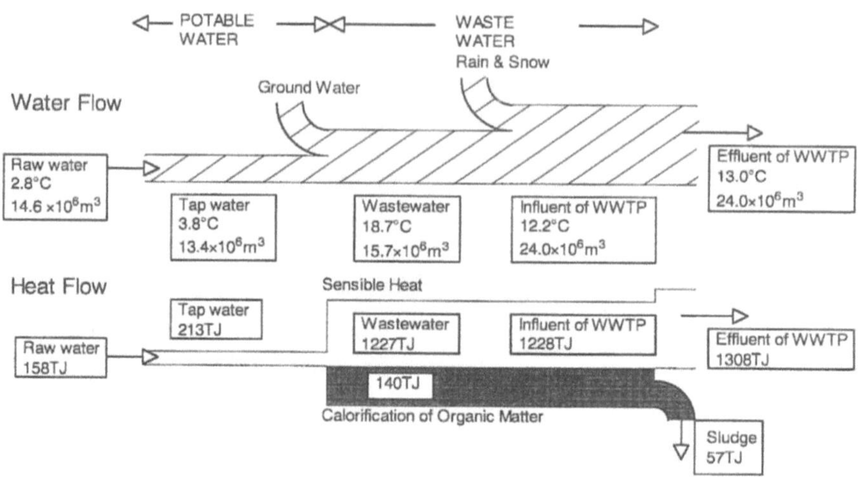

Fig. 3. Heat energy flow through water supply and sewage system of Sapporo in February, 1991

3
Application of the activated sludge model to the reaction basin at low temperatures

3.1
Plant operation and simulation analysis

Using the sewage system to transport and melt snow causes a drop in temperature of wastewater. The predicted temperature of sewage may be in the range of 4 to 8°C. Jansen et al.[4] showed that nitrification capacity of a wastewater treatment plant depends strongly on the temperature according to data from the Søholt treatment plant in Denmark. Chiemchaisri and Yamamoto[5] studied biological nitrogen removal under low temperatures in a membrane separation bioreactor. They reported that nitrogen removal performance decreased from more than 90% at 25°C to 20% at 5°C as a result of inhibition of nitrification at low temperatures. They also reported that increasing the aeration time in an operational cycle improved nitrification. Insufficient nitrification was also observed at the wastewater treatment plant in Paris, and the requirement of high sludge concentration levels was stressed by Gousailles et al.[6] The design manual of a wastewater treatment plant in Japan[7] suggests the minimum required sludge retention time (SRT) related to the wastewater temperature to keep nitrifying bacteria in a plant. To maintain a given SRT, the excess activated sludge produced daily must be wasted.[8] Control of SRT for nitrification causes a change in the amount of sludge production which is the input of sludge treatment processes.

In order to determine the influence of transporting and melting snow, plant performance should be evaluated in terms of the effluent water quality as well as the quality and quantity of sludge produced from the plant. Simulation analysis is one

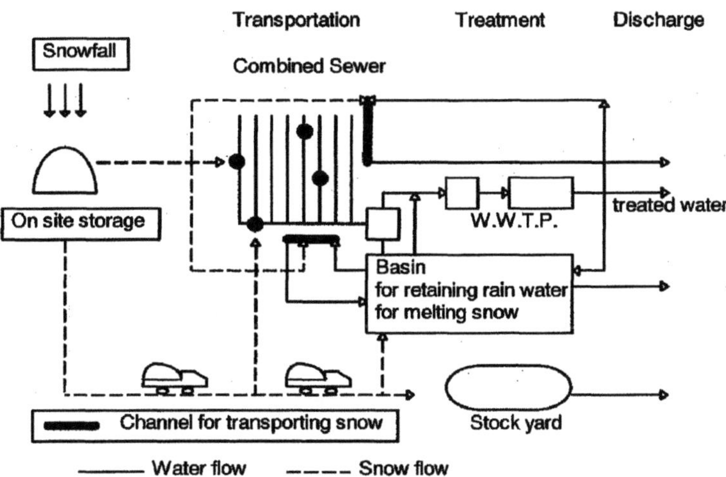

Fig. 4. Sewage system used to transport and melt snow in a snowy region

of the most powerful tools for this evaluation, and we can set adequate conditions of operating biological wastewater treatment plants at low temperatures. This simulation requires a model of all unit processes in the treatment plant: water treatment processes including primary and final clarifier and aeration basin, and sludge treatment processes.

3.2
Modeling of temperature dependence of the biological reaction rate

The Activated Sludge Models 1[9] (ASM1) and 2[10] (ASM2) have been frequently used to predict the performance of wastewater treatment processes, especially of biological reactions. Siegrist and Tschui[11] calibrated the parameters in the activated sludge model with data from municipal wastewater treatment plants in Switzerland. Lesouef et al.[12] demonstrated the on-site calibration techniques, and verified the model on the full-scale plant. It has been well recognized that the Activated Sludge Model has applicability as a prognostic tool.[13]

The ASM1 is the simplest mechanistically based model incorporating carbon oxidation, nitrification and denitrification. ASM1 has 13 components as shown in Table 1, and the fundamental processes selected for inclusion in it are the eight processes listed in Table 2. The first four components in Table 1 characterize organic matters of wastewater. Grady[14] gave us the explanation of these organic matters as follows:

"They are defined operationally, rather than physically. The operational defines are readily biodegradable, slowly biodegradable, and inert. Readily biodegradable organic matter is material that is degraded so easily that the use of electron acceptor responds immediately to a change in its concentration. Slowly biodegradable substrate is material that requires hydrolysis. Inert organic material is any with a reaction rate so low that losses are unlikely to be seen within the time constraints of normal wastewater treatment system."

Table 1. Components in Activated Sludge Model No. 1 (ASM1)

Symbol	Component
S_I	Soluble inert organic matter (g COD m^{-3})
S_s	Readily biodegradable substrate (g COD m^{-3})
X_I	Particulate inert organic matter (g COD m^{-3})
X_S	Slowly biodegradable substrate (g COD m^{-3})
$X_{B,H}$	Active heterotrophic biomass (g COD m^{-3})
$X_{B,A}$	Active autotrophic biomass (g COD m^{-3})
X_P	Particulate products arising from biomass decay (g COD m^{-3})
S_O	Dissolved oxygen (g O_2 m^{-3})
S_{NO}	Nitrate and nitrite nitrogen (g N m^{-3})
S_{NH}	Ammonium plus ammonia nitrogen (g N m^{-3})
S_{ND}	Soluble biodegradable organic nitrogen (g N m^{-3})
X_{ND}	Particulate biodegradable organic nitrogen (g N m^{-3})
S_{ALK}	Alkalinity (mole HCO_3^- m^{-3})

Table 2. Processes in Activated Sludge Model No. 1

1 Aerobic growth of heterotrophs
2 Anoxic growth of heterotrophs
3 Aerobic growth of autotrophs
4 Decay of heterotrophs
5 Decay of autotrophs
6 Ammonification of soluble organic nitrogen
7 Hydrolysis of entrapped organics
8 Hydrolysis of entrapped organic nitrogen

The ASM1 gives a set of stoichiometric coefficients, reaction rate expressions and typical values of parameters. The ASM2 is an extension of the ASM1, and presents a concept for dynamic simulation of combined biological processes for chemical oxygen demand, nitrogen and phosphorous removal. We used the ASM1 in this work because the interested components in the simulation were organic matters and nitrogen components.

In the ASM1, nitrification is assumed to be a single step process and soluble ammonia nitrogen serves as the energy source for growth of nitrifiers and nitrate nitrogen as end products. The rate of aerobic growth of autotrophic biomass, ρ_A, is given as follows:

$$\rho_A = \mu_A \left(\frac{S_{NH}}{K_{NH} + S_{NH}} \right) \left(\frac{S_O}{K_{O,A} + S_O} \right) X_{B,A} \tag{1}$$

where μ_A is the maximum specific growth rate of nitrifying bacteria, and K_{NH} and $K_{O,A}$ are half-saturation coefficients. The expression of the aerobic growth rate of heterotrophic biomass, ρ_H, in the ASM1 has also double Monod type as:

$$\rho_H = \mu_H \left(\frac{S_S}{K_S + S_S} \right) \left(\frac{S_O}{K_{O,H} + S_O} \right) X_{B,H} \tag{2}$$

where μ_H is the maximum specific growth rate of heterotrophic biomass, and K_S and $K_{O,H}$ are half-saturation coefficients.

For modelling and design purposes, the impact of temperature on the maximum growth rate is often described by an Arrhenius type function:[15–17]

$$\mu(T) = \mu(20)\theta^{(T-20)} \tag{3}$$

or

$$\mu(T) = \mu(20)\exp[K_T(T-20)] \tag{4}$$

where $\mu(T)$ is the maximum specific growth rate at temperature $T(°C)$, $\mu(20)$ is the maximum specific growth rate at 20°C, θ (theta) is the simplified Arrhenius temperature coefficient, and K_T is the Arrhenius temperature coefficient. Values of theta reported in the literature are summarized in Table 3 for nitrifying biomass and in Table 4 for heterotrophic biomass.

Table 3. Simplified temperature coefficient for modeling growth of nitrifying autotrophic bacteria

Theta	Conditions	Ref.
1.103	*Nitrosomonas*, T=5–30°C	17
1.103	Nitrification, calculated from parameter values for 10 and 20°C	9
1.028	Nitrification, MLVSS=430 mg l⁻¹, T=4–33°C	16
1.061	Nitrification, MLVSS=1200 mg l⁻¹, T=4–25°C	16
1.129	Nitrification, MLVSS=3200 mg l⁻¹, T=4–25°C	16
1.116	Nitrosomonas, T=6–14°C	18
1.103	Nitrification, T=10°C	11
1.044	Nitrification, T=5–10°C	this work

Theta simplified Arrhenius temperature coefficient, *MLVSS* mixed-liquor volatile suspended solids, *T* temperature

Table 4. Simplified temperature coefficient of heterotrophic bacteria

Theta	Conditions	Ref.
1.080	Anoxic	15
1.096	Anoxic	19
1.03–1.1	Anoxic	20
1.035	Aerobic	21
1.072	Anoxic and aerobic	9
1.00–1.08	Aerobic	1
1.072	Aerobic	this work

Theta simplified Arrhenius temperature coefficient

3.3
Calibration of the activated sludge model with pilot plant data

The pilot plant had a fully mixed aeration tank (1.2 m³) and a secondary clarifier, and it treated the primary effluent 2.6 l min⁻¹ of the Sosei WWTP in Sapporo. Experiments were performed at 5, 6.5, 8 and 10°C. Measured variables were biological oxygen demand (BOD), chemical oxygen demand (COD), nitrogen compounds of influent and effluent of the plant, volatile and non-volatile suspended solids of mixed liquor, and return sludge at steady state.

The model was calibrated with data from operation at 10°C. Only two parameters, the yield coefficient of the heterotrophic biomass Y_H and the maximum specific growth rate for the autotrophic biomass μ_A, were adjusted from the default values.[9] Y_H was 0.715 g cell COD formed g⁻¹ COD oxidized (default value 0.67) and μ_A was 0.29 d⁻¹ (default 0.3). Run 1 in Table 5 shows the computed and observed results. Then, temperature coefficients in the reaction equations were estimated using the data from operation at 5°C. Reducing the simplified temperature coefficient for the growth of autotrophic biomass from 1.103 (default) to 1.044 yielded the results in Run 2. Comparisons of the simulated results with data from operations at other temperatures (Run 3 and Run 4) showed that the ASM1 could be adopted for describing biological reactions in the aeration basin at low temperatures (Table 5).

Table 5. Comparison of computed results with pilot plant data (reprinted with permission from ref. 2)

	Run 1		Run 2		Run 3		Run 4	
Temp. (°C)	10.3		4.9		7.7		6.5	
R_S (%)	30		30		33		31	
R_e (%)	1.28		1.08		2.02		0.77	
	obs	comp	obs	comp	obs	comp	obs	comp
BOD_T (g m⁻³)	7.1	9.5	8.9	12.8	7.4	10.1	6.0	8.7
Total N (g m⁻³)	21	22.8	18.3	17.2	15.9	24.0	15.0	18.0
NH_4-N (g m⁻³)	17.5	19.1	11.5	12.6	12.5	22.6	7.8	7.8
NO_3-N (g m⁻³)	2.5	2.8	3.2	3.6	0.9	0.2	4.8	9.4
MLDO (g m⁻³)	1.3	1.55	3.4	3.9	2.5	4.4	1.9	0.7
MLSS (g m⁻³)	1430	1330	1270	1220	985	980	1680	1720
MLVSS (%)	79.5	78.1	79.2	83.9	82.2	77.8	73.5	75.6
RSSS (g m⁻³)	5530	5520	4620	5050	3910	3700	6880	7050
SRT (d)	5.8	5.4	6.3	6.2	3.2	3.8	9.5	8.8

Rs (return sludge flow)/(sewage flow rate into plant), R_e (waste sludge flow)/(sewage flow rate into plant), *BOD* biological oxygen demand, measured after addition of allylthiourea, *MLDO* dissolved oxygen in the aeration basin, *MLSS* mixed-liquor suspended solids, *MLVSS* mixed-liquor volatile suspended solids, *RSSS* suspended solids in return sludge, *SRT* sludge retention time, *obs* observed, *comp* computed

4
Simulation model of full-scale municipal wastewater treatment plant

4.1
Description of the investigated treatment plant

The Sosei wastewater treatment plant in Sapporo treats about 93,500 m³ d⁻¹ of wastewater during dry weather in winter. Figure 5 shows sludge and water treatment processes in this plant. The Sosei plant has two lanes in its water treatment system, and one sludge treatment system. Each water treatment lane consists of primary clarifiers, aeration basins and final clarifiers. Plant operation data show that nitrification occurs partly in lane 1 and fully in lane 2 in winter. The sludge treatment system has gravity thickeners and pressure filtration units. Eighty percent of the cake from this system is incinerated. Supernatant of thickeners and filters is returned to the water treatment system. All water and sludge flows in Figure 5 were modeled in the simulation.

4.2
Simulation model for the plant

4.2.1
Primary clarifier

Since a primary clarifier has two output flows, effluent and concentrated sludge, we have to predict two values, that is, solid concentrations of effluent and sludge. A Voshel and Sak-type formula[22] was used to simulate solid removal performance. This empirical model shows that solid removal efficiency r is directly proportional to a power function of influent solid concentration X_{in} and inversely proportional to a power function of overflow rate L_p. Regression analysis using data from the plant yielded the following relationships:

$$r=0.133X_{in}^{0.312}L_p^{-0.0265} \quad \text{(Lane 1)}, \qquad r=0.200X_{in}^{0.290}L_p^{-0.0807} \quad \text{(Lane 2)} \tag{5}$$

Figure 6 is a comparison of computed results with the plant operation data, showing that Eq. (1) can express the solid removal efficiency of the primary clari-

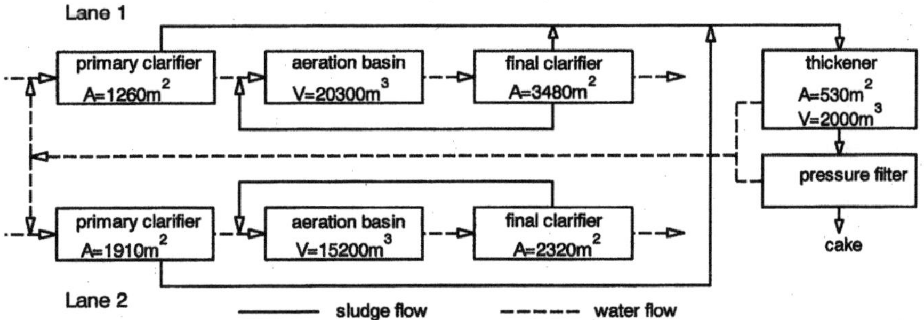

Fig. 5. Sludge and water flows in the investigated plant (reprinted with permission from ref. 2)

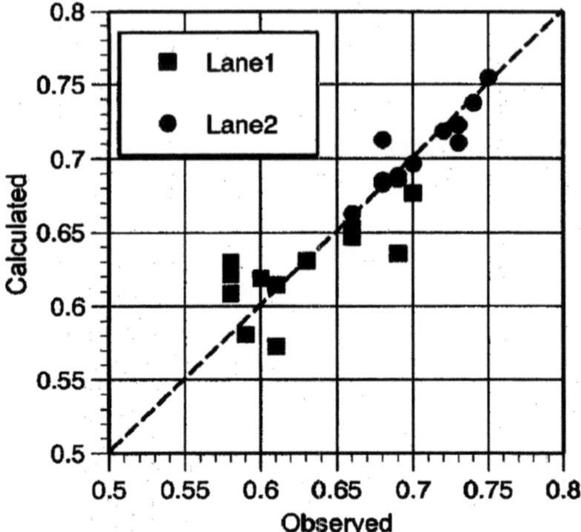

Fig. 6. Comparison of calculated results of solid removal efficiency with plant data (reprinted with permission from ref. 2)

fier in the investigated plant. Solid concentration of sludge from the primary clarifier was estimated by the following flow volume and solid mass balance equation at steady state:

$$Q_{in}=Q_e+Q_u \qquad Q_{in}X_{in}=Q_eX_e+Q_uX_u \tag{6}$$

where Q_{in}, Q_e and Q_u are flow rates of influent, effluent and sludge, respectively, and X_{in}, X_e and X_u are concentrations of suspended solids. In Eq. (6), flow rate of sludge Q_u was fixed at 3% of influent flow rate in the simulation, which was the average value of the plant operation data.

4.2.2
Aeration basin

Simulation of an aeration basin requires two models: a model of biological reactions and a model of mixing conditions of a basin. In this study we used the ASM1 for the biological reactions in the aeration basin. Parameter values of this model were estimated by the pilot plant data. Mixing condition of the basin was approximated with four complete mixing tanks in series.

4.2.3
Final clarifier

A final clarifier has two functions: clarification and thickening. The models of clarification proposed so far relate effluent concentration of suspended solids to both flow rate and concentration of suspended solids of the influent. A typical empirical model has the following form:[23,24]

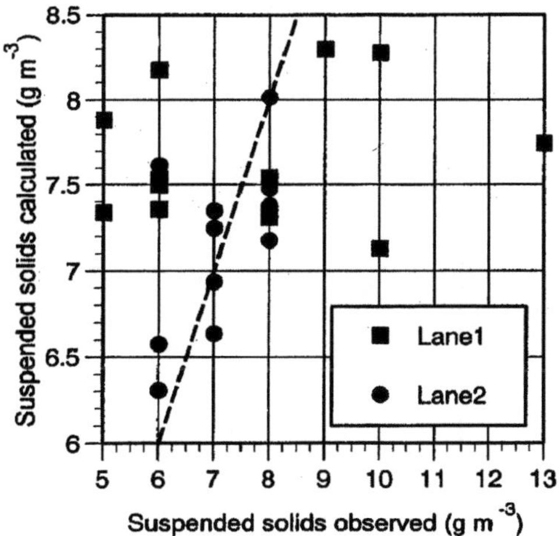

Fig. 7. Results of regression analysis on effluent suspended solids concentration from the final clarifier (reprinted with permission from ref. 2)

$$X_e = a_1 + a_2 X_{in} + a_3 L_P \tag{7}$$

The application of Eq. (7) to data from the plant did not yield a close correlation, as shown in Figure 7. Because the plant data varied randomly, we set a constant value, 7.5 g m^{-3}, for the concentration of suspended solids of the effluent in the simulation. For thickening, the balance equations of flow volume and solid mass at steady state were employed in the same manner as the model for a primary clarifier.

4.2.4
Thickener

Outputs of a thickener are supernatant and concentrated sludge. A solid recovery percentage, $R_{TH,mass}$ ([concentrated solid mass]/[input solid mass]), was adopted as a measure of thickening performance. The data for solid recovery percentage of the plant are shown in Figure 8. The mean value in winter, about 80%, was used in the simulation. The operational condition was specified by the withdrawal percentage $R_{TH,Q}$ ([concentrated sludge flow rate, Q_u]/[input flow rate, Q_{in}]). Plant data showed that $R_{TH,Q}$ had a constant value of about 10%. The balance equations of flow:

$$Q_u = R_{TH,Q} Q_{in} \qquad Q_e = (1 - R_{TH,Q}) Q_{in} \tag{8}$$

yielded flow rate of the supernatant, Q_e, and flow rate of the concentrated sludge, Q_u. The solid mass concentration X_e in supernatant and X_u in concentrated sludge are estimated by the following mass balance equations:

$$X_u Q_u = R_{TH,mass} X_{in} Q_{in} \qquad X_e Q_e = (1 - R_{TH,mass}) X_{in} Q_{in} \tag{9}$$

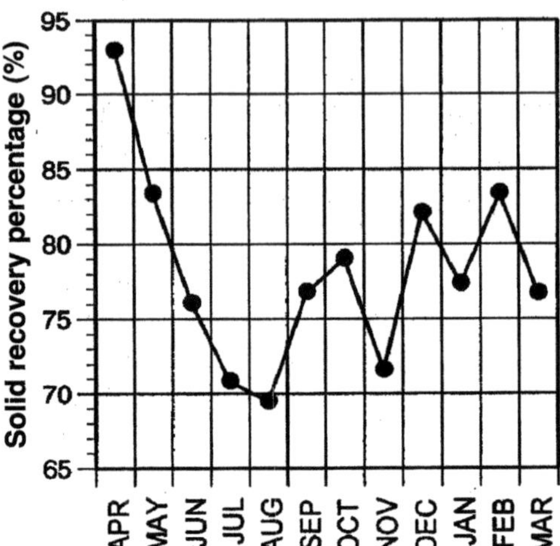

Fig. 8. Monthly variation of solid recovery percentage in the investigated plant's thickener (reprinted with permission from ref. 2)

Concentration of dissolved matter in supernatant was specified by the plant data.

Besides computation using these balance equations, the minimum required cross-sectional area A_{Req} was calculated by the limiting flux theory.[25] According to this area, one can examine the possibility whether the desired concentrated sludge can be obtained by the thickening process under the current operating conditions adopted in the simulation. Since this calculation required information on the solid mass flux, the hindered settling velocity was measured by the settling test with the sludge from the pilot plant operating at 5°C. The analysis of the settling curve in Figure 9 based on Kynch's theory[26] yielded the relationship between the hindered settling velocity and the solids concentration. This relationship was correlated by a logarithmic model, and limiting flux G_L was computed by the following equations:[27]

$$V = V_0 \left(\frac{X}{X_0} \right)^{-n}, \quad \frac{G_L}{V_0 X_0} = (n-1)\left(\frac{n}{n-1} \right)\left(\frac{U_u}{V_0} \right)^{\frac{n-1}{n}} \tag{10}$$

where V is hindered settling velocity of sludge, X is solid concentration of sludge, and U_u is bulk downward flow velocity, $U_u = Q_u/A_{Req}$.

The settling test showed that the parameters in Eq. (10) were $n = 0.4$, $V_0 = 5$ cm min^{-1}, and $C_0 = 1$ g l^{-1}, respectively. Solid mass and volume in the withdrawal flow can be written as

$$A_{Req} G_L = X_{in} Q_{in} R_{TH,mass} \qquad A_{Req} U_u = Q_{in} R_{TH,Q} \tag{11}$$

Substituting Eq. (10) into Eq. (11) and rearranging the resulting expressions yields

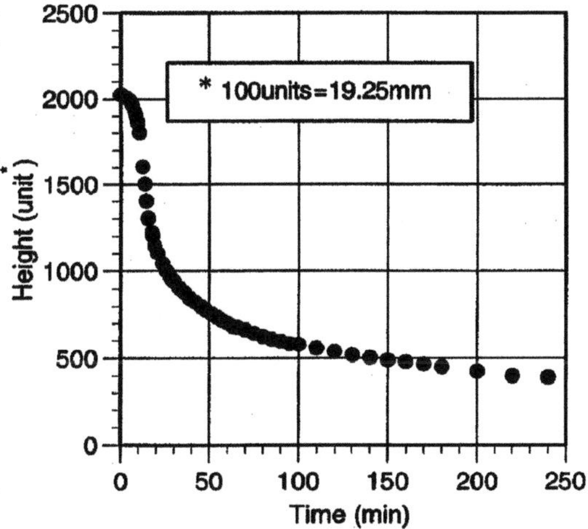

Fig. 9. Settling curve of sludge in pilot plant at 5°C (reprinted with permission from ref. 2)

$$\frac{A_{Req}}{R_{TH,Q}\,Q_{in}/V_0} = (n-1)\left(\frac{n}{n-1}\right)^{\frac{1}{n}}\left(\frac{X_{in}}{X_0}\right)^{n}\left(\frac{R_{TH,mass}}{R_{TH,Q}}\right)^{n} \tag{12}$$

Stokes' law for the settling velocity was assumed, then the temperature dependence of settling velocity was expressed by viscosity, and hence, V_0 at T°C was expressed by

$$V_0(T) = \frac{\eta(10)}{\eta(T)}\,V_0(10) \tag{13}$$

4.2.5
Pressure filter

The purpose of the pressure filter model is the estimation of heating value of sludge cake. This heating value is the measure of burning characteristics of the sludge cake. Coagulants used in the plant are ferric chloride and lime. The average dose rates in winter were about 7 and 31%, respectively. There was a small variation in water content of cake, as shown in Figure 10, and a constant value, 63%, was used in the simulation. The solid recovery percentages were nearly constant, at about 90%. Heating value of cake was calculated by the experimental formula:[28]

$$H_C = (1-w/100)H_{H,C} - 540/100 \tag{14}$$

$$H_{H,C} = 0.93\left(\frac{(58.3X_{CAKE}-193)-353(k_2-0.69k_1)}{1+k_2-0.03k_1}\right) - 33 \tag{15}$$

where H_C is heating value of sludge cake (kcal kg⁻¹ cake), w is water content of cake

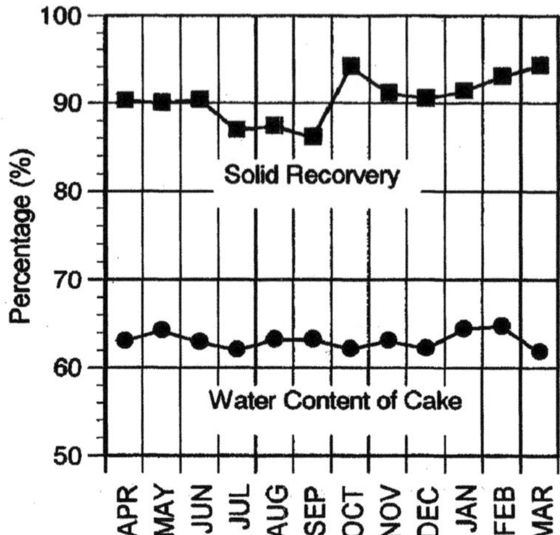

Fig. 10. Solid recovery percentage and water content of cake (reprinted with permission from ref. 2)

Table 6. Influent concentrations used in simulation

Parameter	$(g\ m^{-3})$
Readily biodegradable chemical oxygen demand (COD)	70
Soluble inert COD	10
Slowly biodegradable COD	120
Particulate inert COD	30
Oxygen	5
NH_4-N	16.0
NO_X-N	0.3
Soluble organic nitrogen	4
Particulate organic nitrogen	4
Non-volatile suspended solids	60

(%), X_{CAKE} is organic matter content in cake (%), and k_1 and k_2 are respective addition rates of ferric chloride and lime. Murakami et al.[28] reported that the lower limit of heating value for combustion without fuel depended on the type of incinerator, but 430 kcal kg^{-1}cake was taken for the lower limit in the simulation.

4.3
Comparison of simulated results with plant data

We simulated the average performance in winter by steady state analysis and compared the results with the plant data. Influent concentrations used in the simulation are summarized in Table 6. The comparison in Table 7 shows that the simulation model yields reasonable results.

Table 7. Comparison of simulated results with Sosei Plant data

Parameter	Lane 1		Lane 2	
Flow rate (m³ d⁻¹)	42,700		50,800	
R_S	0.26		0.31	
R_e	0.013		0.0077	
	observed	computed	observed	computed
BOD_T (g m⁻³)	3.4	6.2	3.5	6.2
Total N (g m⁻³)	13.9	14.8	12.1	16.3
NH_4-N (g m⁻³)	6.3	5.4	0	0.6
NO_3-N (g m⁻³)	7.6	9.0	11.7	15.4
MLDO (g m⁻³)	2.8	2.9	5.2	5.2
MLSS (g m⁻³)	1,980	1,660	2,650	2,450
MLVSS (%)	72.0	70.5	71.0	71.3
RSSS (g m⁻³)	8,050	7,700	9,560	10,130
sludge	observed		computed	
Thickener input	24.2 t d⁻¹		27.5 t d⁻¹	
Cake production	24.4 t d⁻¹		25.6 t d⁻¹	
Heating value	585 kcal kg⁻¹ cake		520 kcal kg⁻¹ cake	

For abbreviations see Table 5

5
Plant operation maps at low temperatures

It was estimated that melting and transporting snow would cause influent temperatures to fall to 4–8°C. Operation maps of the plant at 4 and 8°C were drawn by the simulation of steady state. In the investigated wastewater treatment plant, we cannot change reactor volume without construction, and cannot control wastewater flow rate either. We selected the sludge recycle rate R_s and the excess sludge withdrawal rate R_e as the operational variables in this simulation. The goal of operation was specified by the following four variables:

1. nitrate nitrogen concentration in the effluent,
2. minimum required cross-sectional area of thickener,
3. sludge cake production mass rate,
4. heating value of cake.

We prepared four scenarios and studied feasible operating conditions for each case:

Scenario 1: Keep the present level of nitrate nitrogen concentration in effluent. Neither thickener nor incineration plant has an excess capacity.

Scenario 2: Keep the present level of nitrate nitrogen concentration in effluent. Both thickener and incineration plant have 10% excess capacity.

Scenario 3: Keep the present level of nitrate nitrogen concentration in effluent. There is no limitation either in thickener or in incineration.

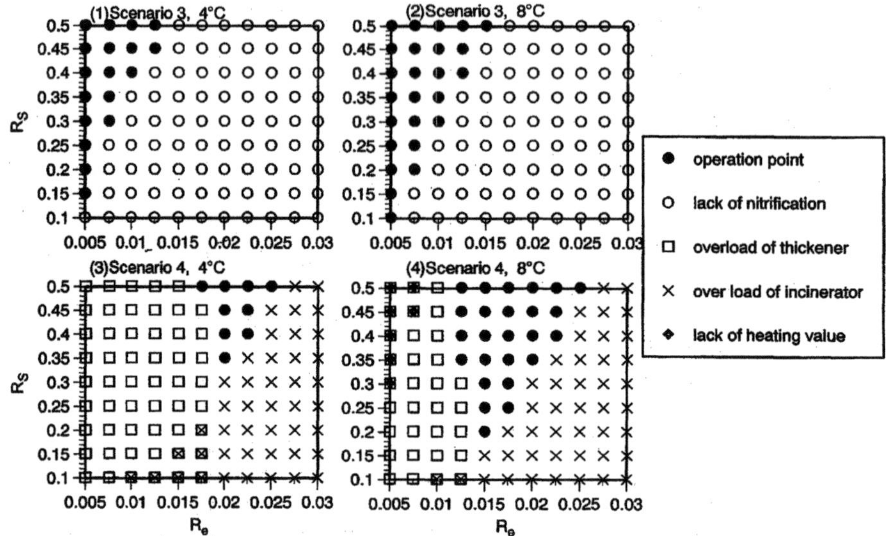

Fig. 11. Plant operation maps (reprinted with permission from ref. 2)
R_e excess sludge withdrawal percentage defined by [waste sludge flow] / [sewage flow rate into plant], R_s sludge recycle percentage defined by [return sludge flow] / [sewage flow rate into plant]

Scenario 4: No limitation on nitrification. Both thickener and incineration plant have 10% excess capacity.

In scenario 1, no available operating point was found at 4 or at 8°C. In scenario 2, only a few operating points were obtained only at 8°C. Figure 11 shows the operation maps of scenarios 3 and 4. In scenario 3, feasible operating points are in the range of small R_e values, i.e., long SRT operation. The operating points in scenario 4 at 8°C are in a wide range including the present operating points. At 4°C in scenario 4, a large R_s is required, and this may cause some troubles in solid–liquid separation at final clarifiers. The difference between the maps of scenarios 3 and 4 shows that it is impossible to keep nitrifying bacteria at the present level in the plant while preventing the excess load for sludge treatment processes. This suggests that the enlargement of the sludge treatment system or the addition of nitrifying-bacteria-holding apparatus to aeration basins is required.

6
Conclusions

The city of Sapporo set a new policy: combined sewer systems convey and treat snow as well as wastewater and stormwater. In accordance with this policy, several multipurpose facilities are being constructed and planned and research works have been performed. Among them, the simulation study for operation of a wastewater treatment plant at low temperatures is presented in this chapter.

Using the sewage system for transporting and melting snow has various advantages:

1. effective use of facilities for controlling combined sewer overflows in a snow season;
2. effective use of heat energy wasted to receiving waters;
3. reduction of energy consumption for melting snow;
4. reduction of pollution load caused by snowmelt.

This multi-purpose usage of facilities, on the other hand, has disadvantages such as the drop in sewage temperature. To assess the influence of the low temperature on the performance of a wastewater treatment plant, the simulation analysis was performed. The simulation model for the full-scale plant at the steady state was developed. Processes included in the model are those of primary clarifier, aeration basin, final clarifier, thickener and pressure filter. The Activated Sludge Model 1 was used for simulation of reactions in the aeration basin. The model was calibrated by data from 5 and 10°C pilot plant experiments. The calibrated model was able to predict the results of other temperature experiments at the pilot plant. The full- scale plant model yielded fairly accurate results on the sludge mass load of thickener, production rate and heat value of sludge cake as well as on nitrification of the plant in winter. Plant operation maps showed that enlargement of the sludge treatment system or the addition of nitrifying-bacteria-holding apparatus to the aera-tion basin is required to maintain the present treatment level when the sewage system is used for transporting and melting snow and the wastewater treatment system is operated at low temperatures.

7
References

1. Tchobanoglous G, Burton FL. Wastewater Engineering–Treatment, Disposal and Reuse. Metcalf & Eddy Inc., 3rd ed. New York: McGraw-Hill, Inc., 1991:372-373.
2. Funamizu N, Takakuwa T. Simulation of the operating conditions of the municipal wastewater treatment plant at low temperatures using a model that includes the IAWPRC Activated Sludge Model. Wat Sci Technol 1994; 30:105-113.
3. Ochifuji K, Nagano K, Nakamura M. Utilization of the waste heat energy in the urban energy system. In: Proceedings of the 8th Conference on the Energy and Economy, 1992:13-18.
4. Jansen JC, Kristensen GH, Laursen KD. Activated sludge nitrification in temperate climate. Wat Sci Technol 1992; 25:177-184.
5. Chiemchaisri C, Yamamoto K. Biological nitrogen removal under low temperature in a membrane separation bioreactor. Wat Sci Technol 1993; 28:325-333.
6. Gousailles M, Rovel JM, Nicol R. Purification of wastewater from the paris conurbation: biological removal of nitrogen at the valenton purification plant. Wat Sci Technol 1991;23: 773-779.
7. Japan Sewage Works Association. The Design Manual of a Wastewater Treatment Plant. Tokyo: Japan Sewage Works Association, 1994.
8. Tchobanoglous G, Burton FL. Wastewater Engineering - Treatment, Disposal and Reuse. Metcalf & Eddy Inc., 3rd ed. New York: McGraw-Hill, Inc., 1991:551-552.
9. Henze M, Grady CP, Gujer W, Marais GR, Matsuo T. Activated Sludge Model No.1. IAWPRC Scientific and Technical Report No.1. London: International Association on Water Pollution Research and Control, 1987.

10. Henze M, Gujer M, Mino T, Matsuo T, Wentzel MC, Marais GR. Activated Sludge Model No.2. IAWQ Scientific and Technical Report No.3. London: International Association on Water Quality, 1995.

11. Siegrist H, Tschui M. Interpretation of experimental data with regard to the activated sludge model No.1 and calibration of the model for municipal waste water treatment plants. Wat Sci Technol 1992; 25:167-183.

12. Lesouef M, Payraudeau M, Rogalla F, Keiber B. Optimizing nitrogen removal reactor configurations by on-site calibration of the IAWPRC Activated Sludge Model. Wat Sci Technol 1992; 25:105-124.

13. Pedersen J, Sinkjaer O. Test of the activated sludge model's capabilities as a prognostic tool on a pilot scale wastewater treatment plant. Wat Sci Technol 1992; 25:185-194.

14. Grady CPL. Dynamic modeling of suspended growth biological wastewater treatment processes. In: Patry G, Chapman D, eds. Dynamic Modeling and Expert Systems in Wastewater Engineering. Chelsea, Michigan: Lewis Publishers Inc., 1989:4-5.

15. Lewandoswki Z. Temperature dependency of biological denitrification with organic materials addition. Water Research 1982; 16:19-22.

16. Shammas NK. Interaction of temperature, pH, and biomass on the nitrification process.J WPCF 1986; 58:52-59.

17. US Environmental Protection Agency. Manual of Nitrogen Control, EPA/625/R-93/010. Washington, 1993.

18. Gujer W. Design of a nitrifying activated sludge process with aid of dynamic simulation. Progress Wat Technol 1977; 9:323-336.

19. Nakajima J, Kaneko K. Practical performance of nitrogen removal in small-scale sewage treatment plants operated in intermittent aeration mode. Wat Sci Technol 1991; 23:709-718.

20. Ekenfelder WW, Argaman Y. Principles of biological and physical/chemical nitrogen removal. In: Sedlak R, ed. Phosphorus and Nitrogen Removal from Municipal Wastewater - Principles and Practice, 2nd ed. Chelsea: Lewis Publishers Inc, 1991:20.

21. Barton DA, McKeown JJ. Evaluation of an aerator control strategy utilizing time varying mathematical model simulations. Wat Sci Technol 1986; 18:189-201.

22. Voshel D, Sak JG. Effect of primary effluent suspended solids and BOD on activated sludge production. J WPCF 1968; 40, part 2, R203-R212.

23. Chapman D. The influence of dynamic loads and process variables on the removal of suspended solids from an activated sludge plant. PhD thesis, University of Alberta, Edmonton, Canada, 1984.

24. Vitasovic Z. Continuous settler operation: A dynamic model. In: Patry G, Chapman D, eds. Dynamic Modeling and Expert Systems in Wastewater Engineering. Chelsea: Lewis Publishers, 1989:59-81.

25. Coe HS, Clevenger GH. Determining thickener areas. Trans AIME 1916; 55:356.

26. Kynch GJ. A theory of sedimentation. Tran Farady Society 1952; 48:166-176.

27. Baskin DE, Suidan MT. Unified analysis of thickening. J of Environ Engin ASCE 1985; 111:10-26.

28. Murakami T, Kudo J, Suzuki K, Sasabe K, Kakuta K. Report on energy saving in sludge incinerations. Japan Sewage Works Agency, Research and Technology Development Division 1986; 62-63.

The potentials of sub-mesophilic and/or psychrophilic anaerobic treatment of low strength wastewaters

G. Lettinga*, S. Rebac, J. van Lier and G. Zeman

Wageningen Agricultural University, Sub-department Environmental Technology, Bomenweg 2, NL-6703 HD Wageningen, The Netherlands

1
Introduction

1.1
Cold low and medium strength wastewaters

Particularly under moderate climate conditions many low and medium strength wastewaters are discharged at lower ambient temperatures, including domestic wastewater and a large variety of industrial wastewaters, e.g. those of bottling, malting, brewery and soft drinks manufacturing. The chemical oxygen demand (COD) concentrations of these wastewaters frequently are lower than 1,500 mg COD dm^{-3} and frequently they contain dissolved oxygen concentration of up to 5 mg O_2 dm^{-3}. The established sanitary wastewater engineering world has so far considered anaerobic wastewater treatment (AnWT) of cold and very low strength wastewaters as unfeasible. Although this opinion may be based mainly on prejudice and a serious lack of sound insight into the anaerobic digestion process and technology, in fact it restrained scientists in the past to start research in this field.

On the other hand, it certainly is true that AnWT of low strength cold wastewaters is indeed not so obvious, i.e. a number of bottlenecks have to be eliminated. So, for instance, the low $COD_{influent}$ will result in extremely low substrate levels inside the reactor, and in a low biogas production rate as well. In conventional anaerobic sludge bed reactors this implies a too low mixing intensity in the reactor and, consequently, in a poor substrate–biomass contact. Another serious problem when treating very low strength wastewaters is that the permissible amount of sludge washout per m^3 wastewater is extremely small, which sets exceptionally high requirements on the sludge retention abilities of the reactor. Therefore the required reactor volume in case of low strength wastewaters generally will be determined by the permissible hydraulic loading rate (HLR) rather than by the organic loading rate (OLR).[1]

---

* Corresponding author

1.2
Anaerobic technology (AnTec)

One of the major successes in the development of AnWT technology has been the introduction of high-rate reactors in which biomass retention and liquid retention are greatly uncoupled.[2-4] This feature comprises a crucial issue for the treatment of low(er) strength wastewaters. For those reactor systems where the sludge retention is based on the settling characteristics of sludge aggregates, as is the case for the well-known upflow anaerobic sludge bed (UASB) reactors, the hydraulic load therefore will become the restrictive factor with respect to the required reactor in case of the treatment of very low strength wastewaters.

As a result of the high sludge concentration of UASB systems, conversion rates exceeding 40–60 kg COD m^{-3} d^{-1} can be easily attained at 30–40°C for medium strength soluble wastewaters.[1] The feature of the high biomass retention in principle also enables the application of AnWT to relatively cold (mainly) soluble wastewaters.[5-8] So far practically all full-scale applications of AnWT are restricted to wastewaters with temperatures exceeding 18°C. The reported maximum organic loading rate achieved at temperatures around 10°C was about 4 g COD dm^{-3} d^{-1}; the achieved treatment efficiency was 80%.[8] However, these very promising results were at that time not sufficiently encouraging to implement AnWT at full scale for the treatment of cold wastewaters (< 18°C). In fact, in the late 1970s high-rate AnWT systems were still not accepted as important, not even for medium strength wastewaters under optimal mesophilic conditions. Because temperature strongly affects the rates of the anaerobic conversion processes, some essential improvements have to be made in the conventional design of high-rate reactors in order to enable their application under 'sub-optimal' temperatures and for very low strength wastewaters. When successful, such a modified (improved) reactor system would represent a major technological break-through, because with this then indeed an efficient bioengineering of bacterial catalysis under sub-optimal temperatures would be possible. A successful application of psychrophilic anaerobic biocatalysis would be also of great economic importance, since generally (depending on the temperature of the wastewaters) a significant amount of energy is required to bring the wastewater temperature into the more optimal mesophilic range (30–40°C).[9]

So far, it is not clear whether high-rate psychrophilic anaerobic wastewater treatment needs the development of psychrophilic or psychrotolerant sub-populations, nor to what extent mesophilic sludges can become psychrotolerant.

1.3
Appropriate reactor technology

Previous experiments in optimizing the sludge–wastewater contact in UASB reactors led to the development of an advanced reactor design (Fig. 1), namely the expanded granular sludge bed (EGSB).[5,10-12] The EGSB system uses exclusively granular sludge, while in anaerobic fluidized bed (FB) or attached film expanded bed (AFEB) reactors inert carrier materials are used for attachment of active bio-

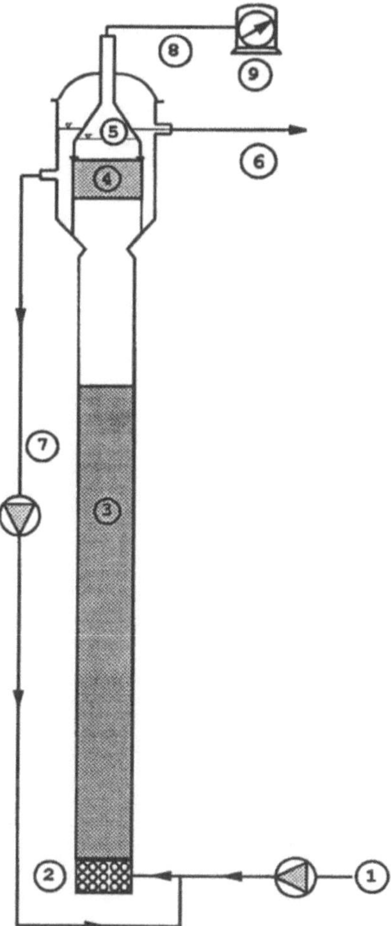

Fig. 1. Schematic diagram of an expanded granular sludge bed (EGSB) reactor system
(1) feed; (2) feed distribution; (3) expanded sludge bed; (4) sieve drum; (5) gas liquid separation;
(6) effluent; (7) effluent recirculation; (8) biogas; (9) wet test gasmeter

mass. The superficial liquid velocities (V_f) which can be applied in the EGSB system are between 4 and 10 m h^{-1}. These high V_f values can be achieved by applying effluent recycle and/or by using tall reactors. The feasibility of high-rate AnWT systems for cold wastewaters depends primarily on:

1. the quality of the seed material used and its development under submesophilic conditions,
2. the types of the organic pollutants in the wastewater,[13]
3. the reactor configuration, especially its capacity to retain viable sludge. Single and multi-compartment (moduled) reactors can be applied. To accomplish the highest possible overall treatment efficiency, especially in case of multi-

component wastewaters, moduled reactors, namely the staged sludge anaerobic reactor (SSAR), offer significantly better potentials than single compartment reactors. In these SSAR reactors sludge developing in the separate compartments is prevented from being mixed up. They look particularly attractive for wastewaters containing a variety of soluble and insoluble and/or non-acidified compounds.[14–17] The system was found to offer great potential for thermophilic treatment.[18–20]

2
Psychrophilic anaerobic treatment of wastewaters in an EGSB reactor system

2.1
Single module EGSB system

Recent results obtained with single module EGSB reactors (Fig. 2) clearly reveal the extraordinary potentials of the expanded bed concept as a 'high-rate' treatment system for low strength soluble wastewaters under psychrophilic conditions (10–12°C). COD removal efficiencies over 90% were achieved with a volatile fatty acids (VFA) mixture (190 mg COD dm^{-3} for C_2, 270 mg COD dm^{-3} for C_3 and 330 mg COD dm^{-3} for C_4) as feed at OLRs of up to 12 kg COD m^{-3} d^{-1} and at hydraulic retention times (HRTs) as low as 1.6 h. The seed sludge used in EGSB

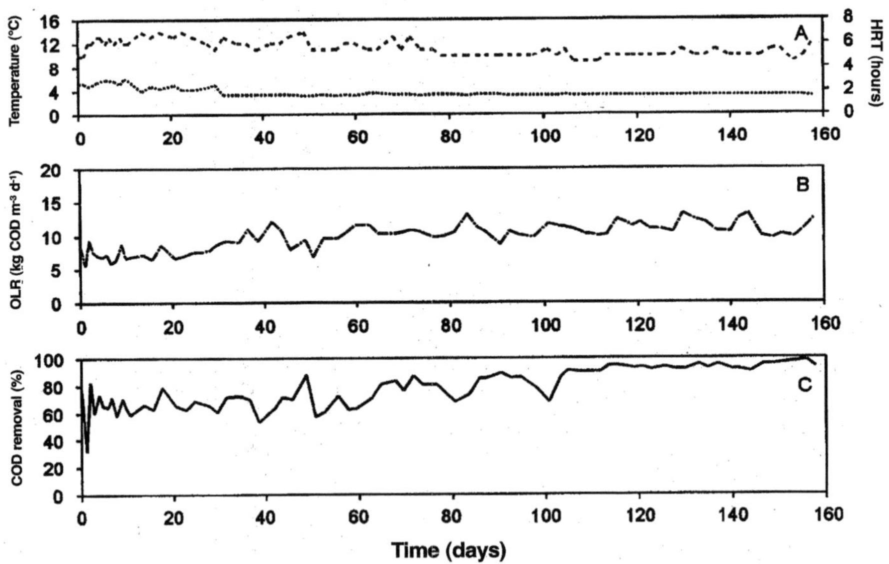

Fig. 2. Operational conditions and efficiency of the expanded granular sludge bed reactor system fed with low strength wastewater
(**A**) Temperature (---) and hydraulic retention time (HRT, ·····); (**B**) Organic loading rate (OLR, —·—·—); (**C**) COD removal efficiency, (——) (adapted from ref. 21)

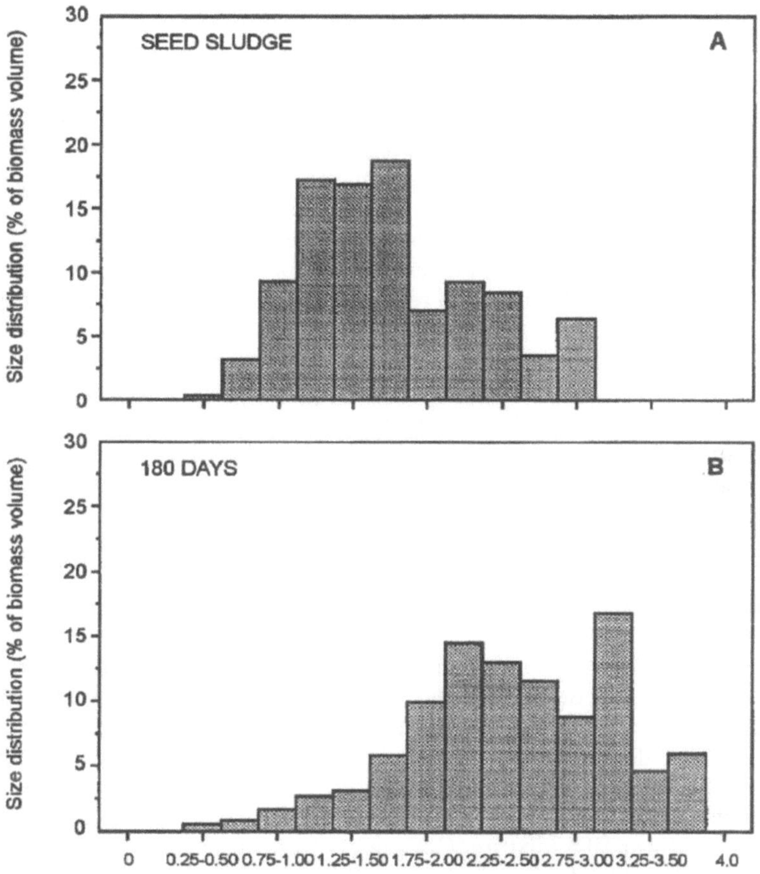

Fig. 3. Size distribution of granular sludge at the start of experiments (**A**) and after 180d (**B**) (adapted from ref. 21)

reactor in these experiments, i.e. 30 kg VSS m^{-3} reactor, consisted of a mesophilic granular sludge from a full-scale UASB reactor treating brewery wastewater. The reactor was operated at up-flow velocities up to 10 m h^{-1}, which provided a sufficient expansion of the sludge bed in the reactor. The results in Figure 2 show that immediately from the start-up, despite the low temperatures of 10–15°C applied, the COD-removal efficiencies were in the range of 40–60% at an imposed OLR of 8 kg COD m^{-3} d^{-1}. Furthermore, within 3 months the system could accommodate OLRs as high as 12 kg COD m^{-3} d^{-1} full at treatment efficiencies exceeding 95%. Moreover, it turned out (Fig. 3) that also the growth of granular sludge proceeded satisfactorily under low temperature conditions, although, on the other hand, the results give rise to some concern that too large granules might develop at due term upon continuing the operation under the prevailing conditions. The extraordinary high treatment efficiencies achieved are the more surprising in view of the low influent VFA concentrations of 500–800 mg dm^{-3} applied. This implies that the VFA concentrations prevailing in the reactor medium are in the range of 30–40 mg dm^{-3}, which indicates an extremely high substrate affinity of the sludge at 10°C. The estimated values for the half saturation constant in the Michaelis-Menten equation (K_m) for acetate, propionate and butyrate amounted to 0.04, 0.01 and 0.14 g COD dm^{-3}, respectively. The results demonstrate the importance of adequate hydraulic mixing in anaerobic systems for lowering the apparent K_m.[12,20,21]

It will be obvious that the above findings with EGSB systems represent a very important step forward in the applicability of anaerobic treatment in practice. Nevertheless, further significant improvements are possible. Results with additional experiments in single module EGSB reactors[21,22] revealed that propionate degradation is the limiting factor in treating pre-acidified wastewaters. As under mesophilic and thermophilic conditions, apparently also under psychrophilic conditions it is rather difficult to develop a balanced methanogenic sludge, i.e. a sludge exerting a high C$_3$-degrading capacity. This likely can be attributed to the presence of too high H$_2$ or acetate concentration in a single module reactor; on the basis of thermodynamic consideration it can be shown that growth of C$_3$-degrading organisms then does not proceed satisfactorily.[23]

2.2
Two module EGSB system

As for mesophilic and thermophilic AnWT systems,[18,19] also the psychrophilic AnWT process can be distinctly improved by applying sequentially operated modular reactors. This is clearly demonstrated by results obtained in recent experiments with a two module EGSB reactor set-up. The results shown in Figure 4 clearly reveal the feasibility of such a concept, i.e. concerning experiments consisting of two sequentially operated EGSB reactors, at extremely low temperatures, i.e. even down to 2°C. In treating a VFA mixture (C$_2$:C$_3$:nC$_4$ = 1:1.5:1.8 based on COD ratio), COD removal efficiencies exceeding 90% were achieved at 8 and 4°C at organic loading rates of 12 and 5 kg COD m^{-3} d^{-1} and at HRTs of 2 and 4 h, respectively. Such a two module EGSB reactor is capable of accommodating 3–5 times higher OLRs at 90% COD$_{vfa}$ removal efficiency than reported ever before for psychrophilic AnWT (Table 1).

The enhancement of the biodegradation process in a properly designed and operated staged reactor system can be attributed to the development of a balanced micro-ecosystem in the sludge in the various separate reactor compartments. In order to achieve this, two conditions must be met:

1. mixing up of the sludge present in the various reactor modules should be prevented,
2. the biogas evolving in the separate reactor compartment should be released from these compartments (there should be no mixup of the biogas).

As a result of these measures, the extent of product inhibition, e.g. in the conversion of propionate, can be reduced to a minimum. Consequently, in a two module (staged) reactor, a compound like propionate will be degraded mainly in the sec-

Table 1. Anaerobic wastewater treatment under psychrophilic conditions

Reactor type	Influent	Concentration (g COD dm^{-3})	OLR (kg COD m^{-3} d^{-1})	Temperature (°C)	HRT (h)	Efficiency (%)	Reference
AAFEB	Glucose	0.2–0.6	4–16	10	1–6	40–80	8
UASB	Vinasse	0.2–0.4	0.7–6.5	8	1.5–14	32–65	5
EGSB-S	VFA	2.6	2.0	12	32	50	5
ASF	Peptone	0.2*	0.64	5–10	7.5	27–35	24
EGSB-S	Domestic	0.3	4.5	9–11	2.1	20–48	25
EGSB-S	VFA	0.5–0.8	8–12	10–12	1.6–2.5	90	21
UASB	Beef consommé	1.4–7.0	2–10	10	16	49–80	26
ASBR	Dry milk	0.6	0.6–2.4	5–10	6	65–85	27
EGSB-T	VFA	0.5–0.9	5–12	4 - 8	2–4	90	unpubl. data

AAFEB anaerobic attached film expanded bed reactor, *UASB* upflow anaerobic sludge blanket reactor; *EGSB-S* single module expanded granular sludge bed reactor, *ASF* anaerobic submerged filter tank, *ASBR* anaerobic sequencing batch reactor, *EGSB-T* two module expanded granular sludge bed reactor system, *VFA* volatile fatty acids, *COD* chemical oxygen demand, *OLR* organic loading rate, *HRT* hydraulic retention time
* g BOD (biological oxygen demand) dm^{-3}

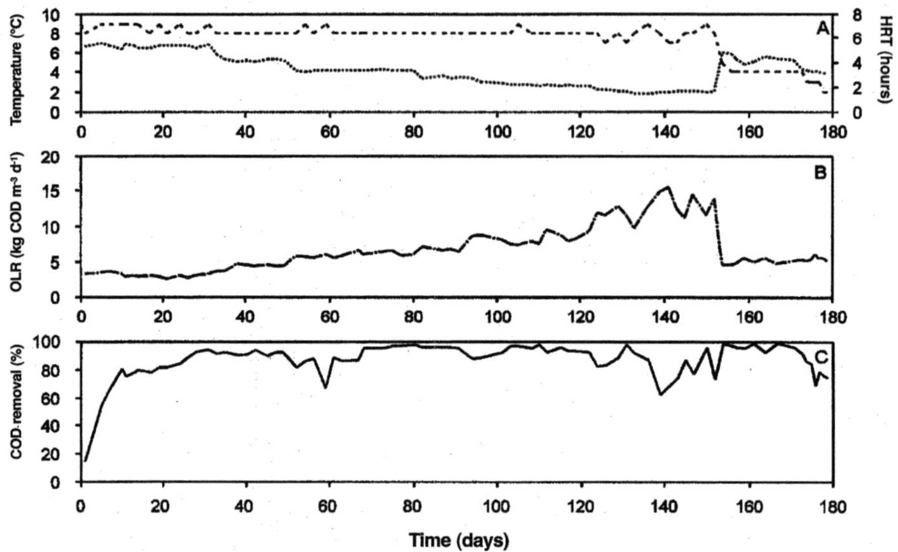

Fig. 4. Operational conditions and efficiency of the two module expanded granular sludge bed reactor system fed with low strength wastewater
(A) Temperature (---) and hydraulic retention time (HRT, ····); (B) Organic loading rate (OLR, ·—·—·—·); (C) COD removal efficiency (——)

ond module. Here, the concentration of H_2 and acetate will be relatively low and, consequently, the conditions for propionate degradation more optimal; in fact, it becomes a self-regulating process. As a consequence of moduling, a sludge with an exceptional high specific acetogenic and methanogenic activity will develop in the second module, leading to a substantial increase in the organic loading potentials of the system.

As the wash-out of viable sludge from the EGSB reactors used in the experiments could be kept at extremely low values, the amount of in-growing viable biomass under all conditions applied considerably exceeded the amount of viable biomass rinsing out from the reactor.

2.3
Development of granular 'psychrophilic' biomass

The long-term operation of lab-scale and pilot-scale EGSB reactors resulted in the development of a methanogenic sludge, well adapted to the low temperature conditions and with a high activity. Although they are very likely present, so far the prevalence of psychrophilic bacteria in the sludge has not been demonstrated. The assessed temperature response curves of the psychrophilically (2–12°C) grown sludge still merely reveal an optimum temperature in the mesophilic range, i.e. between 30 and 40°C (Tables 2 and 3). Presumably the psychrophilic homologues in the sludge cannot manifest in these temperature-activity profiles, due to the

Table 2.   Temperature dependence of the maximum specific degrading activities of mesophilic seed granular sludge from full scale UASB reactor with various substrates. Standard deviation is given in parentheses (adapted from ref. 21)

Temperature	Maximum specific activity (g COD g^{-1} VSS d^{-1})		
(°C)	Acetate	Propionate	Butyrate
10	0.090(0.000)	0.050(0.002)	0.050(0.010)
20	0.380(0.010)	0.301(0.020)	0.172(0.013)
30	0.980(0.120)	0.551(0.031)	0.331(0.042)

Table 3.   Temperature dependence of the maximum specific degrading activities of granular sludge cultivated at 10°C for 300 d with various substrates. Standard deviation is given in parentheses (adapted from ref. 28)

Temperature	Maximum specific activity (g COD g^{-1} VSS d^{-1})				
(°C)	Hydrogen[a]	Hydrogen[b]	Acetate	Propionate	Butyrate
10	1.744(0.374)	0.296(0.009)	0.331(0.003)	0.112(0.009)	0.228(0.002)
20	8.064(0.624)	1.020(0.379)	1.057(0.004)	0.328(0.010)	0.530(0.002)
30	18.024(1.170)	2.732(0.076)	2.204(0.011)	0.663(0.002)	0.915(0.025)

[a] Homoacetogenic activity.
[b] Hydrogenotrophic activity.

presence of (still) much higher amounts of mesophiles. These mesophiles apparently still could grow prosperously under low temperature conditions, because the specific methanogenic activity of the seed sludge improved significantly during the course of the experiment (Tables 2 and 3). This can even be the case, despite their low growth rates under these conditions; the activity growth rates assessed at 10°C amounted to 0.04 and 0.016 d^{-1} for acetate and propionate degraders, respectively.[28] High propionate degrading activities were also found at 5°C (unpubl. data), indicating satisfactory growth and enrichment of these particular acetogens at such a low temperature. The results in Table 3 reveal very high substrate degrading activities of the sludge at 30°C; the values found even exceed those assessed for a typical mesophilic granular sludge,[12,29] and they approach those found for a sludge from thermophilic 55–65°C anaerobic reactors.[30]

An observation of considerable practical importance, furthermore, comprises the ability of the methanogenic sludge to preserve its achieved methanogenic activity.[31] A 6-month storage period at 4°C did not affect the methanogenic capacity of the sludge. Apparently, the starvation rate of methanogens when grown at the low temperatures is extremely low. In practice this means that psychrophilically grown sludge will enable a good and fast start-up of a new psychrophilic reactor system.

The most abundant methanogens present in the psychrophilic sludge were the acetate-consuming *Methanosaeta* and the hydrogenotrophic *Methanobrevibacter* species (or relatives).[20] *Methanosarcina* sp. was found to represent less than 1% of the total methanogenic 16S rRNA, suggesting that this bacteria did not play an important role in methanogenic acetate removal.[20] The low level of *Methanosarcina* sp. in the psychrophilic granular sludge might be related to the low acetate concentrations in the reactor system over a long period of time. Owing to the fact that K_m values for acetate for *Methanosaeta* sp. are 5–10 times lower than for *Methanosarcina* sp.,[32] the latter were outcompeted during the study of the EGSB reactors.

2.4
Psychrophilic treatment of more complex low strength wastewater

In order to assess the feasibility of psychrophilic AnWT for more complex wastewaters, comprehensive pilot plant experiments were conducted with malting wastewater originating from the batch steep process of the Bavaria B.V. malting factory, Wageningen, The Netherlands. The concentrations of soluble and total COD in the wastewater were between 230–1,800 and 320–4,450 mg dm^{-3}, respectively. The strength of the malting wastewater fluctuates strongly, depending on the type of barley used and its growth conditions prior to harvesting. The anaerobically biodegradable COD of the wastewater amounted to about 73%; these values were found using batch bioassays conducted at 15°C. The biodegradability of the malting wastewater varies slightly over time, depending on its composition. The wastewater contained anaerobically completely biodegradable compounds, such as different kinds of sugars, lactic acid, glycerol, ethanol and volatile fatty acids, but also compounds which are only partially or poorly (or slowly) biodegraded at low temperature, such as fats, proteins, tannin, cellulose and barley grains suspended solids. The COD of the settleable and the colloidal suspended solids in the malting wastewater ranged from 20–231 to 0–176 (mg dm^{-3}), respectively.[22,31]

In the AnWT of malting wastewater using single module EGSB reactor at 16°C, the achieved COD removal efficiencies averaged around 56% at imposed OLRs ranging between 4.4 and 8.8 kg COD m^{-3} d^{-1} and HRTs of approximately 2.4 h. At 20°C the removal efficiencies achieved amounted to approximately 66% and 72%, respectively, at imposed OLRs of 8.8 and 14.6 kg COD m^{-3} d^{-1}, and corresponding HRTs of 2.4 and 1.5 h.[22] The lower values of COD removal efficiency coincided with high H_2 concentrations in the biogas. In these situations an imbalance between the acidification rate and methanogenesis in the reactor very likely led to higher H_2 concentrations with well negative effects for the conversion of, e.g., propionic acid.

With regard to the removal of the non-soluble COD fraction, a psychrophilic single EGSB reactor system clearly does not comprise the proper system; the superficial velocities prevailing in the reactor are too high. This has already been demonstrated with earlier EGSB experiments with settled domestic sewage.[25] Moreover, even in cases removed by some 'entrapment' mechanism, little conversion of these solids can be expected due to the very low hydrolytic activity under the prevailing low temperature conditions.[25,33]

Of particular importance is the fractional content of the wastewater of non-acidified organic matter, especially for as design criteria. The reason for this is that granular sludge easily may become over-grown by acidogenic biomass. As this will lead to a serious deterioration of the granular sludge, this should be prevented when using single module reactors.[20]

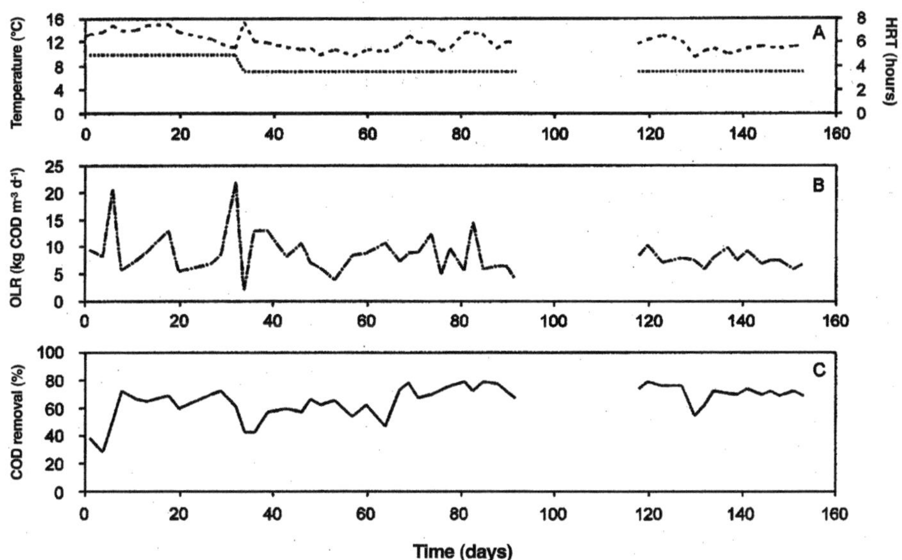

Fig. 5. Operational conditions and efficiency of the two module pilot-scale expanded granular sludge bed reactor system fed with low strength malting wastewater
(A) Temperature (---) and hydraulic retention time (HRT, ····); (B) Organic loading rate (OLR, ·—·—·); (C) COD removal efficiency (——) (adapted from ref. 31)

The problems will be less serious in a two module EGSB reactor.[20,31] The formation of a layer of acidifying sludge around granules then can be restricted to the first module of the system, which then should serve as a high-rate acidification step. It was observed that the formation of layers of acidifying biomass around the granular seed sludge present here may lead to gas entrapment in the granule and subsequent flotation of these granules.[31] This obviously represents operational problems in the 'high-rate' acidogenic reactor module. Also, particularly at low temperatures, these problems are very serious, because the acidifiers have – compared to mesophilic conditions[29] – a very high growth yield (0.22 g VSS-COD g^{-1} COD$_{removed}$),[20] but their starvation rate then is extremely low, i.e. around 1.57×10^{-5} h^{-1} at 10°C.[20] Moreover, also the presence of oxygen in the wastewater can contribute to the growth of facultative anaerobic and/or aerobic organisms; they have at least a 10-times higher growth yield than methanogens.[34–36] Despite these peculiar problems, a two module EGSB system appeared to be capable of accommodating wastewaters containing up to 10% of non-acidified substrate-COD at temperatures as low as 8°C, provided the oxygen level in the wastewater was maintained below 2 mg O$_2$ dm^{-3}.[20,31] Moduling the EGSB system markedly improved the performance and stability of the system compared with a one-step process.

When applying a two module pilot-scale EGSB system (Fig. 5), anaerobic treatment of acidified malting wastewater in the temperature range 8–12°C can be accomplished at an OLR of up to 12 kg COD m^{-3} d^{-1} and at a HRT of 3.5 h. The system provided a remarkable long-term (6-month) performance stability at low temperatures even when imposing strong variations in OLR between 3 and 12 kg COD m^{-3} d^{-1}. The COD removal efficiencies were very high and comparable to the maximum efficiency of 85% found during the mesophilic anaerobic treatment of brewery wastewaters at 30°C.[37] Similar results were even found at 37°C where an efficiency of 80% was obtained in a three-stage system.[38] The removal efficiencies found in this study are comparable to those found in a thermophilic (55°C) anaerobic fluidized bed reactor.[39]

3
Conclusions

Psychrophilic AnWT using well designed one and two module EGSB systems is now available for full-scale application to low strength non- or partially acidified wastewaters, including various types of industrial wastewaters and settled domestic sewage. Implementation of these systems and the mesophilic and thermophlic AnWT systems as well, combined with the physical–chemical resource recovery post-treatment methods, will lead to a very significant decrease in the operational costs of wastewater treatment, in many cases even to an economically profitable operation for the concerned industry or municipality. In view of its highly sustainable features, high-rate AnWT-concepts should replace the conventional – principally non-sustainable – aerobic treatment systems so far used in the traditional (and very expensive) sanitary engineering environmental protection approaches.

4
References

1. Lettinga G, Hulshoff Pol LW. UASB-process design for various types of wastewaters. Wat Sci Technol 1991; 24:87-107.
2. Lettinga G. Anaerobic digestion and wastewater treatment systems. Antonie van Leeuwenhoek. 1995; 67:3-28.
3. Iza J, Colleran E, Paris JM, Wu WM. International Workshop on anaerobic treatment technology for municipal and industrial wastewaters: summary paper. Water Sci Technol 1991; 24:1-16.
4. McCarty PL One hundred years of anaerobic treatment. In: Hughes DE, Stafford DA, Wheatly BI, Baader W, Lettinga G, Nyns EJ, Verstraete W, eds. Anaerobic Digestion 1981. Amsterdam: Elsevier, 1982:3-22.
5. de Man AWA, van der Last ARM, Lettinga G. The use of EGSB and UASB anaerobic systems for low strength soluble and complex wastewaters at temperatures ranging from 8 to 30°C. In: Hall ER, Hobson PN, eds. Proceedings of the Fifth International Symposium on Anaerobic Digestion. Bologna, Italy, 1988:197-209.
6. Lin CY, Noike T, Sato K, Matsumoto J. Temperature characteristics of the methanogenesis process in anaerobic digestion. Water Sci Technol, 1987; 19:299-310.
7. Jewell WJ, Morris JW. Influence of varying temperature, flow rate and substrate concentration on the anaerobic attached film expanded bed process. In: Proceedings of the 36th Industrial Waste Conference. Purdue University, 1981:1-24.
8. Switzenbaum MS, Jewell WJ. Anaerobic attached film expanded bed reactor treatment of

dilute organics. In: Proceedings of the 51st Annual WPCF Conference. Anaheim, California, 1978:1-164.

9. Mills PJ. Minimisation of energy input requirements of an anaerobic digestor. Agric Wastes 1979; 1:57-66.

10. Frankin RJ, Koevoets AA, van Gils WMA, van der Pas A. Application of the BIOBED upflow fluidized bed process for anaerobic waste water treatment. Water Sci Technol 1992; 25:373-382.

11. Rinzema A, van Veen H, Lettinga G. Anaerobic digestion of triglyceride emulsions in expanded granular sludge bed upflow reactors with modified sludge separators. Environ Technol 1993; 14:423-432.

12. Kato TM, Field JA, Versteeg P, Lettinga G. Feasibility of expanded granular sludge bed reactors for the anaerobic treatment of low strength soluble wastewaters. Biotechnol Bioengin 1994; 44:469-479.

13. Koster IW, Lettinga G. Application of the upflow anaerobic sludge bed (UASB) process for treatment of complex wastewaters at low temperatures. Biotechnol Bioengin 1985; 27:1411-1417.

14. Weber H, Kulbe K D, Chmiel H, Trösch W. Microbial acetate conversion to methane: kinetics, yields and pathways in a two-step digestion process. Appl Microbiol Biotechnol 1984; 19:224-228.

15. Cohen A, Breure AM, Van Andel JG, Van Deursen A. Significance of partial pre-acidification of glucose for methanogenesis in an anaerobic digestion process. Appl Microbiol Biotechnol 1985; 21:404-408.

16. Dinopoulou G, Lester JN. Optimization of a two-phase anaerobic digestion system treating a complex wastewater. Environ Technol Lett 1989; 10:799-814.

17. Komatsu T, Hanaki K, Matsuo T. Prevention of lipid inhibition in anaerobic processes by introducing a two-phase system. Water Sci Technol 1991; 23:1189-1200.

18. Wiegant WM, Hennik M, Lettinga G. Separation of the propionate degradation to improve the efficiency of thermophilic anaerobic treatment of acidified wastewaters. Water Res 1986; 20:517-524.

19. Lier van JB, Boersma F, Debets MMWH, Lettinga G. High-rate thermophilic anaerobic wastewater treatment in compartmentalized upflow reactors. Water Sci Technol 1994; 30:251-261.

20. Lier van JB, Rebac S, Lens P, Bijnen van F, Oude Elferink SJWH, Stams AJM, Lettinga G. Anaerobic treatment of partly acidified wastewater in a two-stage expanded granular sludge bed (EGSB) system at 8°C. Water Sci Technol 1997; 36:317-324.

21. Rebac S, Ruskova J, Gerbens S, van Lier JB, Stams AJM, Lettinga G. High-rate anaerobic treatment of wastewater under psychrophilic conditions. J Ferment Bioengin 1995; 5:15-22.

22. Rebac S, van Lier JB, Janssen MGJ, Dekkers F, Swinkels KTM, Lettinga G. High-rate anaerobic treatment of malting waste water in a pilot-scale EGSB system under psychrophilic conditions. J Chem Technol Biotechnol 1997; 68:135-146.

23. Stams AJM. Metabolic interactions between anaerobic bacteria in methanogenic environments. Antonie van Leeuwenhoek 1994; 66:271-294.

24. Matsushige K, Inamori Y, Mizuochi M, Hosomi M, Sudo R. The effects of temperature on anaerobic filter treatment for low-strength organic wastewater. Environ Technol 1990; 11:899-910.

25. Last van der ARM, Lettinga G. Anaerobic treatment of domestic sewage under moderate climatic (Dutch) conditions using upflow reactors at increased superficial velocities. Water Sci Technol 1992; 25:167-178.

26. Grant S, Lin KC. Effects of temperature and organic loading on the performance of upflow anaerobic sludge blanket reactors. Can J Civ Engin 1995; 22:143-149.

27. Banik GC, Dague RR. ASBR treatment of dilute wastewater at psychrophilic temperatures. In: Proceedings of 69th Annual Water Environmental Conference. Dallas, Texas, USA, 1996:235-246.

28. Rebac S, Gerbens S, Lens P, van Lier JB, Stams AJM, Lettinga G Kinetics of fatty acid degra-

dation by psychrophilically cultivated anaerobic granular sludge. Bioresource Technol 1998 (in press).

29. Alphenaar A. Anaerobic granular sludge: characterization, and factors affecting its functioning. PhD Thesis, Agricultural University Wageningen, The Netherlands, 1994:93-112.

30. Lier van JB, Groeneveld N, Lettinga G Development of thermophilic methanogenic sludge in compartmentalized upflow reactors. Biotechnol Bioengin 1996; 50:115-124.

31. Rebac S, van Lier JB, Lens P, van Cappellen J, Vermeulen M, Stams AJM, Dekkers F, Swinkels KTM, Lettinga G. Psychrophilic (6-15°C) high-rate anaerobic treatment of malting wastewater in a two module EGSB system. Biotechnol Prog 1998:14 (in press).

32. Jetten MSM, Stams AJM, Zehnder AJB. Methanogenesis from acetate: A comparison of the metabolism in *Methanothrox soehngenii* and *Methanosarcina* spp. FEMS Microbiol Rev 1992; 88:181-198.

33. Zeeman G, Sanders WTM, Wang KY, Lettinga G. Anaerobic treatment of complex wastewater and waste activated sludge. In: Proceedings of the Conference on Advanced Wastewater Treatment. Amsterdam, The Netherlands, 1996:225-232.

34. Shen CF, Guiot SR. Long-term impact of dissolved O_2 on the activity of anaerobic granules. Biotechnol Bioengin 1996; 49:611-620.

35. Kato MT, Field JA, Lettinga G. Methanogenesis in granular sludge exposed to oxygen. FEMS Microbiol Lett 1993; 11:317-324.

36. Gerritse J, Gottschal JC. Oxic and anoxic growth of a new *Citrobacter* species on amino acids. Arch Microbiol 1993; 160:51-61.

37. Pereboom JHF. Methanogenic granule development in full scale internal circulation reactors. Water Sci Technol 1994; 30:211-221.

38. Stadlbauer EA, Oey LN, Weber B, Jansen K, Weidle R, Löhr H, Ohme W, Döll G. Anaerobic purification of brewery wastewater in biofilm reactors with and without a methanation cascade. Water Sci Technol 1994; 30:395-404

39. Péres M, Romero LI, Sales D. Thermophilic anaerobic degradation of distillery wastewater in continuous-flow fluidized bed bioreactors. Biotechnol Prog 1997; 13:33-38

Effectiveness of wastewater lagoons in cold regions

D. W. Smith* and K. M. E. Emde

Department of Civil and Environmental Engineering, 304 Environmental Engineering Building, University of Alberta, Edmonton, Alberta, Canada T6G 2G7

1
Introduction

Development in the world's cold regions has lead to increasing environmental concerns, including the proper disposal of domestic wastes generated by burgeoning northern cities, towns, villages, hamlets, and work camps.[1,2] Growth in many of the northern areas tends to be the result of economic boom periods, rather than a gradual increase.[3] A smaller percentage of the communities in these regions have sewage treatment facilities, such as found in more southern areas.[4] Wastewater treatment lagoons are the most common, and economical, method used to treat domestic sewage for the cold regions of North America.[4]

As with other wastewater treatment processes, lagoons are designed to ensure that there is sufficient opportunity, and time, to allow necessary physical, chemical and biological treatment resulting in effluents with a significant reduction of organic matter, with reduced levels of suspended solid matter, a reduction in nutrients (primarily nitrogen, phosphorus), and a reduction in indicator and pathogenic organisms.

Wastewater treatment lagoons have also been called sewage treatment lagoons, sewage lagoons, waste stabilization ponds, oxidation ponds, stabilization ponds, and waste ponds.[5] All of these terms refer to a natural or constructed cell, or series of cells, whereby a major portion of organic matter entering this system is decomposed, or "stabilized", by indigenous microorganisms.[6] Today lagoons can be used, if properly designed, for treatment of some types of industrial, as well as, domestic wastes. Figure 1 illustrates common design options for cold regions lagoon systems.[7] The major benefits of using lagoons for biological treatment of wastewater includes:

- low capital and operating costs,
- low maintenance requirements,
- minimal energy requirements,
- minimal knowledge needed for operators,

---

* Corresponding author

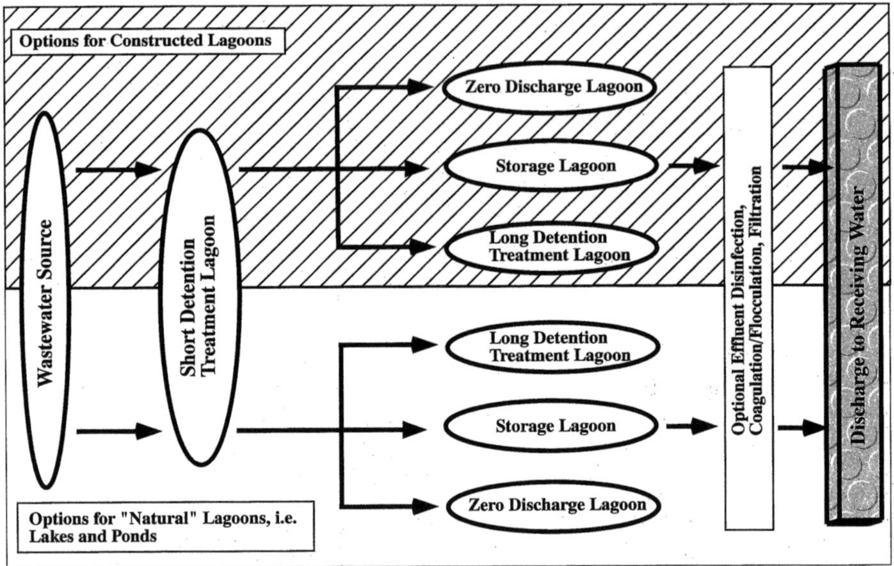

Fig. 1. Typical cold climates lagoon design options (after ref. 7)

– land costs are low where there are large expanses of land available, and
– ability to adequately serve the needs of smaller populations.[9]

2

Description of cold regions

Depending on the definition of cold regions used, the North American regions include the state of Alaska, the Canadian territories of Yukon and the Northwest Territories, the northern portions of certain Canadian provinces (British Columbia, Alberta, Saskatchewan, Manitoba, Ontario, Quebec, and Newfoundland), and some areas of the continental United States.[3] Also included are the arctic coasts of Europe (Norway, Sweden, Finland, Russia, and the Republics of Chukotia, Koryakia and Taymyria (formerly part of the USSR), and the coastal areas of Iceland and Greenland.[8]

These regions can be distinguished on the basis of three general climatic types: polar climate, tundra climate and taiga climate.[8] These climate types not only influence the design and operation of wastewater lagoons, but also the influence of effluents on the receiving environments. As the ecosystem in the cold regions is, in some respects, rather fragile relative to more southern climates, the environmental impacts of any anthropogenic discharges are more keenly felt, and more difficult to remediate. Table 1 summarizes some typical northern sewage flow rates.[4] Depending on the type of community and water/wastewater system available, the wastewater generated can become extremely concentrated as a result of low water use, or

Table 1. Typical northern community sewage flow rates (after ref. 4)

Source	Average amount (l person^{-1} d^{-1})
Communities (1,000 population, or more) with:	
truck-haul water and conventional internal plumbing	140
truck-haul water, low flush toilets	90
truck-haul water, home water tanks, "honey-bucket" toilet	1.5
Work Camps	220
Remote military camps with limited access to water	130

extremely dilute when water is allowed to flow at a constant, low rate to prevent the potable water piping from totally freezing during the winter months.[9]

Environmental conditions in northern receiving waters vary significantly with the time of year and geographic location. Typically winter river flows are very low, due to frozen sources, whereas spring run-off and summer flows may be quite high, with high concentrations of sediments.[4]

The extreme polar regions are considered arid, based on a mean annual precipitation of less than 100 mm. The land is ice, or snow-covered, much of the year and there is little or no vegetation. These regions are considered to be a "polar desert" and as such, not suitable for conventional, long-term habitation. There is little human habitation in the extreme polar region except for military outposts, and off-shore oil exploration in the Arctic Ocean. In these regions lagoon treatment of wastes is extremely slow, therefore requiring very long retention times. In the frozen interior of Greenland, lagoons are not a suitable waste treatment alternative.

Between the approximate latitudes of 65° and 75° in North America lies a climatic region characterized by permafrost and scant vegetation. This region is known as the tundra region. Summer temperatures commonly remain below 10°C, with mean temperatures typically staying below 0°C for 5 to 8 months of the year. Precipitation, usually in the form of snow, is generally less than 380 mm annually. These regions, for a portion of the year, may be suitable for lagoon disposal, if the lagoons are properly designed and operated.

The taiga region is characterized by low winter temperatures and a relatively wide range of summer temperatures, as high as 35°C. Precipitation is typically less than 500 mm annually, with the majority as rainfall during the spring and summer months. Vegetation consists primarily of stands of spruce and aspen poplar, as well as other vegetation. This region comprises most of the northern part of selected Canadian provinces, Alaska, the northern areas of Norway, Sweden, Finland, and the northern regions of the former USSR. The majority of cold region wastewater lagoons are found in this area, as well as the northern boreal forest and alpine areas.

3
Physical limnology of cold regions lagoons

Cold regions wastewater lagoons operate under ice-cover for a portion of the year. This influences the design of the facility, as well as the ecology of the lagoon. Typically the lagoons will be designed to operate with intermittent discharge, to avoid discharging effluents during the winter when the effluents are of low quality, and receiving waters have low assimilative capacities due to minimal water flow and ice cover.[10–12]

Lagoons in permafrost areas have been shown to cause a geothermal disturbance on the surrounding permafrost soils. If the soil contains ice lenses, their melting most likely will lead to consolidation of the soil and settling of the bottom and sides of the lagoon.[13,14] Liners are commonly used to reduce the rate of seepage from the lagoon, and to minimize the adverse impacts on the surrounding environment.[15] Wastewater lagoon liners may be multi-layered, and consist of a thick packed clay layer, low permeability polymeric membranes, or some combination of both.[15]

Lagoons are influenced by other environmental factors, such as wind, amount of sunlight, and ambient precipitation/evaporation. Wind will promote aeration of the lagoon surface, facilitate mixing of the cell during periods of no ice-cover, but may be one factor in short-circuiting of the treatment process. Sunlight aids algal photosynthesis and provides some microorganism reduction as a result of ultraviolet light penetration into the cell. Ambient precipitation and evaporation affects the water balance of the cell, that is, water can enter the cell due to precipitation and collection system run-off, causing dilution in the treatment cell. During periods of no ice-cover warm daytime ambient temperatures result in a degree of water evaporation, causing some concentration of nutrients in the cell.

Stratification, and the resulting lack of mixing, significantly affects algal dynamics in wastewater lagoons.[16,17] Mixing within the lagoon helps to balance nutrient and oxygen levels, as well as to moderate the water temperature. Mixing also helps transport non-motile algae closer to the epilimnion region of the lagoon. If the mixing process had been short-circuited, then photosynthetic, non-motile organisms in the metalimnion and hypolimnion will die and settle out, resulting in an increase in biological oxygen demand (BOD) and a decrease in dissolved oxygen levels.[16,18] Algae trapped in the epilimnion by lack of water mixing, may encyst or die off due to the increased water temperatures in this region.[16] To facilitate the mixing processes in cold regions lagoons, a deep sub-surface outlet is typically part of the engineered lagoon design.[4]

4
Types of cold regions lagoons

Three types of lagoon cells are commonly used, each having different treatment objectives. These are aerobic, anaerobic, and facultative cells.

Anaerobic cells, also known as short detention cells, are used for sedimentation

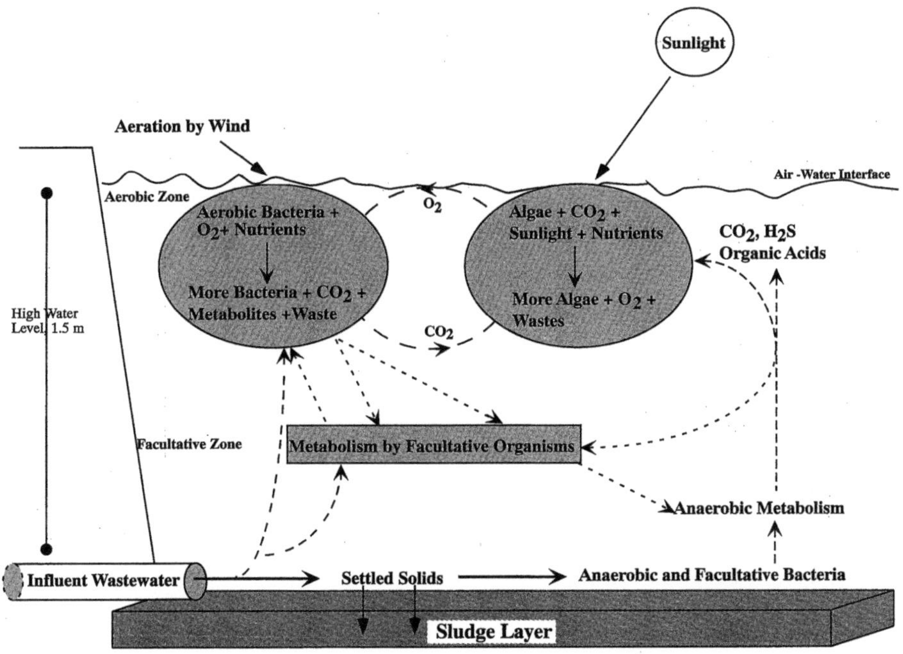

Fig. 2. Typical algal-bacterial symbiosis in a northern region lagoon (after refs. 2 and 16)

of wastes, similar to primary sedimentation in a conventional wastewater treatment plant. High loading rates to the cell ensures that oxygen is limiting, and is used faster than could be replaced by atmospheric diffusion or algal production. Detention times are short, and are typically 2–3 d. Biological processes are primarily anaerobic, as a result of the high loading of nutrients and solids to the cell.

Three major groups of microorganisms are found in these lagoon cells, and they often are referred to as the consortium of anaerobic waste treatment organisms. These are the: acid-formers (e.g. *Clostridium* species, *Propionibacterium* species, *Bacteroides* species), methanogens, and sulfate reducing bacteria. The "acid-formers" use the processes of hydrolysis and fermentation of in-coming, complex waste materials, such as large carbohydrates, proteins and fats, to produce smaller, simpler organic molecules. These include alcohols, aldehydes, and fatty organic acids.[19] There is relatively little stabilization of organic waste accomplished by methanogens.[20]

The methanogens perform the bulk of waste stabilization in anaerobic lagoons, through the metabolism of simple organics, generated by "acid-formers", to carbon dioxide (CO_2) and methane (CH_4). The degree of waste stabilization can be measured by the amount of CH_4 generated from this type of lagoon.[20] As methanogens are quite temperature sensitive,[21] the rate of waste stabilization decreases significantly with decreasing lagoon temperature.[20] Often during the winter months, the temperature of the incoming wastewater will often be higher than the mean tem-

Table 2. Role of selected aerobic microorganisms in lagoon wastewater treatment

Organism	Function
Acinetobacter	Phosphorus removal
Achromobacter	Denitrification
Arthrobacter	Degradation of carbohydrates
Bacillus	Degradation of proteins
Microthrix	Degradation of fats and lipids
Nitrosomonas	Nitrification
Nitrobacter	Nitrification
Pseudomonas	Degradation of carbohydrates, denitrification

perature of the lagoon.[4] The ideal design for optimum, anaerobic, nuisance free treatment of wastes by anaerobic lagoons would favor methanogens.[21] This can include the use of floating covers for anaerobic lagoons located in cold regions to maintain a higher lagoon water temperature.[22] Floating covers, man-made or the result of the natural buoyancy of certain wastewater constituents such as animal fats, can also help control objectionable odors, including hydrogen sulfide (H_2S) rising from the lagoon during spring break-up.[23]

In conjunction with methanogens, anaerobic sulfide-generating organisms will be found in most cold regions anaerobic lagoons. These are often considered to be more of a nuisance, due to the considerable production of H_2S resulting in odor complaints from near-by residents or facilities. These types of odors are associated with the spring time loss of ice cover on anaerobic cells.

Facultative, or treatment, cells have aerobic conditions near the air-water interface, with conditions becoming facultatively anaerobic with increasing water depth. Figure 2 illustrates a simplified version of the algal-bacterial symbiosis that operates in facultative lagoons.[2,16] Algae commonly isolated from cold regions lagoons include species of *Euglena, Chlorella, Chlamydomonas* and *Oscillatoria.*[14,17] Algae are able to use CO_2, NH_4^+ and PO_4^- generated, or released, by bacterial metabolism as a primary source of nitrogen, phosphorus and carbon. Dominant algal species, such as *Chlamydomonas* and *Chlorella*, are often motile, and can move up and down the water column to exploit available light.[17] The microbial population in facultative lagoons is extremely diverse and complex. This population is composed of indigenous organisms, as well as, those entering from the incoming wastewater that are able to successfully adapt to the environment of the facultative lagoon. Table 2 summarizes the role of some selected microorganisms in lagoon wastewater treatment.[19] Detention times are long, and are typically 2–3 months at the design hydraulic loading. During periods of ice-cover, facultative lagoons become anaerobic.

Storage cells, also known as maturation, finishing, or polishing lagoons, operate similar to facultative cells. The organic loading to the cell is low, as a result of substrate removal in previous treatment cells. There is a great diversity of micro- and macroorganisms. The primary purpose of these cells is nutrient (nitrogen and phosphorus) removal, primarily by algae, reduction of pathogens and indicator organisms, and removal of additional suspended solids. Detention times are long,

Table 3. Typical raw sewage temperatures at selected northern North American treatment facilities (after refs. 4 and 25)

Location	Temperature range (°C)	
	Summer	Winter
Fairbanks, Alaska	3–11	0–11
Juneau, Alaska	9	2
Kenai, Alaska	10–4	8
Clinton Creek, Yukon Territory	22	17
Hay River, NWT	10–15	
Inuvik, NWT	18–30.5	0–1
Yellowknife, NWT	6–11	3–5
Sutherland, Saskatchewan		10–14
Lacombe, Alberta	10–18	0–2
Stettler, Alberta	15–22	1.5–8

Table 4. Typical northern primary wastewater characteristics (after ref. 4)

Parameter	Units	Conventionally concentrated wastewater[26]	Dilute wastewater[27]
COD[a]	(mg l[-1])	81,000–134,000	40–900 (measured as TOC)
Total N	(mg l[-1])	7,300–9,500	8
Suspended solids	(mg l[-1])	66,000–85,000	40–2,000
Phosphorus	(mg l[-1])	1,200	9
Calculated flow	(l person[-1] d[-1])	1.2	310

COD chemical oxygen demand, *TOC* total organic carbon

usually totaling 12 months at the design hydraulic loading. Storage may be in several cells, which together hold the treated wastewater until an appropriate time for discharge, ideally in the fall. Effluents, meeting all regulatory requirements, are discharged to receiving waters during periods of no ice cover.

The microbial population of lagoons is directly influenced by seasonal changes in temperature and the resulting ice and snow cover. A large portion of the lagoon microbial population is facultatively psychrophilic. During winter months typical influent wastewater temperatures range between 10 to 0.5°C.[24] Table 3 summarizes some raw sewage temperatures measured at a number of cold region treatment facilities.[4,25] This influences metabolism of solids, as well as accumulation of sludge in the lagoon. Typical primary wastewater sludge characteristics are given in Table 4.[4,26,27]

Although there may be partially aerobic conditions found under moderate ice and snow-covered lagoons, the rapid drop in temperature in the late fall kills the bulk of the algal population present.[28] As a result, many cold region lagoons operate under predominately anaerobic conditions during the winter months. During this period lagoons typically provide the equivalent of primary treatment. Slaughter et al. reported that the winter discharge of a northern Ontario lagoon had an effluent quality of BOD$_5$ of 90 mg l[-1], an ammonia concentration of 16 mg l[-1],

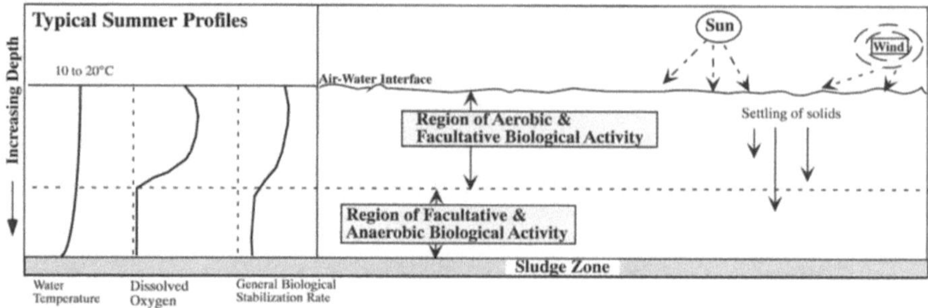

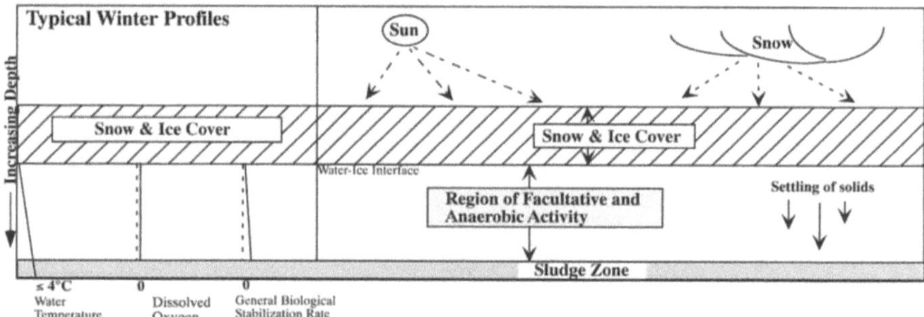

Fig. 3. Expected process conditions found during summer and winter in a cold regions lagoon (after ref. 2)

and a hydrogen sulfide concentration of 26 mg l^{-1}, resulting in significant fish mortality in the receiving waters.[29] Figure 3 illustrates the expected process conditions in a cold region lagoon during the summer and winter.[2,7]

5
Factors influencing design of cold regions lagoons

There are a number of key factors to consider when planning and designing lagoons for wastewater treatment in cold regions. Generally, only facultative and anaerobic lagoons are practical for northern communities. As a result of long periods of ice-cover, cold region lagoons operate under aerobic conditions only after spring ice break-up and during the summer months. Aerated lagoons, where the oxygen is supplied by a turbine or compressed air system, are only applicable for selected applications where the community is able to provide proper support for operation and maintenance, as well as supply sufficient electricity.

The design of cold region wastewater lagoons is commonly based on hydraulic detention times, and in some jurisdictions, on organic loading.[4] Both of these parameters are directly related to the size of community, including any industrial wastewater flows, being served. The design must also consider public health impacts, including reduction of opportunistic and direct pathogens which may be present in the sewage. Pathogen removal is most often measured relative to the reduction in fecal coliform levels during the course of the wastewater retention time in the lagoon.

Organic and/or hydraulic loading influences sludge accumulation in lagoons. When the bottom lagoon water temperatures are less than 19°C, conditions favoring sludge accumulation are created.[21] Sludge accumulates at a somewhat greater rate as the result of reduced rate of microbial metabolism of solids during cold months of the year.[24] Sludge disposal in cold region lagoons tends to be infrequent, occurring typically once every 6–12 years.[4] The removed sludge is generally stored in a separate holding area, allowed to freeze dry, and covered with 0.5–1.0 m of soil.[4]

6
Reduction of organisms

Reduction of microorganisms, including indicator organisms and pathogens, relies on a number of "natural disinfection" processes. These include: adsorption of organisms and consequent sedimentation in the lagoon; inability of organisms to successfully compete with the adapted and indigenous microbial population for nutrients; predation by other microorganisms, phages, or higher organisms; and reduction by the penetration of the sun's ultraviolet rays into the lagoon.

Chlorine disinfection of wastewater lagoon effluents has been proposed as one method to reduce microbial levels in treated effluents.[30,31] One study found that significant factors for determining the proper chlorine dose for disinfection included the lagoon water temperature, sulfide concentration, and the total chemical oxygen demand.[31] Cold water temperatures, less than 5°C, required 83% more applied chlorine to achieve the same level of disinfection if the water temperature was 15°C.[31] Unfiltered effluents exerted a greater chlorine demand than did filtered effluents.[31] However, high dosages of chlorine, when combined with the levels of organic matter still present in lagoon effluents, could lead to unacceptable chlorine residuals and the possible formation of a number of disinfection by-products being released to the receiving waters.

An alternative method to reduce levels of microorganisms from wastewater lagoons is the use of coagulation and flocculation. This method will also assist with the removal of additional suspended solids and phosphorus, and reduce effluent BOD.[32] A study on a cold region lagoon in Alberta Canada found that fecal coliform levels could be reduced by 99.9%, together with a reduction of 90% in suspended solids, a 97% reduction in phosphorus, and a 35% reduction in BOD.[32] However, there are additional problems created due to increased accumulation of sludge in the lagoon, together with the sludge dewatering, and disposal processes needed for cold climates.[33]

Sand filtration has also been used to polish some cold region lagoon effluents prior to discharge.[33–35] Larger organisms can be removed easily, but without coagulation-flocculation, smaller organisms such algae and others are not readily removed.[33] Direct filtration of lagoon effluents resulted in algal removal ranging from 14–90%, depending on the filter loading rate and the sand size used in the filter.[33] In contrast intermittent sand filtration, which involves the periodic application of wastewater lagoon effluent to a natural or constructed sand bed, has been found to be more effective for the removal of algae, BOD, phosphorus, bacteria and suspended solids.[35] With both types of filtration there may be certain operational problems in the more northern cold regions, that is in the arctic and sub-arctic regions, due to the relatively harsh and long winter conditions found there. However, polishing filtration of selected cold regions lagoon effluents may prove useful in the control of encysted pathogens, such as *Giardia* and *Cryptosporidium,* or helminth eggs as opposed to disinfection, and the potential formation and release of wastewater disinfection by-products into the receiving environment.

7
Discharge timing

The time of effluent discharge is critical in cold regions lagoons. This is to not only ensure that proper treatment of wastes contained within the lagoon has occurred, but also to minimize adverse impacts on the receiving environment. Winter discharge is undesirable, as cold regions lagoons only provide primary treatment of wastes.

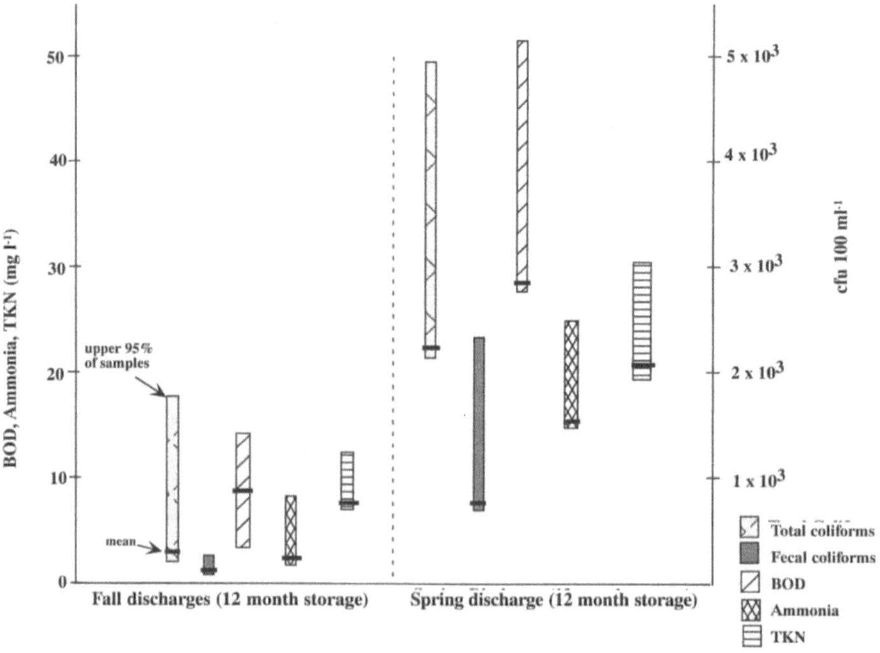

Fig. 4. Changes in effluent quality as a function of season of discharge (after refs. 10 and 11) *BOD* biological oxygen demand, *TKN* total kieldahl nitrogen

Early spring discharge of lagoon effluents is generally not desirable as the anaerobic conditions over the winter typically result in increased hydrogen sulfide emissions to the surrounding environment.[17] It has also been found that in spring many lagoons exhibit a significant decrease in BOD and an increase in total suspended solids (TSS), reflecting the decrease in biological treatment over the winter months. Also ammonia released by microbial activity during the winter will add to the effluent toxicity. Figure 4 illustrates changes in effluent BOD and TSS, as a function of the season of discharge.[10–12] These data were collected in Alberta during 1983 to 1986.

Although the algal population in spring is usually low, there can be very rapid algal blooms of *Euglena* and *Chlamydomonas* species, even under partial ice cover.[32] These spring algal blooms can result in unacceptably high TSS levels in the effluent. As a result of the low efficiency of winter waste treatment, levels of indicator organisms, and possibly wastewater pathogens, are typically high in effluents discharged in spring.[10–12,32]

Fall discharges tend to be of a much better bacteriological quality than spring effluents.[10–12,36] This is due to the increase in microbial metabolism and resulting treatment processes occurring during the summer months. Table 5 summarizes differences between fall and spring discharge for total and fecal coliform levels.[10] The data in Table 5 clearly demonstrate that fall discharges are preferable to spring discharges, in terms of total and fecal coliform levels.[10]

In addition to concerns about the bacteriological quality of lagoon effluents, there are serious concerns about the toxicity of the final lagoon effluent on the receiving body of water. Specifically, effluent toxicity is most often thought of in terms of the toxic effects of nitrogen, phosphorus, and in some cases sulfide and metals, on the receiving environment. Toxicity can be reduced as the result of algal and bacterial assimilation or biotransformation, sedimentation, and air stripping of volatile forms of certain chemicals, such as ammonia.[37] Gaseous ammonia stripping is dependent on the pH of the lagoon, water and ambient temperature and mixing conditions in the lagoon.[37] Biological nitrification and denitrification are known to play significant roles in ammonia and nitrate reduction cycles.[19]

One study evaluating the toxicity potential of a number of wastewater storage lagoons in Alberta and Yukon, Canada, found that approximately 92% of lagoon systems studied were able to produce an effluent that met the regulatory 100% LC_{50} requirement (mean lethal concentration of a compound that will kill 50% of the population exposed), and that the risk factor for construction of a storage lagoon system with annual fall discharge was less than 10%.[37]

High levels of nitrogen and phosphorus have been directly linked with an increase in growth of algae and aquatic plants resulting in depletion of dissolved oxygen and possible eutrophication of the water body.[19] High ammonia ($\geq$ 5 mg l^{-1}), as well as high sulfide concentrations are toxic to fish in the receiving environment. Metals, including copper, iron, lead, mercury and zinc, are commonly found in wastewater lagoons receiving industrial and commercial discharges. Lagoons tend to be less effective than conventional wastewater treatment for removal of metals, and lagoon sludge can have high concentrations of metals due to long retention times and infrequent sludge removal.[4]

Table 5. Differences between fall and spring effluent levels of total and fecal coliforms in Alberta (after ref. 12)

Lagoon type	No. of short and long term storage cells		Fall effluent quality			Spring effluent quality		
			n	Total coliform (cfu 100 ml^{-1})	Fecal coliform (cfu 100 ml^{-1})	n	Total coliform (cfu 100 ml^{-1})	Fecal coliform (cfu 100 ml^{-1})
A	4 short, 2 long (or more)	mean value	23	6.0×10^1	2.0×10^1	30	1.6×10^3	6.7×10^2
		upper 95%	23	2.1×10^2	4.3×10^1	30	6.1×10^3	3.1×10^3
B	2 short, 2 long (or more)	mean value	10	2.3×10^2	7.4×10^1	12	5.4×10^3	1.6×10^3
		upper 95%	10	2.4×10^4	1.0×10^3	12	9.2×10^3	1.6×10^4
C	0 short, 1 long (or more)	mean value	17	3.3×10^3	2.1×10^2	52	1.2×10^3	6.8×10^2
		upper 95%	17	4.5×10^5	1.0×10^3	52	1.9×10^4	8.3×10^3

n total number of samples taken

8
Effects on pathogen removal and survival and persistence of pathogens

It has been found that shallow wastewater lagoons, having constant low average water temperatures, did not significantly reduce pathogen levels.[38] In fact, persistence of some pathogens, and indicator organisms, appears to be enhanced by the cold temperatures.[21,39–41] In one study of lagoon wastewater treatment in two communities, Slanetz et al. found that while there were marked seasonal differences in total and fecal coliform reduction, the reductions in levels of salmonellae and enteric viruses remained relatively constant, regardless of season or water temperature.[39] A study by Dutka and Bell showed that viable *Salmonella* species were recovered in the absence of high levels of total coliform (< 10 cfu 100 ml^{-1}) from the St. Lawrence River, indicating that low levels of indicator organisms are not always a guarantee of the complete absence of pathogens.[43] *Salmonella* species, as well as total coliforms have been shown to be capable of survival in cold, subarctic waters for at least 7 d or longer, at temperatures near 0°C, and under total ice-cover.[40,41] Fecal coliform survival in these cold regions receiving waters can range between 2–6 times higher than for a similar wastewater discharge in a temperate climate.[41] In a study by Putz et al., it was found that the cold water temperatures slowed the microbial death rate and likely the predation by higher organisms, while ice cover eliminated the lethal effects of the sun's ultraviolet rays.[42] Given and Smith found that increased wastewater strength was positively correlated with increased bacterial survival in ozonated wastewater effluents from Whitehorse, Yukon Territory, Canada.[44] It was also found that the ability to detect *Salmonella* species in the final wastewater effluent did not show a definate relationship to the levels of fecal coliforms measured. *Salmonella enteridis* and *Salmonella montevideo* were the most frequently isolated species of *Salmonella* in this study.[44]

Many studies of survival and persistence of facultatively psychrophilic microorganisms in wastewater often do not consider the special laboratory conditions needed to resuscitate these adapted, and possibly injured, organisms. Routine regulatory monitoring does not consider this factor at all. Standard Methods for the Analysis of Water and Wastewater, 19th edition, and similar laboratory guidance manuals published in other countries, requires total coliform testing to be conducted at 35°C, and fecal coliform testing at 44.5°C.[45] If an organism, falling into one of these two groups, has adapted well to colder temperatures, the incubation temperature used in the laboratory may now be at the extreme upper temperature limit for growth. The laboratory results generated are directly dependent on the ability of the organism to quickly adapt to the new environment, and length of lag period of growth of the organism being studied. As a result, some reported levels may be artificially low due laboratory handling, and thus not truly representative of actual populations present.

Table 6 summarizes the effect of recovery media and incubation temperature on recovery of total heterotrophs from the Inuvik, NWT sewage lagoon.[46] Table 7 summarizes survival times in temperate waters for some commonly excreted human pathogens.[47] Many of the organisms listed in Table 7 are common agents of disease in northern regions, can be isolated from wastewater influents, as well as from

Table 6.   Effect of recovery medium and incubation temperature on microorganism recovery from the Inuvik, NWT sewage lagoon (after ref. 46)

Recovery medium	Incubation temperature	Organisms recovered (cfu ml^{-1})
Nutrient agar	55°C	2×10^2
Nutrient agar	37°C	5.1×10^4
Nutrient agar	22°C	1.9×10^5
Nutrient agar	2°C	19×10^4
Desoxycholate agar	37°C	3.8×10^3
Sabouraud agar[a]	22°C	1.0×10^1
Thornton agar[b]	22°C	7.4×10^4

Sewage samples were collected in August, with a mean lagoon temperature of 16°C
[a] Used for cultivation of yeasts and molds.
[b] Used for recovery of total heterotrophs (ref. 47).

Table 7.   Transfer of human pathogens in wastewater, via surface waters, to contaminated drinking water (after ref. 50)

Organism	Typical amount excreted by an infected adult (cfu g^{-1} feces)	Average survival in surface water[a] (d)	Typical infective dose for "normally healthy" adult[b] (cfu ingested)
Bacteria			
Toxigenic *E. coli*	10^8	90	10^9
Salmonella species	10^6	60 to 90	10^6 to 10^7
Shigella species	10^6	30	10^2
Campylobacter species	10^7	7	10^6
Vibrio species	10^5	30	10^8
Yersinia enterocolitica	10^5	90	10^8
Viruses			
Enterovirus	10^7	90	10^2
Hepatitis A	10^6	5 to 30	—
Rotavirus	10^6 to 10^9	5 to 30	<10
Norwalk virus	—	5 to 30	?
Protozoans			
Entamoeba	10^7	20 to 30	10 to 10^2
Giardia lamblia	10^5	20 to longer (?)	5 to 10^2
Cryptosporidium parvum		20 to longer (?)	
Helminths			
Ascaris	10^3	365 to longer	2 to 5
Taenia	10^3	?	

[a] Survival time may be increased/decreased by factors such as water temperature, ice cover, biological oxygen demand, hydraulic flow and others.
[b] Infectious dose found to cause clinical symptons in 50% of individuals tested.

Table 8. Commonly excreted pathogens in wastewater and the efficiency of lagoon removal (after ref. 7)

Type	Agent	Typical disease caused	Lagoon (% removal)
Virus	Coxsackii	Various	99.99 to 100
	Hepatitis A	Infectious hepatitis	
	Norwalk	Gastroenteritis	
	Poliovirus	Poliomyelitis	
	Rotavirus	Gastroenteritis	
Bacteria	*Campylobacter fetus*	Enteritis	
	Escherichia coli	Gastroenteritis	>99.99
	Salmonella species	Salmonellosis	
	Salmonella typhi	Typhoid fever	
	Shigella species	Shigellosis	99.99 to 100
	Vibrio cholerae	Cholera	100
	Yersinia enterocolitica	Gastroenteritis	
Protozoa	*Entamoeba histolytica*	Amebic dysentry	100
	Giardia lamblia	Giardiasis	
	Cryptosporidium	Cryptosporidiosis	
Helminths	*Ascaris lumbricoides*	Ascariosis	80 to 100
	Taenia sp	Tapeworm	
	Enterobius vermicularis	Pinworm	
	Diphyllobothrium latum	Broad or fish tapeworm of humans	

some wastewater effluents and receiving waters.[26,40,44,47] Table 8 illustrates some commonly excreted pathogens found in wastewaters entering cold regions lagoons, and the removal efficiencies.[7]

The apparent ability of indicator and pathogenic organisms to survive for extended periods of time in the receiving water becomes extremely important for management of drinking water supplies and public health in many northern communities. In the Canadian and Alaskan arctic and sub-arctic regions many smaller communities and aboriginal settlements derive their drinking water directly from streams, rivers, lakes and ponds, which may have been also used for discharge of treated wastewater.[48] In some cases the drinking water supply is not adequately treated prior to consumption.[48] These receiving waters also are used for recreational purposes in the summer, as well as, providing a source of food in the form of fish and in certain types of waters, shellfish. In addition, there may be significant reservoirs of opportunistic and direct pathogens harbored in wild animals consuming water from these sources.[49,50] Thus the efficacy of lagoon treatment is critical for the control of waterborne disease, as well as for the maintenance of surface water quality in cold regions.[7,8]

An additional concern with high levels of microorganisms in lagoon effluents is the presence of plasmid-transmitted antimicrobial resistance (R) factors, and the potential for transfer of properties such as antibiotic resistance, to indigenous organisms in receiving waters.[51,52] Gene transfer, in the presence of a disinfectant residual, has been demonstrated in water and may prove significant in the transfer of properties aiding the survival and persistence of pathogens.[53] Bell et al. found that approximately 7% of fecal coliforms isolated from the Slave River, NWT Canada, demonstrated antibiotic resistance to three or more antibiotics.[51,52] In contrast fecal coliforms isolated from the Red River in Manitoba, which receives wastewater effluents from the City of Winnipeg, approximately 53% of fecal coliforms isolated demonstrated resistance to multiple antibiotics.[51,52] This was believed to be a direct consequence of increased population size and as a result, antibiotic use, in larger communities.

Study of a long retention wastewater lagoon in Michigan, USA, demonstrated that levels of total coliforms do not always decrease significantly over the winter months, but remained stable at $1x10^7$ cfu 100 ml^{-1}.[54] In addition, as effluent passed from one cell to the next at this site, the levels of antibiotic resistant coliforms increased, with more than 60% of total coliforms in the final effluent discharged demonstrating resistance to one or more antibiotics.[54]

A survey of sewage applied to land in northern Canada (Fort Simpson, Norman Wells and Tuktoyaktuk, NWT) found that cysts/eggs of a number of intestinal protozoan parasites could be recovered, including *Diphyllobothrium latum*, *Entamoeba coli*, and *Enterobius vermicularis*.[55] Studies by Fitzgerald demonstrated that it was possible to have transmission of eggs of *Ascaris suum* to swine pastured on land that had been treated with anaerobically digested sewage sludge for a period of years.[56] Similar results were noted for the transmission of *Moniezia* eggs, eggs of trichostronylinate nematodes and *Trichinella spiralis* to cattle.[56] Conventional sewage treatment was unlikely to kill helminth eggs in any reasonable period of time, nor would conventional sewage treatment be very effective in the control of sewageborne parasitic infections.[57] This problem could be exacerbated in cold regions

Table 9. Wastewater treatment lagoon use in Canada (as of 1994; adapted from ref. 59)

Location	No. of lagoons	Wastewater treatment facilities (%)
Alberta	278	84
British Columbia	34	31
Manitoba	127	85
New Brunswick	58	63
Newfoundland	1	2
Northwest Territories	15	71
Nova Scotia	14	16
Ontario	128	33
Prince Edward Island	17	9
Quebec	59	21
Saskatchewan	129	92
Yukon Territory	8	70
Canadian Total[a]	868	4

[a] This table illustrates that lagoons are a significant method of wastewater treatment in Canada, especially in provinces with significant portions of the province or territory falling within the definition of "cold regions". This includes Alberta, British Columbia, Manitoba, Northwest Territories, Ontario, Quebec, Saskatchewan and Yukon Territory.

Table 10. Design criteria for selected Canadian cold regions lagoons (after ref. 60)

Province	Loading rate (kg BOD ha^{-1} d^{-})	Type of discharge	Detention time (d)	Type and no. of cells	Type of liner and minimum thickness	Cell depth (minimum)
Alberta	N/A	Controlled	Anaerobic 2 d Facultative 60 d Storage 365 d	0A-1F-1S,[a] flow <70 m^3 2A-1F-1S, flow <250 m^3 4A-1F-1S, flow >250 m^3	Natural *insitu* liners – 0.9 m Compacted clay – 0.6 m Meet seepage requirements	Anaerobic: 3 to 3.5 m Facultative: 1.5 m Storage: 0.5 m, max. 2.5 m
Saskatchewan	Primary cell: 30 kg	Controlled	Combined storage: 180 d	Minimum of 2	Meet seepage requirements	Treatment cell mimimum of 0.6 m, max. of 1.5 m Storage cell minimum of 0.3 m, maximum of 2.1 m
Manitoba	Primary cell: 44.6 kg	Controlled	Nov 1 to May 15: 197 d	Not specified	N/A	Minimum of 0.3 m, maximum of 1.5 m
Ontario	19.6 kg	Site specific requirements	Site specific requirements	Minimum of 2 of maximum size of 8 ha	Site specific	Treatment cell: 1.8 m Storage cell: 2.7 m
New Brunswick	35 kg 500 people ha^{-1}	Controlled, if ice cover expected	Sufficient for reduction in BOD	Minimum of 2	Meet seepage requirements	Treatment cell: 1.5 to 1.8 m Storage cell: 2.0 m Minimum depth: 0.6 m
Nova Scotia	22 kg 250 people ha^{-1}	Site specific requirements	Sufficient for reduction in BOD	Minimum of 2 Small community exception = 1 cell Maximum size of 5 ha	Minimum of 0.5 m compacted clay	Treatment cell: 1.5 to 1.8 m Storage cells: 2.0 m Minimum depth: 0.6 m
Newfoundland	20 to 30 kg	Site specific requirements	Controlled discharge: 180 d Flow-through discharge: 90 to 120 d	Minimum of 3 Small community exception = 2 cells	Meet seepage requirements	Minimum depth: 0.6 m Maximum depth: 1.8 m

[a] *A* anaerobic, *F* facultative, *S* storage cell.

Table 11. Wastewater effluent quality standards for the Northwest Territories, Canada (after ref. 61)

Wastewater flow (l person^{-1} d^{-1})	Regulatory parameter	Units	Stream, river or estuary dilution factor[a]		Lake residence time[b]		Marine mixing condition	
			>10:1 and <100:1	>100:1 and <1000:1	Treatment > 5yr	Treatment < 5yr	Open coastline	Bay or fjord
Summer								
<150	BOD	mg l^{-1}	30	80	30	80	360	100
	Suspended solids	mg l^{-1}	35	100	35	100	300	120
	Phosphorus	mg l^{-1}	10	—	—	—	—	—
	Fecal coliforms[c]	cfu 100 ml^{-1}	1x10^3	1x10^4	1x10^3	1x10^4	*	*
150–600	BOD	mg l^{-1}	30	40	30	40	120	120
	Suspended solids	mg l^{-1}	35	60	35	60	180	180
	Phosphorus	mg l^{-1}	9	—	—	—	—	—
	Fecal coliforms[c]	cfu 100 ml^{-1}	1x10^4	1x10^4	1x10^4	1x10^4	*	*
<150	BOD	mg l^{-1}	No discharge	Special permit	30	80	30	80
	Suspended solids	mg l^{-1}	No discharge	Special permit	35	100	35	100
	Phosphorus	mg l^{-1}	No discharge	Special permit	—	—	—	—
	Fecal coliforms[c]	cfu 100 ml^{-1}	No discharge	Special permit	1x10^3	1x10^4	1x10^3	1x10^4
150–600	BOD	mg l^{-1}	No discharge	Special permit	30	80	30	80
	Suspended solids	mg l^{-1}	No discharge	Special permit	35	100	35	100
	Phosphorus	mg l^{-1}	No discharge	Special permit	—	—	—	—
	Fecal coliforms[c]	cfu 100 ml^{-1}	No discharge	Special permit	1x10^3	1x10^4	1x10^3	1x10^4

[a] Dilution calculated by equation: Dilution factor = (minimum average monthly stream flow) + (average daily wastewater flow).

[b] Lake residence time calculated by equation: Residence time = (Volume of lake, m^3) + (Annual outflow from lake, m^3 yr^{-1}).

[c] Geometric mean of fecal coliform density outside the initial mixing zone should be ≤100 cfu ml^{-1}

* If marine bay/fjord areas are open and well flushed, then bacteriological standards are only of concern in areas used for fishing and shellfish harvesting, or for water contact sports.

lagoons, especially if retention time is inadequate. Any evaluation of health risk from cold regions lagoons, or other waste treatment processes, or land application of sludge is directly related to the levels and types of enteric disease(s) found in a given community. Types of human parasites most frequently found in Canada that are associated with sewage are: *Entamoeba histolytica, Giardia lamblia,* with lesser occurrences of *Balantidium coli, Taenia saginata, Ascaris, Trichuris trichiura, Strongyloides, Toxacara* species, *Enterobius vermicularis, Diphyllobothrium,* and *Toxoplasma gondii.*[58]

9
Standards for lagoon design, operation and effluent quality in selected Canadian cold regions areas

As an appendix, wastewater treatment lagoon use in Canada (Table 9), design criteria for cold regions lagoons (Table 10), and effluent quality standards for Canada (Table 11) are presented.

10
Conclusions

Lagoons can be an effective, and relatively inexpensive wastewater treatment alternative for communities in many cold regions of the world. If properly designed and managed, lagoons can provide significant reductions in total suspended solids, nutrients, and indicator and pathogenic microorganisms, so that the effluent discharged to the receiving water will not create adverse environmental impacts. Physical-chemical treatment processes, such as coagulation-flocculation or disinfection, can be used to enhance the final effluent quality.

Treatment processes in cold regions lagoons involve the metabolic activities of a consortium of indigenous and adapted wastewater microorganisms. Periods of ice cover change the lagoon ecology such that facultatively anaerobic and anaerobic organisms as well as facultatively psychrophilic and psychrophilic organisms are favored. Wastewater treatment during ice cover is minimal, and is comparable to primary treatment processes. Fall discharges are typically of better quality than spring discharge, as a result of the degree of treatment received in the lagoon environment.

Future challenges for cold regions lagoon and conventional wastewater treatment processes will need to focus on the efficacy of the treatment processes in reduction of encysted, wastewater pathogens such as *Giardia* and *Cryptosporidium,* as well as on better recovery and detection techniques for injured, but viable indicator and pathogenic microorganisms that may still be present in the final effluent.

11
References

1. Alter AJ. Cold region environmental concerns. Critical Rev Environ Control 1990; 20: 257-298.
2. Heinke GW, Smith DW, Finch GR. Guidelines for the planning and design of wastewater lagoon systems in cold climates. Can J Civil Engineering, 1991; 18:556-567.
3. Smith DW. Introduction — cold regions utilities. In: Smith DW, ed. Cold Regions Utilities Monograph, 3rd ed. New York, NY: American Society of Civil Engineers, 1996:1.1-1.6.
4. Smith DW, Crum JA. Wastewater treatment. In: Smith DW, ed. Cold Regions Utilities Monograph, 3rd ed. New York, NY: American Society of Civil Engineers, 1996:10.1-10.42.
5. Allum MO, Carl CD. The role of ponds in wastewater treatment. In: McKinney R, ed. Second International Symposium on Wastewater Treatment Lagoons. Kansas City: 1970:7-10.
6. Gloyna EF. Waste Stabilization Ponds. Geneva: World Health Organization, 1971.
7. Heinke GW, Smith DW, Finch GR. Guidelines for the planning, design, operation and maintenance of wastewater lagoon systems in the Northwest Territories, vol. 1 — Planning and design. Yellowknife, NWT: Department of Municipal and Community Affairs, Government of the Northwest Territories, 1988.
8. Fair GM. Sanitary Engineering in Polar Regions. Medicine and Public Health in the Arctic and Antarctic. Geneva: World Health Organization, 1963:116-137.
9. Smith DW, Finch GR. A Critical Evaluation of the Operation and Performance of Lagoons in Cold Climates. Ottawa: Report for Environment Canada, 1983.
10. Prince DS, Smith DW, Stanley SJ. Performance and factors affecting performance of lagoons in cold regions. In: Smith DW, Sego DC, eds. Cold Regions Engineering - A Global Perspective. Proceedings 7th International Cold Regions Engineering Specialty Conference. Edmonton: Canadian Society for Civil Engineering, 1994:539-556.
11. Prince D., Smith DW, Stanley SJ. Performance of lagoons experiencing seasonal ice cover. WEF 1995; 67:318-326.
12. Prince DS, Smith DW, Stanley SJ. Intermittent discharge lagoons for use in cold regions. J Cold Regions Engin, American Society of Civil Engineers 1995; 9:184-194.
13. Scher RL. Geotechnical considerations. In: Smith DW, ed. Cold Regions Utilities Monograph, 3rd ed. New York, NY. American Society of Civil Engineers, 1996:3.1-3.115.
14. Likens GE, Johnson PL. Limnological reconnaissance in interior Alaska. Hanover, New Hampshire. Cold Regions Research and Engineering Laboratory. Research Report 239, 1968.
15. Komex Consultants Ltd. Design and construction of liners for municipal wastewater stabilization ponds. Calgary, AB. Prepared for the Municipal Engineering Branch, Alberta Environment, 1983.
16. Marais G vR. Dynamic behaviour of oxidation ponds. In: McKinney RE, ed. 2nd International Symposium for Waste Treatment Lagoons. Kansas City: University of Kansas, 1970:15-45.
17. McKinney RE. Algal based wastewater treatment systems. Edmonton, AB. Conf Can Soc for Civil Eng, 1982:41.
18. Hartley WR, Weiss CM. Light intensity and vertical distribution of algae in tertiary oxidation ponds. Chapel Hill. ESE Publ. No. 199. School of Public Health, Univ of North Carolina, 1968.
19. Atlas RM, Bartha R. Microbial Ecology: Fundamentals and Applications, 3rd ed. Redwood City, CA: The Benjamin/Cummings Publishing Company Inc., 1993.
20. Dornbush JN. State of the art – anaerobic lagoons. In: McKinney RE, ed. 2nd International Symposium for Waste Treatment Lagoons. Kansas City, University of Kansas, 1970: 382-404.
21. Oswald WJ. Advances in anaerobic pond systems design. Advances in Water Quality Improvement, Water Quality Symposium No. 1, Austin, TX: Univ Texas Press, 1968.
22. McIntoch GH, McGeorge GG. Year round lagoon operation. Food Processing 1964; 25:82-86
23. Coerver JF. Anaerobic and aerobic ponds for packinghouse wastes treatment in Louisiana.

Proc 19th Ind Waste Conf Purdue Univ Ext Serv, 1964:117,200.

24. Schneiter RW. Cold Region Wastewater Lagoon Sludge: Accumulation, Characterization, and Digestion. PhD Thesis. Logan, Utah: Department of Engineering, Utah State University, 1982.

25. Dawson RN. Lagoon Sewage Treatment in the MacKenzie District, Northwest Territories. Edmonton, AB. Division of Public Health Engineering, Department of National Health and Welfare 1967.

26. Heinke GW. Disposal of human waste in northern areas. In: Some Problems of Solid and Liquid Waste Disposal in the Northern Environment. Ottawa: Environment Canada Report EPS4-NW-76-1, 1973:87-140.

27. Hrudey SE, Raniga S. Greywater management for isolated northern communities. New York. In: Proceedings, The Northern Community: A Search for a Quality Environment. American Society of Civil Engineers, 1981:471-481.

28. Sparling AB. Winter operation of sewage lagoons in Manitoba. Winnipeg, Dept of Health, Environmental Sanitation Section, 1967.

29. Slaughter R, Boland B, Thornley S. Quality of sewage lagoon discharges during winter and impact on stream water quality and biota. Canadian Hydrology Symposium — Cold Climate Hydrology Proceedings, 1979:547-558.

30. Hom LW. Chlorination of waste pond effluents. In: McKinney RE, ed. 2nd International Symposium for Waste Treatment Lagoons. Kansas City: University of Kansas, 1970:151-159.

31. Johnson BA, Wight JL, Bowles D, Reynolds JH, Middelbrooks EJ. Waste Stabilization Lagoon Microorganism Removal Efficiency and Effluent Disinfection with Chlorine. EPA-600/2-79-018. Cincinnati, OH. US Environmental Protection Agency, 1979:360.

32. Finch GR, Smith DW. Effect of Chemical Coagulation on the Removal of Fecal Coliforms from a Seasonal Discharge Sewage Lagoon. Environmental Engineering Technical Report 84-1, Edmonton, Department of Civil Engineering, University of Alberta, 1984:124.

33. Caldwell DH, Parker DS, Uhte WR, Stenquist RJ. Upgrading Lagoons. EPA Technology Transfer Seminar Publication. Cincinnati. US Environmental Protection Agency, 1977.

34. Middlebrooks EJ, Falkenborg DH, Lewis RF, Ehreth DJ. Upgrading Wastewater Stabilization Ponds to Meet New Discharge Standards. PRWG150-1. Logan, Utah. Utah Water Research Laboratory, Utah State University, 1974:244.

35. Tupyi B, Filip DS, Reynolds JH, Middlebrooks EJ. Separation of Algal Cells From Wastewater Lagoon Effluents, vol. II: Effect of Sand Size on the Performance of Intermittent Sand Filters. EPA-600/2-79-152. Cincinnati, OH: Municipal Environmental Research Laboratory, US Environmental Protection Agency, 1979.

36. Beier AG. Lagoon performance in Alberta. In: Smith DW, Tilsworth T, eds. Cold Regions Environmental Engineering: Proceedings of the Second International Conference, Edmonton. Kitchner, ON, TekTran International, 1987:239-249.

37. Klohn Leonoff Yukon Consulting Engineers, NovaTec Consultants Inc. Toxicity Assessment of Alberta Sewage Effluent Storage Lagoons. Final Report. Whitehorse, Yukon Territory, Canada. Yukon Community and Transportation Services, Whitehorse, 1991.

38. Clark SE, Coutts HJ, Christianson C. Biological Waste Treatment in the Far North. Report 1610-06/70. Alaska. Federal Water Quality Administration, Department of the Interior, Alaska Water Laboratory, 1970.

39. Slanetz LW, Bartley C, Metcalf TG, Nesman R. Survival of enteric bacteria and viruses in municipal sewage lagoons. In: McKinney RE, ed. 2nd International Symposium for Waste Treatment Lagoons. Kansas City: University of Kansas, 1970:132-141.

40. Van Donsel DJ, Gordon RC, Davenport CV. Preliminary Study of Winter Survival of Fecal Bacteria in a Subarctic River. US Environmental Protection Agency, EPA Working Paper No. 28, 1974.

41. Gordon RC. Winter Survival of Fecal Indicator Bacteria in a Sub-arctic Alaskan River. EPA-R2-72-013. Corvallis, OR. US Environmental Protection Agency, 1972.

42. Putz G, Smith DW, Gerard R. Microorganism survival in an ice-covered river. Can J Civ

Engin 1984; 11:177-186.

43. Dutka BJ, Bell JB. Isolation of Salmonella from moderately polluted waters. J Water Poll Control Fed 1973; 45:316.

44. Given PW, Smith DW. Disinfection of dilute, low-temperature wastewater using ozone. Ozone Sci Engin 1979; 1:91-106.

45. APHA, AWWA, WEF. Standard Methods for the Analysis of Water and Wastewater, 19th ed. Washington, DC, American Public Health Association, 1995.

46. Boyd WL, Boyd JW. Microbiological studies of the aquatic habitats of the area of Inuvik, Northwest Territories. Arctic 1967; 20: 27-41.

47. Allen ON. Experiments in Soil Bacteriology. Minneapolis, Minn. Burgess Publishing Co., 1951:117.

48. Facey RM, Smith DW. A comprehensive domestic water quality study in the Northwest Territories. Proceedings of the 7th International Specialty Conference on Cold Regions Engineering. Edmonton, AB. Canadian Society for Civil Engineering, 1994:461-480.

49. Geldreich EE. Waterborne pathogens. In: Mitchell R, ed. Water Pollution Microbiology. New York: John Wiley, 1972: 207-241.

50. Geldreich EE. Microbial water quality concerns for water supply use. Environ Toxicol Water Qual 1991; 6:209-223.

51. Bell JB, Macrae WR, Elliott GE. Incidence of R-factors in coliform, fecal coliform, and Salmonella populations of the Red River in Canada. Appl Environ Microbiol 1980; 40:486-491.

52. Bell JB, Macrae WR, Elliott GE. R-factors in coliform – fecal coliform sewage flora of the Prairies and the NWT. of Canada. Appl Environ Microbiol 1981; 42:204-210.

53. Lisle JT, Rose JB. Gene exchange in drinking water and biofilms by natural transformation. Water Sci Technol 1995; 31:41-46.

54. Kish AJ, Lampky JR. Survival incidence of antibiotic resistant coliforms in a lagoon system. J Water Poll Control Fed 1983; 55:506-512.

55. Bienek G. Parasitological Implications of Land Sewage Disposal in the Arctic and Subarctic. Edmonton, AB. Report for Environment Canada – Environmental Protection Service, 1976.

56. Fitzgerald PR. Natural transmission of parasites from anaerobically digested sludge to animals. In: Wallis PM, Lehmann DL, eds. Biological Health Risks of Sludge Disposal to Land in Cold Climates. Calgary, AB. Kananaskis Centre for Environmental Research: The University of Calgary Press, 1983:179-191.

57. Crewe W, Owen RR. The parasitiology of sewage sludge in cold climates, with special reference to the United Kingdom. In: Wallis PM, Lehmann DL, eds. Biological Health Risks of Sludge Disposal to Land in Cold Climates. Calgary, AB. 1983. Kananaskis Centre for Environmental Research: The University of Calgary Press, 1983:193-211.

58. Sekla L, Lehmann DL, Wallis PM. Parasites and the land application of sewage: A literature review and an assessment of the health risk in Canada. In: Wallis PM, Lehmann DL, eds. Biological Health Risks of Sludge Disposal to Land in Cold Climates. Calgary, AB. Kananaskis Centre for Environmental Research: The University of Calgary Press, 1983:231-240.

59. Smith DW. Workshop on sewage lagoon processes. Winnipeg, Manitoba, February, 1994.

60. Technical Services Branch, Environmental Protection Service. Sewage Lagoons in Cold Climates, EPS4/NR/1, Ottawa, Environment Canada, 1985.

61. Northwest Territories Water Board. Guidelines for the discharge of treated municipal wastewater in the Northwest Territories. Yellowknife, NWT: Government of the Northwest Territories, 1992.

Low temperature anaerobic treatment of swine manure

D. I. Massé

Dairy and Swine Research and Development Centre, Agriculture and Agri-Food Canada
P.O. Box 90, 2000 Route 108 East, Lennoxville, Québec, Canada J1M 1Z3

1
Introduction

Over the last 30 years or so, agriculture has changed from being subsistent to being very specialized and intensive. As a result, many agricultural practices have experienced major changes. These changes took place in order to increase the animal production but unfortunately not enough effort was made to develop adequate practices for sound animal manure management. Animal manure management practices are often detrimental to the environment and also represent a potential hazard to human and animal health.

Anaerobic treatment is a natural process that has several ecological benefits and great potential to treat animal manure. Conventional anaerobic digestion of animal manure in farm scale digesters was tried at several locations across Canada during 1975–1985 and found not to be successful for several reasons.[1] Digesters were designed to operate at mesophilic or thermophilic temperatures. Because of subfreezing winter temperatures in Canada, digesters operated during the winter used not only most of the gas they produced, but sometimes required supplementary heating to maintain the digester temperature. Anaerobic digesters were not cost effective because they were designed to produce electricity which made them capital intensive, and they were not practical for farm use because their control and maintenance required skilled operators, increased labor input, and sometimes changes in farm operational procedures. Digesters were unstable and difficult to control because they were pushed to the limit to achieve maximum gas production.

2
Treatment objectives

The farm industry will be interested in animal manure treatment only if the process can be achieved at low cost, is very stable, easy to operate, requires minimum skill and does not interfere with regular farm operations. Psychrophilic anaerobic digestion (PAD) at temperatures ranging between 5 and 25°C holds promise for success under Canada's cool climatic conditions compared to mesophilic and thermophilic anaerobic processes previously studied. The purpose of this chapter is to provide an overview of the feasibility of using PAD at 20°C in intermittently fed sequencing batch reactors (SBRs) to reduce the pollution potential, recover usable energy and reduce odors of swine manure slurry on both small and large farm operations.

3
Literature review

Anaerobic digestion of municipal waste water and animal manures at low (psychrophilic) temperature has been reported in previous studies by several authors.[2-13] These studies have demonstrated that PAD can be used successfully to produce methane from animal manure and other organic wastes. But there was a large variation in PAD process performance for unexplained reasons. The energy or fibre content of the diet of the animals and the presence of antibiotics or food additives were not indicated. Also, several reports did not provide information on the age of the manure or its characteristics. Most of the studies concentrated on biogas production while little consideration was given to odor reduction, waste stabilization or increase in availability of plant nutrients. Additional research was therefore necessary to evaluate precisely the feasibility of PAD in SBRs.

4
Process description

An SBR is a simple operating system. It consists of a tank where the following five consecutive steps take place: 1) fill; 2) react; 3) settle; 4) draw; and 5) idle. During the fill period the organic waste is loaded into the SBR. When the SBR is full, the react period starts and its length should be sufficient to meet the treatment objectives. During the settle period no mixing is provided and quiescent settling conditions prevail to allow treated liquid to be separated from the solids and to retain bacteria in the system. During the draw period the treated liquid is removed and finally the idle period offers the flexibility of coordinating simultaneous operation of two or more SBRs.

According to Harper and Suidan,[14] the microbial activities and ecosystem of anaerobic digestion are affected by the digester design as well as by the environmental and operational conditions. The SBR is highly suitable for the treatment of animal manure at ambient temperatures because it offers optimum conditions to retain a high concentration of slow growing microorganisms in the tank. Dague et al.[15] noted that with anaerobic SBR the food to microorganism ratio is high after the fill period and low just prior to the settling period. They indicated that the above operating conditions result in efficient bioflocculation and solids separation, and that with SBR the partial pressure of CO_2 is maintained in the reactor during the settling period. As a result no significant quantity of CO_2 is transferred to the head space. This reduces the suspension of particulates in the supernatant that occurs when CO_2 is transferred from the liquid to the gas phase. The long biomass retention time in the SBR may allow PAD to adapt to environmental changes such as temperature variations, changes in organic loading rate, and presence of inhibitory elements.

5
Experimental design

5.1
Operating strategy

For results to be applicable to farm conditions, laboratory tests should simulate as closely as possible the actual farm operation. At a typical farm, manure is generally removed from the barn 1–3 times a week. Therefore the SBR should be intermittently fed 1–3 times a week. The fill cycle should not be longer than a month in order to limit the volume of the SBR. The react period should be long enough to produce almost odorless effluent with reduced pollution potential and increased fertilizer value. For the PAD in SBR to be cost effective, it is very important that the operational cost is kept very low. The operation of SBR at ambient temperatures and the reduction or elimination of mechanical mixing would substantially reduce the energy input and increase the energy efficiency of the SBR because all the energy produced will be available for on-farm utilization.

Table 1 gives the operating conditions that were used to simulate current pig manure slurry managements at the farm and to minimize the interference of the SBR operation with regular farm activities.

5.2
Experimental setup and procedure

Experiments were carried out in laboratory scale digesters located in a controlled temperature room. All the tests were carried out at a temperature of 20°C. The SBRs were operated without mixing. Figure 1 is a schematic diagram of the bench scale SBRs and feeding system used in this study.

Manure slurry was obtained from gutters under a partially slatted floor in a growing-finishing barn at a commercial swine operation. The manure was up to 4 d old at the time of collection. Its composition was determined using standard methods[16] and is given in Table 2. SBRs were fed 3 times a week and the feed and react periods lasted 4 weeks each. Digesters 1, 2, 5 and 6 were initially started using

Table 1. Sequencing batch reactor operating conditions (after ref. 22)

Digester no.	Loading rate (g COD l^{-1} d^{-1})	Fill period (weeks)	React period (weeks)
1–2	0.72	4	4
3–4	0.72	4	4
5–6	1.20	4	4
7–8	1.20	4	4

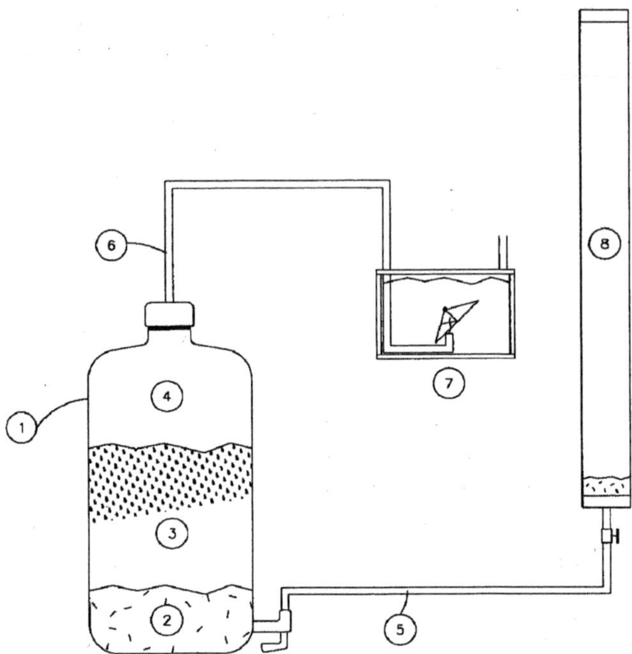

Fig. 1. Schematic of laboratory scale sequencing batch reactor (after ref. 22)
1 25 l nalgene digester; **2** sludge bed zone, 7.5 l; **3** variabale volume zone, 12 l; **4** head space zone, 5.5 l; **5** influent line; **6** gas outlet; **7** gas meter; **8** feeder tube

7.5 l of anaerobic granular sludge obtained from the Agropur Co-Operative Dairy anaerobic wastewater treatment plant at Notre-Dame du Bon Conseil, Quebec, Canada. Digesters 3, 4, 7 and 8 each received a mixture of sludge (5.9 l of Agropur Sludge and 1.6 l of anaerobic non-granulated sludge obtained from the Robert O. Pickard Environmental Centre, Ottawa, Ontario. The Agropur sludge substrate consisted mainly of fats and proteins. The anaerobic municipal sludge substrate came from both primary and secondary clarifiers. Some characteristics of the sludges are given in Table 3.

6
Process performance

6.1
Process stability

Figures 2 and 3 give the cumulative biogas production as a function of time, loading rate and inoculum type. The shapes of cumulative biogas production curves are similar for the four treatments considered. The rate of gas production was low during the fill period and increased during the react period. The lag phase in biogas

Table 2. Composition of swine manure slurry (after ref. 22)

Constituent	Concentration
Total solids (%)	4.8
Total suspended solids (%)	3.6
Volatile solids (%)	3.0
Volatile suspended solids (%)	2.6
Soluble COD (g l^{-1})	39
Total COD (g l^{-1})	84
Total Kjeldahl nitrogen (g l^{-1})	7.5
NH_4-N (g l^{-1})	5.8
pH	7.4
Alkalinity (g $CaCO_3$ l^{-1})	19.0
Acetic acid (g l^{-1})	6.3
Propionic acid (g l^{-1})	1.9
Butyric acid (g l^{-1})	2.5
Cellulose (% TS)	2.43
Hemicellulose (% TS)	4.15
Lignin (% TS)	1.31
Calcium (mg kg^{-1} TS)	54790
Copper (mg kg^{-1} TS)	957
Magnesium (mg kg^{-1} TS)	8643
Lead (mg kg^{-1} TS)	96
Potassium (mg kg^{-1} TS)	42760
Sodium (mg kg^{-1} TS)	13900
Zinc (mg kg^{-1} TS)	4470

TS total solids, *COD* chemical oxygen demand

production lasted about 40 d. It was probably due to the acclimatization of microorganisms to the decrease in operating temperature from 35 to 20°C and to the new substrate (swine manure). During the react period the biogas production rate increased exponentially until the end of the react period, when it started to decline as the availability of substrate became the limiting factor.

The digesters with combined sludge produced the highest amount of biogas. The cumulative biogas production was 30 and 70% higher in these digesters at organic loading rates of 0.72 and 1.2 g chemical oxygen demand (COD) l^{-1} d^{-1}, respectively. The higher biogas production may have been caused by an increased hydrolysis rate. The combined sludge contained anaerobic sludge from municipal wastewater treatment plants. This sludge is already acclimatized to compounds such as cellulose, hemicellulose and lignin. The Agropur sludge was only acclimatized to proteins and fats which are the major constituents of dairy wastewater. Another possible reason for the higher biogas production could be that the activity of the municipal sludge was higher than the activity of the Agropur sludge. Actual sludge activities were not measured in this study.

Figures 4, 5 and 6 illustrate the acetic, propionic and butyric acid concentrations as a function of time respectively. They also indicate the cumulative feeding concentration of each individual total volatile acid (VA). Figure 4 indicates that acetic acid accumulated rapidly from 0 to 5,500 mg l^{-1} during the fill period. This accu-

Table 3. Inocula characteristics (after ref. 22)

Constituent	Agropur sludge	Municipal sludge
Total solids (%)	11.0	2.6
Total suspended solids (%)	10.7	2.3
Volatile solids (%)	5.6	1.3
Volatile suspended solids (%)	5.4	1.2
Carbon (% VS)	48.41	55.9
Nitrogen (% VS)	9.64	8.4
Hydrogen (% VS)	7.54	10.6
Oxygen (% VS)	34.41	25.1
Soluble COD (g l^{-1})	10.0	3.0
Total COD (g l^{-1})	73.0	8.2
NH_4-N (g l^{-1})	1.3	1.0
Total Kjeldahl nitrogen (g l^{-1})	7.9	1.8
Cellulose (% TS)	0.70	0.84
Hemicellulose (% TS)	0.73	3.98
Lignin (% TS)	1.56	2.9
pH	7.6	7.3
Alkalinity (g $CaCO_3$ l^{-1})	16.0	6.0
Operating temperature (°C)	35.0	35.0
Sludge residence time (weeks)	26.0	2.0

TS total solids, *COD* chemical oxygen demand

mulation is about 4 times larger than the amount of acetic acid fed to the digesters. Figure 5 shows that propionic acid was accumulating faster in digesters with the Agropur sludge than in digesters with combined sludge during the fill period. For digesters with combined sludge the propionic acid accumulations were equal to the cumulative concentration fed. Figure 6 shows that butyric acid was not accumulating during the fill period, but rather was consumed because the concentrations of butyric acid in the digesters were substantially lower than the cumulative concentration fed.

The rapid increase in acetic acid concentration during the fill period shows that hydrolysis and acidification were occurring. It also indicates that during the fill period the utilization of acetic acid by the methane formers was the rate limiting step. The rapid increase in acetic acid is usually due to the faster growth rate of acid formers or inhibition of methane formers by an increase in concentration of VAs or other compounds. By comparing Figures 2 and 3 with Figure 4 it is obvious that methane formers were not inhibited by the increase in VAs concentration. These figures indicate that during the period of increased VAs concentration the methane production rate is also increased. Therefore the increase in VAs is more probably due to the faster growth rate of acid formers. The large increase in VAs did not affect the process stability. This was because the alkalinity in the SBRs was very high (16,000 mg $CaCO_3$ l^{-1}, the large increase in VAs caused only a small drop in pH); furthermore, the pH was maintained between 7.5 and 7.8, the unionized VAs concentration was always low ($\leq$ 6 mg l^{-1}). The inoculum type did not have much effect on acetic acid concentrations in SBR. The SBR with the combined sludge inoculum

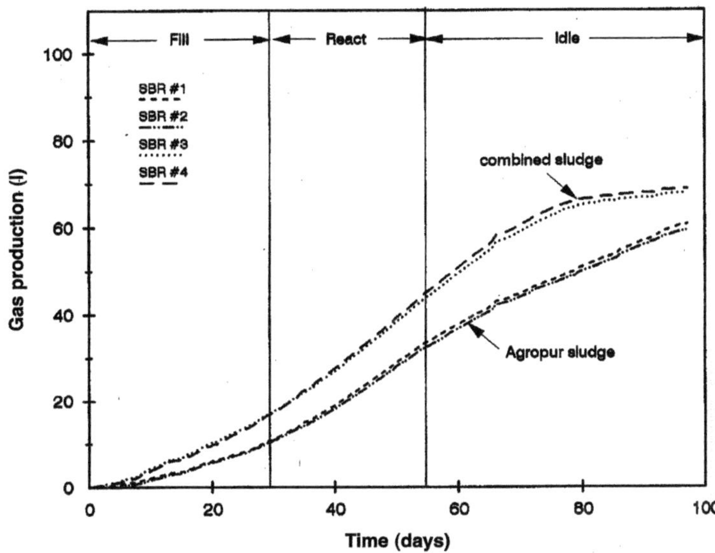

Fig. 2. Cumulative biogas production as a function of time for sequencing batch reactors (SBR) with an organic loading rate of 0.72 g COD l^{-1} d^{-1} (after ref. 22)

had higher CH_4 production and lower propionic and butyric acid concentrations at any time. This indicates that SBRs were more stable with combined sludge than with Agropur sludge.

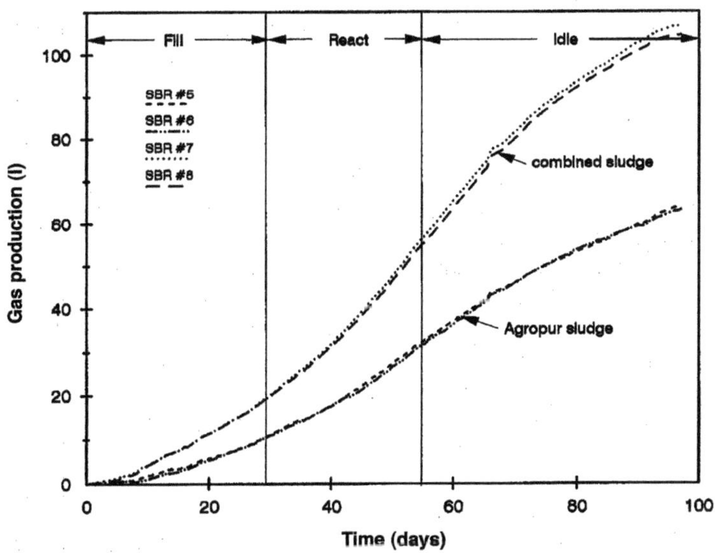

Fig. 3. Cumulative biogas production as a function of time for sequencing batch reactors (SBR) with an organic loading rate of 1.20 g COD l^{-1} d^{-1} (after ref. 22)

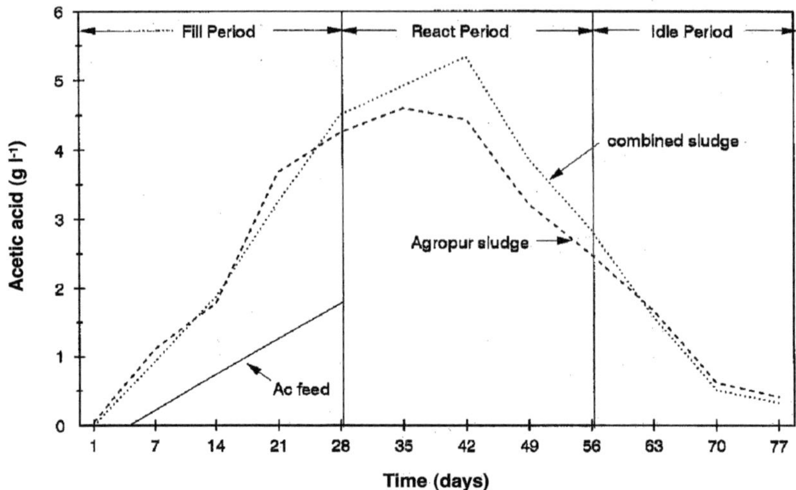

Fig. 4. Average acetic acid concentration as a function of time for sequencing batch reactors (SBR) with a loading rate of 1.20 g COD l^{-1} d^{-1} in SBRs 5–6 (······) and SBRs 7–8 (——) (after ref. 22)

Figure 7 gives the ammonia concentrations as a function of time for the SBR with an organic loading rate of 1.2 g COD l^{-1} d^{-1}. A similar curve was obtained at the lower organic loading rate of 0.7 g COD l^{-1} d^{-1}. The high concentration of ammonia nitrogen did not inhibit the methane formers in the SBR, because both the methane production and the ammonia-nitrogen concentration increased concomitantly. Kroecker et al.[17] found that ammonia is inhibitory to the methanogenic

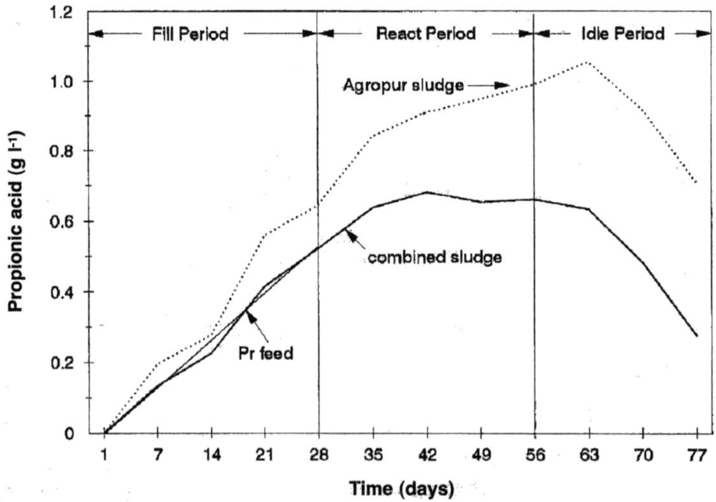

Fig. 5. Average propionic acid concentration as a function of time for sequencing batch reactors (SBR) with a loading rate of 1.20 g COD l^{-1} d^{-1} in SBRs 5–6 (······) and SBRs 7–8 (——) (after ref. 22)

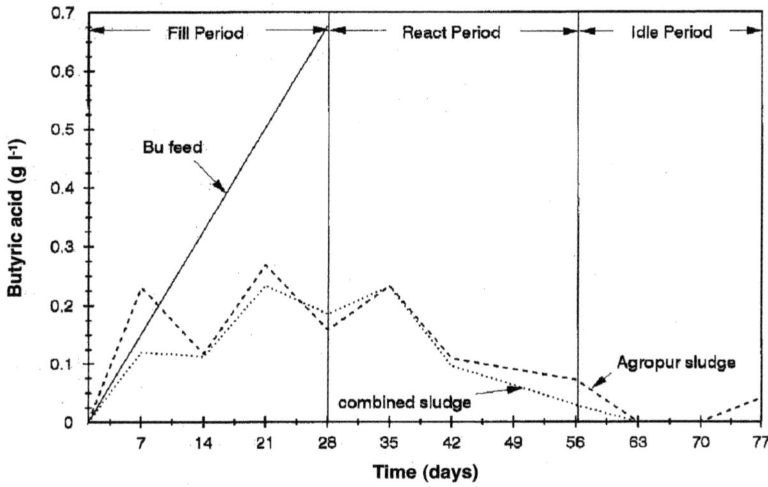

Fig. 6. Average butyric acid concentration as a function of time for sequencing batch reactors (SBR) with a loading rate of 1.20 g COD l⁻¹ d⁻¹ in SBRs 5–6 (⋯⋯) and SBRs 7–8 (——) (after ref. 22)

bacteria when its concentration exceeds 2,000 mg l⁻¹. Melbinger and Donnellon[18] found that ammonia is toxic only when its concentration exceeds the threshold limit of 1,700–1,800 mg l⁻¹ and is increasing faster than the acclimatization of the methanogenic bacteria. An ammonia nitrogen concentration exceeding 3,000 mg l⁻¹ is toxic to the anaerobic bacteria regardless of pH.[19] Dissolved ammonia gas is substantially more toxic than ammonium ions to anaerobic bacteria; a dissolved ammonia gas concentration ranging between 100 and 200 mg l⁻¹ should have an

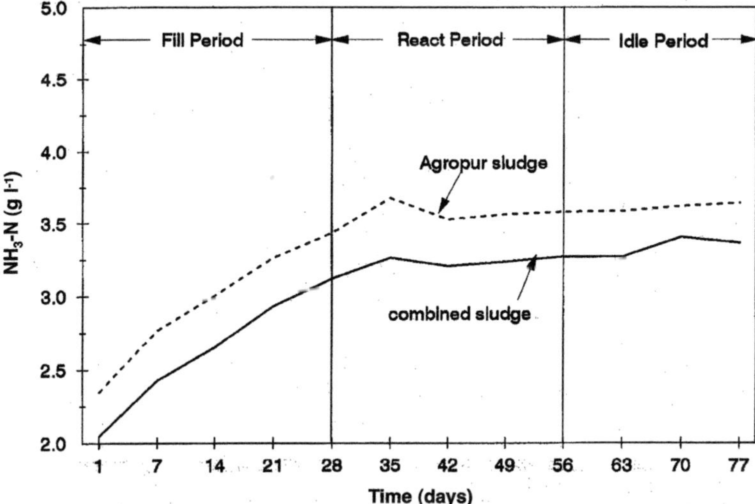

Fig. 7. NH₃-N concentration as a function of time for for sequencing batch reactors (SBR) with a loading rate of 1.20 g COD l⁻¹ d⁻¹ in SBRs 5–6 (⋯⋯) and SBRs 7–8 (——) (after ref. 22)

inhibitory effect on the anaerobic process. In this test, the total ammonia nitrogen concentration (3,700 mg l^{-1}) represents the sum of ammonium ions (3,550 mg NH_4^+ l^{-1}) and dissolved ammonia gas (150 mg NH_3 l^{-1}).

Inhibition by ammonia-nitrogen was not observed in this work. This is likely due to the long hydraulic and solids residence times provided in this test, and allowed the microorganisms to increase their tolerance to high concentrations of ammonia-nitrogen. PAD in SBR appears to be suitable to treat wastewaters with high nitrogen content.

6.2
Energy recovery

The biogas produced in four test runs was of high quality with a methane concentration between 75 and 80%. Table 4 gives the methane production as a function of unit mass of volatile solids (VS) fed to the digester. The CH_4 production ranged from 0.30 to 0.66 l g^{-1} VS for most of the experimental runs. Methane productions obtained in this study were higher than methane production from swine manure obtained by digestion at 35°C in continuous flow digesters by Kroecker et al.[17] who reported methane production of 0.45 l CH_4 g^{-1} VS added for a loading rate of 2.5 kg VS m^{-3} d^{-1}, and by Hashimoto[21] who reported 0.42 l CH_4 g^{-1} VS added for a loading rate of 2.5 kg VS m^{-3} d^{-1}. The higher methane production g^{-1} VS fed to the SBRs obtained in this study could be due to the lower organic loading rate and longer HRT. Other possible reasons could be the lower operating temperature and the absence of mixing; higher concentrations of hydrogen and carbon dioxide maintained in the liquid phase. As a result more carbon dioxide can be converted to methane by the hydrogen utilizing methanogens. Also, with the continuous flow anaerobic processes previously experimented, some CO_2, H_2 and CH_4 were lost with the digester effluents.

A high rate of methane production was not the main objective of this work, but it is very useful to assess the system performance and stability. The steady production of methane per unit mass of VS fed indicates that the anaerobic digestion of swine manure at 20°C in the laboratory-scale SBR digesters was a stable process. Other interesting findings were that PAD in SBR process does not require mixing. Because of this and that it operates at ambient temperature, this process might not require added energy and most of the energy produced should be available for utilization on the farm.

6.3
Treatment efficiency

Table 4 gives the average level of removal of total chemical oxygen demand (TCOD), soluble chemical oxygen demand (SCOD) and VS of two replicate tests for all runs. The total COD removal ranged from 58 to 73% and the VS removal ranged from 27 to 74%. Results for VS and total COD were highly variable due to sampling variation caused by rapid settling of heavy particulates. Some samples had less solids than others. This affected the VS and TCOD determination as well

Table 4. Average methane production per unit of volatile solids (VS) fed to the digesters (anaerobic sequencing batch reactors, ASBR) and reduction in total chemical oxygen demand (COD), soluble chemical oxygen demand (SCOD) and volatile solids (after ref. 22)

ASBR no.	Loading rate ($g\ COD\ l^{-1}\ d^{-1}$)	Fill period (d)	React period (d)	CH_4 production ($l\ g^{-1}\ VS$ after 56 d)	Removal (%)		
					TCOD	SCOD	VS
1–2	0.72	4	4	0.50	60	90	29
3–4	0.72	4	4	0.66	70	96	74
5–6	1.20	4	4	0.30	58	85	27
7–8	1.20	4	4	0.52	73	91	56

as the calculated methane production g^{-1} VS. The soluble COD test results were consistent. High SCOD removal was achieved during most of the experimental runs. Its removal ranged from 85 to 96%.

Reduction in swine manure slurry odors was one of the objectives of this study. The major volatile compounds that produce odors in animal manure slurries are VAs, amines, carbonyls, esters, hydrogen sulphide and ammonia. In this study, test runs, that achieved complete removal of VAs and 85–96% removal of SCOD, produced treated manure that was relatively odorless compared to raw manure. A large reduction in SCOD may result in complete utilization of amines, carbonyls and esters.

The anaerobic sludge had excellent settling characteristics. A clear interface between the liquid and sludge bed zones started to occur near the end of the react period when the biogas production was very low and the demarcation between the liquid and solids was even more evident. A thick layer of sludge was observable at the bottom of the digesters. The SBR provides excellent settling conditions to retain the slow growing microorganisms.

7

Conclusions

The primary objective of this study was to evaluate the feasibility of PAD in SBR to treat swine manure slurry in order to reduce its pollution potential and odor and to recover energy. Experiments were carried out in eight 25 l SBRs maintained at 20°C under different operating conditions. Experimental results indicated that PAD of swine manure slurry at 20°C in intermittently fed SBR reduced the pollution potential of swine manure slurry by removing 85–95% of the soluble COD; produced important quantities of biogas (0.48–0.66 l CH_4 g^{-1}VS fed); and was very successful to remove odors, the treated manure was almost odorless when compared to raw manure. PAD of swine manure slurry in SBR was found very stable. It was never affected by large concentrations of VAs and ammonia nitrogen even when their concentrations exceeded 6,500 and 3,500 mg l^{-1}, respectively. The experimental results also indicated that this process does not need continuous feeding. This process will therefore minimize the interference with regular farm operations because it can be fed during regular manure removal from the barn, and also the farmer will deal with the digester effluent only once a month or every 2 months. Based on this study, PAD in SBR represent a promising process to treat swine manure slurry on small and large farm operations.

8
References

1. Van Die P. An Assessment of Agriculture Canada's Anaerobic Digestion Program. Engineering and Statistical Research Centre. Contribution No. I-933, Agriculture and Agri-Food Canada, Ottawa, Ontario, K1A 0C6, 1987.
2. O'Rourke JT. Kinetics of anaerobic waste treatment at reduced temperature. PhD thesis, Stanford University, California, US, 1968.
3. Maly J, Fadrus H. Influence of temperature on anaerobic digestion. J Water Poll Control Fed 1971; 43:641-650.
4. Stevens MA, Schulte DD. Low temperature anaerobic digestion of swine manure. Am Soc Agric Engin, Paper 77-1013, St. Joseph, MI, 1977.
5. Wellinger A, Kaufmann R. Psychrophilic methane production from pig manure. Process Biochem 1982; 17:26-30.
6. Chandler JA, Hermes SK, Smith KD. A low cost 75 kW covered lagoon biogas system. In: Proceedings of the Symposium on Energy from Biomass and Waste VII, 1983:627-646.
7. Cullimore RR, Maule A, Mansui N. Ambient temperature methanogenesis from pig manure waste lagoons. Thermal gradient incubator studies. Agric Waste 1985; 12:147-157.
8. Lo KV, Liao PH. Psychrophilic anaerobic digestion of screened dairy manure. Energy Agric 1986; 5:339-345.
9. Sutter K, Wellinger A. ACF-System: A new low temperature biogas digester. In: Szabokcs I, ed. Agricultural Waste Management and Environmental Protection. Proceedings of the 4th International CIEC Symposium. Braunschweig-Volkenrode, Germany: International Scientific Centre of Fertilizers, 1987.
10. Balsari P, Bozza E. Fertilizers and biogas recovery installation in a slurry lagoon. In: Welte E, Szabokcs I, eds. Proceedings of the 5th International CIEC Symposium. Balantonfured, Hungary: International Scientific Centre of Fertilizers, 1988:71-80.
11. Zeeman G, Sutter K, Vens T, Koster M, Wellinger A. Psychrophilic digestion of dairy cattle and pig manure: Start-up procedure of batch, fed-batch and CSTR-type digester. Biol Wastes 1988; 26:15-31.
12. Safley LM, Westerman PW. Performance of a dairy manure anaerobic lagoon. Bioresource Technol 1992; 42:43-52.
13. Safley LM, Westerman PW. Low temperature digestion of dairy and swine manure. Bioresource Technol 1994; 47:165-171.
14. Harper SR, Suidan MT. Anaerobic treatment kinetics. Water Sci Technol 1991; 24:61-78.
15. Dague RR, Habben CE, Pidaparti SR. Initial studies on the anaerobic sequencing batch reactor. Water Sci Technol 1992; 26:2429-2432.
16. APHA. Standard Method for the Examination of Water and Wastewater, 18th ed. Washington, DC: American Public Health Association, 1992.
17. Kroeker E.J, Schulte DD, Sparling AB, Lapp HM. Anaerobic treatment process stability. J Water Poll Control Fed 1979; 51:718-27.
18. Melbinger NR, Donnellon J. Toxic effect of ammonia nitrogen in high rate digestion. J Water Poll Control Fed 1971; 43:1658-1670.
19. McCarty PL. Anaerobic waste treatment fundamentals, Part III, toxic material and their control, process design. Public Works J for October 1964:91-94.
20. Henze M, Harremoes P. Anaerobic treatment of wastewater in fixed film reactors — a literature review. Water Sci Technol 1983; 15:1-101.

21. Hashimoto AG. Thermophilic and mesophilic anaerobic fermentation of swine manure. Agric Wastes 1983; 6:175-191.
22. Massé DI, Droste RL, Kennedy KJ, Patni NK. Potential for the psychrophilic anaerobic treatment of swine manure slurry in sequencing batch reactors. Can J Agric Engin 1997; 39:25-33.

Biodegradation of organic pollutants at low temperatures

R. Margesin* and F. Schinner

Institute of Microbiology (N.F.), University of Innsbruck, Technikerstrasse 25, A-6020 Innsbruck, Austria

1
Introduction

Organic pollutants appear in all climates. The range of contaminating compounds is extremely wide and includes industrial wastes, military wastes, domestic wastes and continuous atmospheric precipitation.[1-3] There is an increasing demand for the clean-up of contaminated sites. Cleaning procedures include chemical, physical (for a review see ref. 4) and biological treatments[2,3] of contaminated soils and ground and surface waters; the biological treatment became a valuable alternative with the increasing attention towards the preservation of the environment.

When organic pollutants enter an ecosystem, they are exposed to a number of physical, chemical and biological factors which contribute to loss or alteration of some of the components:[5] evaporation of volatile compounds; chemical transformations; sorption of the contaminants to colloids such as clay minerals or humus particles in soils or to detritus in aqueous systems; photochemical reactions in surface slicks and in water; formation of emulsions in marine systems due to wind and wave action; microbial metabolization and cometabilization of some of the organic compounds. The prior pollution history of the ecosystem is also of importance, since chronically polluted systems are generally enriched in hydrocarbon-utilizing organisms.[5,6]

Already in 1946, ZoBell[7] reported that a broad spectrum of microorganisms is able to utilize hydrocarbons as the sole source of carbon and energy and that such microorganisms are widely distributed in nature where contaminants may serve as organic carbon sources. The intensity of biodegradation is influenced by several environmental factors such as involved microorganisms, nutrient availability, oxygen, pH, temperature, water content, quality, quantity and bioavailability of contaminants; additional factors in soils are soil characteristics. In many contaminated areas the conditions for microbial degradation may be distinctly limited. In order to accelerate natural biodegradation, these limiting conditions can be optimized by using bioremediation techniques. Bioremediation has been defined as "the use of

---

* Corresponding author

living organisms to reduce or eliminate environmental hazards resulting from accumulation of toxic chemicals and other hazardous wastes"[8] and is a cost-effective, versatile and ecologically acceptable alternative to treat many contaminated sites.[2,3,9]

Bioremediation may involve biostimulation of the indigenous microorganisms by nutrient addition or manipulation of the contaminated media using techniques such as aeration, pH and temperature control.[10] The input of large quantities of organic carbon sources tends to result in a rapid depletion of available inorganic nutrients such as nitrogen and phosphorus. Consequently, rates of biodegradation are limited. Nutrient-amendment considerably increased biodegradation rates in both marine and terrestrial ecosystems.[2,3,11]

Another bioremediation option is the bioaugmentation of contaminated areas with microbial degraders. This method has been tested in both aquatic[12,13] and terrestrial[14,15] environments. It has been repeatedly reported that, regardless of temperature, inoculation had hardly any positive effect on biodegradation rates.[1,11,12] Usually the contaminated environments harbor enough hydrocarbon-degrading microorganisms that are able to metabolize the contaminants more effectively than any introduced microorganisms. The inocula will be replaced by the indigenous biodegraders with time. The bioaugmentation of contaminated soils frequently requires a short-term approach, since, even under best conditions, the introduced organisms will not survive for extensive periods.[16]

One important factor governing microbial activity, and therefore also biodegradation, is temperature. Temperature affects the rates of biodegradation by its effects on the microbial metabolism (Q_{10} effect) and on the physical nature and chemical composition of the contaminants.[2] The effects of temperature are interactive with other factors, such as the quality and quantity of the hydrocarbon mixture.[6] Most studies on the biodegradation of organic pollutants have been conducted at temperatures between 20 and 35°C – therefore biological decontamination under mesophilic conditions is well known – although actual environmental temperatures are usually much lower. Little is known about biodegradation processes in cold environments. In permanently cold climatic regions, but also in subsoils and groundwaters of temperate climates, average temperatures rarely exceed 10°C. Since the optimum process temperature of mesophilic microbial systems, which are normally used to treat wastewaters and contaminated soils, is 30°C, the degradation rate at such temperatures can be severely limited. The use of cold-adapted microorganisms, which have adapted their metabolism to low temperatures[17] and are able to grow and degrade organics over the entire range of seasonal temperature fluctuations, might be adavantageous. Especially the broad growth range of psychrotrophs (0 to >20°C)[18] coincides with the ambient temperature often found in polluted environments. Therefore, these microorganisms may play significant roles in the *in-situ* decontamination and in bioremediation processes at low temperatures.[17,19–23]

2
Biodegradation of petroleum hydrocarbons at low temperatures

Oil spills represent a widespread problem and are caused by tanker accidents, storage tank ruptures, pipeline leaks and transport accidents.[1–3] Petroleum hydrocarbons are a complex mixture[6] of alkanes, cycloalkanes, aromatic hydrcarbons and asphaltic components. The recognition of oil as a largely biodegradable mixture of hydrocarbons and the knowledge that microbial degraders can be enriched in most types of environments has contributed greatly to the development of bioremediation methods for petroleum-contaminated environments.[1,2,6] Hydrocarbon-utilizers comprise less than 0.1% of the microbial community in unpolluted environments but can constitute up to 100% of the viable microorganisms in oil-polluted ecosystems.[6] In most environments, enrichments of oil-degrading microbial communities occur soon after oil contamination.

The biodegradation of petroleum hydrocarbons at low temperatures has been reported in soil, water and marine systems; some examples will be reported below. Studies were conducted on the fate of oil hydrocarbons in arctic,[10,24–27] antarctic[28–30] and alpine[11,14] environments. Some authors discussed the potential use of bioremediation and addressed the effect of temperature on hydrocarbon biodegradation in these systems. At low temperatures, the viscosity of oil increases, the volatilization of toxic short-chain alkanes is reduced and their water solubility is increased, thus delaying the onset of biodegradation. Low winter temperatures can limit rates of hydrocarbon biodegradation, increasing the residence time of oil pollutants.[2] Besides affecting the rate of oil-degradation, changes in temperature may also indirectly alter the pattern of hydrocarbon utilization; under cold conditions the precipitation from crude oil of certain alkanes, such as waxes, would greatly diminish their availability to oil-degrading organisms.[31]

Changes in temperature may affect the utilization of substrates within a hydrocarbon mixture. Fedorak and Westlake[32] challenged marine water samples from three environments with Prudhoe crude oil; after 27 d at 8°C, both the aromatic and saturate fractions were extensively degraded by the indigenous microorganisms when supplemented with nitrogen and phosphorus. Simple aromatics were more readily degraded than the n-alkanes. Without nutrients, the aromatics were more readily attacked than the saturates. Oil quality and temperature of incubation affected the generic composition of crude oil utilizing bacteria. The isoprenoids, phytane and pristane, while readily used at 30°C, were more resistant to bacterial metabolism at 4°C.[33] Similarly, the aromatic fractions of some crude oils were metabolized at 30°C, but not at 4°C.[13,34] Degradation of crude oil was greater at 3°C than at 22°C, with mixed microbial cultures in beach sand samples.[35]

The capability of psychrotrophic biodegraders to mineralize petroleum hydrocarbons over a broad temperature range – e.g., at 5 and 23–25°C[21,36] or in the range of 4 to 30°C[37] – indicates their potential for the *in-situ* decontamination of oil-contaminated sites. A study on the biodegradation of four northern and mid-arctic crude oils under psychrophilic (4°C) and mesophilic (30°C) conditions demonstrated that the utilization of oil was only slightly faster at 30 than at 4°C. Populations obtained at 4°C metabolized crude oils also under mesophilic conditions, whereas mesophilic populations showed only a limited degradation capability under psychrophilic conditions.[33]

2.1
Terrestrial ecosystems

Evidence for biodegradation of petroleum hydrocarbons was found in several cold terrestrial ecosystems.

Westlake and al.[12] monitored the *in-situ* degradation of oil in a soil in Northern Canada for over 3 years. Where fertilizer (urea phosphate, 27:27:0) was applied, there was a rapid increase in bacterial numbers, followed by a rapid disappearance of n-alkanes, isoprenoids, and a continuous loss in weight of saturate compounds in recovered oil. The inoculation of oil-degrading bacteria did not have any effect on the composition of recovered oil, the rate at which chemical changes in recovered oil occurred was not accelerated; this was attributed to the presence of indigenous oil-degrading bacteria in these soils.[12]

Between 1976 and 1978, during the years of operation of the Naval Arctic Research Laboratory near Barrow, Alaska, accidental release of approximately 1,300 m^3 of various types of fuel (gasoline, diesel fuel, jet fuel) occurred in an area adjacent to a drinking water source. Substantial contamination of soils and groundwater near the lake still persists.[27] Groundwater flux in this area is small due to shallow permafrost, which restricts the cross-sectional area available for flow, and to the short annual thaw season which is about 90 d. Permafrost in this region is continuous and extends from near to the surface to depths of up to 300 m. Mean monthly air temperatures in the study area were above freezing only during the months of June–August. These conditions favor the persistence of pure-phase hydrocarbons and high dissolved concentrations.[26] The indigenous microbial community degraded petroleum hydrocarbons also under such severe conditions. Field measurements of dissolved oxygen, nitrate, ferrous iron, and sulfide in groundwater provided evidence that biodegradation of petroleum hydrocarbons is occurring *in-situ*. Even in 1996, nearly 20 years after major spills, the population of hydrocarbon degraders (determined at 10°C) was still higher at contaminated sites.[26]

Diesel oil, a distillate fraction of crude oil, is one of the major pollutants of soils and groundwater near petrol stations. The presence of cold-adapted diesel-oil degrading microorganisms was demonstrated in alpine soils where temperatures above 10°C are reached only during the period of high solar irradiation and on hot summer days.[15,38] An evaluation of 29 uncontaminated and contaminated alpine habitats (soils and glaciers) showed that not only oil-polluted samples but also samples from uncontaminated sites contained a remarkable potential of indigenous microorganisms that degraded diesel oil efficiently at 4 and 10°C.[39] The addition of inorganic nutrients (C:N:P=100:10:2) enhanced biodegradation significantly in both experimentally and chronically contaminated alpine subsoils.[14,15] Comparable results were obtained in a field experiment in an alpine glacier ski resort in 3,000 m a.s.l.[11] The decontamination of alpine subsoils, obtained in laboratory experiments at 10°C, was comparable to those reported at 25–30°C with soils from temperate climate.[40–42]

Biostimulation enhanced diesel-oil biodegradation in alpine soils to a significantly greater degree than bioaugmentation with a cold-adapted hydrocarbon-degrading inoculum.[14,38] Inoculation resulted only in a small increase (5–7%) of the hydrocarbon loss in five unfertilized alpine subsoils, bioaugmentation of fertilized soils was without any effect.[14] All investigated soils harbored enough hydrocarbon-degrading microorganisms that were able to metabolize diesel oil at low temperature more effectively than introduced oil-degraders. Studies on the degradation behavior of a cold-adapted oil-degrading yeast *Yarrowia lipolytica* in liquid culture and in soil demonstrated that the success of bioaugmentation cannot be predicted from liquid culture experiments.[37]

In Antarctica the annual average temperature is -18°C and averages -3°C for the summer month of January. Despite these cold and oligotrophic conditions, high densities of bacterial hydrocarbon degraders were reported in antarctic soils (McMurdo Station), contaminated with fuel oil at 2°C[29] and at 4°C[43]; active petroleum degraders were also found in soils and sediments at an average temperature of -2°C.[44]

In antarctic fertilized mineral soils, artificially contaminated with distillate, significant petroleum degradation was recorded only in plots that had been treated with fertilizer solutions containing nitrogen, phosphorus and potassium (C:N=61:1, C:P=607:1). The hydrocarbon content was reduced from 20,000 to less than 7,000 mg hydrocarbons per kg soil within one year; the fertilized soils had the highest level of microbial activity relative to untreated plots.[28]

In landfarming, contaminated soils or sludges are spread over an impermeable liner, nutrients and/or water can easily be added to the system, as well as any necessary chemical additions to regulate pH.[45] Landfarming was found to be a suitable method for the disposal of oily sludge under South-Norwegian conditions (the average air temperature at 1 p.m. was below 6°C for 5 months of the year).[46] In plots with the highest nutrient addition (600 kg N ha^{-1} year^{-1}) about 80% of the oil was degraded in 33 months. The optimum temperature for oil degradation was about 18°C, approximately 2/3 of the optimum activity was retained at 12°C; minimal degradation was observed at or below 6°C. Studies on the application of land farming and bioremediation in Fairbanks, Alaska, and at McMurdo Station, Antarctica, demonstrated that the extremely low temperatures and consequently low availability of water required special attention to design.[45]

2.2
Marine ecosystems

The rates of biodegradation in marine systems are generally lower compared with the speed and ease of mechanical removal, and are strongly dependent on whether the beaches are sandy or rocky;[1] nutrients are one of the major factors limiting biodegradation.

Cold-adapted petroleum-degrading microorganisms were isolated from water and sediment samples of two areas of Chesapeake Bay in the mid-Atlantic at environmental temperatures of 0, 5 and 10°C.[47] In this region, the water temperature is ≤ 10°C for 6 months of the year. Using a model petroleum incubated with estuarine water from this area collected during winter, when the temperature was 0.7–11.2°C, Walker and Colwell[47] found, that slower but more extensive biodegradation occurred at 0°C than at higher temperatures (5 and 10°C) and assumed a decreased toxicity at lower temperatures, as a consequence of the lower solubility of some hydrocarbons.

In oil-polluted sea ice, in which the mechanical removal of oil and other cleaning technologies are not feasible, bioremediation is the only alternative. Delille et al. investigated the long-term effects of artificial diesel fuel and crude oil contamination on microbial communities in land-fast ice in the Terre Adelie area in Antarctica.[30] During the first weeks of contamination, inhibiting effects of the water-soluble oil fraction on psychrophilic and psychrotrophic bacteria were observed in interface seawater. An enrichment in oil degraders was generally observed from less than 0.001% of the microbial community in uncontaminated samples to 10% after 30 weeks of contamination. Fertilization with Inipol EAP22 enhanced both saprophytic and hydrocarbon-utilizing communities. The natural winter decline of saprophytic bacterial abundance was totally absent in sea ice treated with diesel fuel and fertilizer.

In August 1974, the tanker Metula grounded in the Straits of Magellan; approximately 46,000 metric tons of oil contaminated the cold marine environment.[6] Oil-biodegradation under *in-situ* conditions proceeded slowly[35] due to the low concentrations of nitrogen and phosphorus available in seawater, as well as to restricted accessibility to degradable compounds within aggregated oils. Temperature did not seem to be a limiting factor for petroleum degradation by the indigenous microorganisms in this cold marine environment. Biodegradation of crude oil was greater at 3°C than at 22°C with mixed microbial cultures in beach sand samples.[35] Microbial action may have contributed significantly to the formation of polar material and contributed to the extensive removal of aliphatic hydrocarbons. Microbial degradation was not effective in attacking buried oil or oil that had formed asphalt layers on beaches.[6]

In March 1977, about 406,600 litres of Bunker C fuel oil were lost from the Potomac into the ice-laden waters of Melville Bay in western Greenland. Biodegradation of the oil at the low water temperature proceeded very slowly, if at all. There was no significant increase in hydrocarbon utilizers within a few weeks after the spill.[6]

On January 28, 1989, the Argentine resupply and tourist ship Bahia Paraiso ran aground and sank near the U.S. Research Base Palmer Station, Antarctica. At least 680,000 litres of diesel fuel arctic (a blend of diesel and jet fuel) were released during the initial spill into the semi-enclosed Arthur Harbor and deposited in the nearby intertidal regions.[48] Microbial hydrocarbon oxidation potential was detected throughout both the oil-impacted and control regions. Mineralization rates were extremely low (0.13–1.21 pmol hexadecane g^{-1} dry sediment d^{-1}) in samples without nutrient amendments and incubated at 1°C, microbiological turnover time exceeded 2 years. The low hydrocarbon oxidation rates were explained by the low chronical hydrocarbon pollution in this environment and the low ambient temperatures (perennially ≤ 1°C in Arthur Harbor).[48] On March 24, 1989, the tanker Exxon Valdez grounded in Prince Williams Sound, Alaska, creating the largest spill ever in U.S. territorial waters. The contamination consisted of approximately 41 milion litres of Prudhoe Bay crude oil, an oil that had been the focal point of several previous biodegradation studies in cold water environments.[32,33,49] Despite the cold water temperatures, a significant enrichment of oil-degrading microorganisms was expected and found in the beach material following exposure to the oil.[15] About two months after the spill, concentrations of oil degraders (determined at 16°C) averaged 10^6 g^{-1} oiled beach material, representing a 10,000-fold increase in the number of oil-degrading microorganisms relative to uncontaminated beaches.[50,51] Elevated mineralization potentials, coupled with increased numbers of hydrocarbon degraders, suggested that the cold temperatures (10–16°C) were not overly restrictive to the degradative capabilities of the indigenous microflora.[50] The effectiveness of natural biodegradation was limited by the availability of nitrogen and phosphorus.[25] An oleophilic fertilizer enhanced oil-degradation by the indigenous microorganisms by 2-fold relative to untreated controls, as measured by both changes in oil composition and oil residue weights.[25,50] Biodegradation rates were shown to depend mainly on the concentration of nitrogenous nutrients in sediment pore waters, the oil loading, and the extent to which natural biodegradation had already taken place. Nutrient availability will no longer be the limiting factor, once the content of polar components of the oil residue reaches 60–70% of the total mass; thus bioremediation is not likely to be effective on extensively degraded oil hydrocarbons.[52]

3
Biodegradation of non-halogenated aromatic compounds at low temperatures

3.1
Phenol

Phenol is used as a general disinfectant and for the manufacture of colorless or light-colored artificial resins. Many industrial and medical organic compounds and dyes contain phenol. Biodegradation pathways of phenol under aerobic and anaerobic conditions have been summarized by Allard and Neilson.[1]

Kotturi et al.[20] demonstrated low temperature degradation of phenol (14–1,000 mg l^{-1}) by a psychrotrophic *Pseudomonas putida*. The growth and degradation rate of the strain at 10°C was only about ⅓ to ⅕ of the rates reported for mesophilic pseudomonads at 30°C. However, the use of psychrotrophs may still be economically competitive or even advantageous for wastewaster treatment processes in cold climatic regions, and in cases where influent wastewater temperatures exhibit seasonal variation in the range of 10–30°C, due to reduced energy requirements for heating low temperature effluents as compared to the costs associated with the heating of water prior to mesophilic bioremediation of wastewater and groundwater.[20]

3.2
Toluene

Toluene is of considerable importance as a chemical contaminant. It is a water-soluble component of gasoline, and it is often found in industrial effluents.

Chablain et al.[23] isolated psychrotrophic toluene-degrading strains from a toluene-polluted soil, thereby demonstrating that toluene degradation at low temperature occurs in nature. A *Pseudomonas putida* degraded two aromatic compounds (toluene and benzoate), most likely through different pathways. Studies on the influence of temperature on the growth characteristics on different substrates showed that not only cardinal temperatures but also temperature characteristics deduced from the Arrhenius plot of maximal growth rates differed when the different substrates were used as the sole carbon source. Variations in the optimal temperature were observed in media differing only in the carbon source. The particular temperature of 17°C was considered as critical, due to the metabolic changes occurring at that temperature.

3.3
Polycyclic aromatic hydrocarbons

Polycyclic aromatic hydrocarbons (PAHs) are frequently found in industrial chemical wastes. Many industrial sites, such as coal coking plants, municipial gas plants, and wood treatment facilities are heavily contaminated with PAHs at levels that may severely limit the potential redevelopment and future use of these sites. The concept of bioremediation provides a potentially cheap alternative to traditional disposal techniques. Although PAH components with less than four rings are probably biodegradable, those with more than four rings seem to be persistent.[1] Commercial bioremediation companies identified the low temperature as a major limiting factor in the success of bioremediation strategies employed in the field. There is very little knowledge about the biodegradation of PAHs at low temperatures. Recently, studies conducted in Scotland have demonstrated that microorganisms degrade PAHs at low temperatures; degradation of naphthalene and phenanthrene at 2 and 16°C, respectively, has been reported. All bacteria isolated belonged to the genus *Pseudomonas*. The use of combinations of bacteria in consortia has been identified as a potentially useful strategy in achieving PAH degradation.[53]

3.4
Phthalate esters

Phthalate esters are important industrial chemicals. They are mainly used as plasticizers for vinyl and cellulose polymers. Additionally, they can be present in lubricating oil, cosmetics and defoaming agents in paper manufacture. Phthalate esters can be leached from plastics and hence the environment may be exposed to relatively high levels of phthalate esters from landfill leachates. Phthalate esters can be biodegraded to non-toxic products in various aerobic environments.[54] The environmental persistence of di-n-butyl-phthalate (DBP), one of the most widely used plasticizers, is caused by its low solubility and consequently low microbial degradation, and by its unsuitability as an anaerobic energy source.[55]

Nevertheless, it has been demonstrated that unidentified bacterial stains transformed diethyl phthalate at 5 and 15°C under anaerobic conditions.[56] Simulating a Canadian groundwater ecosystem, Chauret et al.[57] demonstrated that DBP is at least partially biodegradable under anaerobic conditions at 10°C. A psychrotrophic denitrifying *Pseudomonas fluorescens*, isolated from subsurface sediment, transformed DBP during 43 d at 10°C in a chemically defined medium under both aerobic and anerobic conditions using nitrate as the terminal electron acceptor. Biotransformation of DBP by the cold-adapted strain appeared to take place only in the stationary and decline phases of growth. Butanol was produced and utilized by transforming cells. The pseudomonad did not grow with phthalic acid as the sole carbon source, indicating that DBP was not mineralized by this bacterium, but biotransformed. Biotransformation was considered to be a secondary substrate utilization, but not a cometabolism.

Biodegradation of DBP was also investigated in microcosms simulating a Canadian subsurface environment under various redox conditions and at ambient temperature (10°C).[58] Biotransformation of DBP was observed under aerobic, nitrate-reducing, Fe(III)-reducing and sulfate-reducing conditions. DBP-biotransformation was significantly decreased as the redox potential was lowered, especially under sulfate-reducing conditions. It was discussed that other factors, such as loss of essential nutrients and building of toxic intermediates, might have affected the biotransformation rates. The highest biotransformation rate of 0.56 µg DBP g^{-1} sediment d^{-1} occurred under aerobic conditions, the lowest rate (0.05 µg DBP g^{-1} sediment d^{-1}) was observed under sulfate-reducing conditions.

4
Biodegradation of chlorinated aromatic compounds at low temperatures

4.1
Chlorophenols

Chlorophenols and particularly pentachlorophenol have been widely used as wood-preservatives and have polluted various sites. Chlorophenols can be degraded under both aerobic and anerobic conditions,[1] nevertheless they are considered as recalcitrant.

Biodegradation of chlorophenols at low temperatures has been observed in soil, sediment and water in long-term batch experiments. Slow and partial aerobic biodegradation of pentachlorophenol has been shown in batch assays at 10°C,[59] at 4°C[60] and at 0°C.[61]

The first high-rate chlorophenol mineralization from groundwater in an aerobic fluidized-bed system at low temperatures was reported by Järvinen et al.[62] Laboratory scale, continuous-flow reactors were inoculated with non-acclimated activated sludge, the groundwater was amended with inorganic nutrients and phosphate buffer, and continuous groundwater feed was started at 14–17°C. Chlorophenol concentrations in groundwater were 7–11 mg 2,4,6-trichlorophenol l^{-1}, 32–36 mg

2,3,4,6-tetrachlorophenol l⁻¹, and 1.8–2.3 mg pentachlorophenol l⁻¹. The treatment temperature was decreased to the ambient groundwater temperature of 7°C and further to 4°C. At 5–7°C, over 99.9% chlorophenol-biodegradation was achieved at a loading rate of 740 mg l⁻¹ d⁻¹. Mineralization was evidenced by stoichiometric inorganic chloride release and organic carbon removal. The effluent quality was close to drinking water standards. Rapid development and stable maintenance of psychrotrophic chlorophenol-degrading microorganisms from activated sludge in the fluidized-bed system was achieved.

These results suggested a significant potential of the described system for cold-adapted bioremediation of chlorophenols, because the system can be operated and maintained at actual groundwater temperatures and thus avoids the heating expenses. Treatment temperature is crucial for the costs of groundwater bioremediation. At pilot scale, the major costs were labor (43%) and heating (42%) when the temperature had to be increased by less than 10°C. An increase of the process temperature from 7 to 25°C increases annual operating costs by a factor of 2.5. In addition, the heating requirement significally increases plant investment costs.[62]

4.2
Polychlorinated biphenyls

Polychlorinated biphenyls (PCBs) are major pollutants and were used in a wide variety of applications for more than 50 years. The environmental fate is important because PCBs accumulate in biota and have been associated with potential health effects.[63] PCB contamination can enter the marine and terrestrial food chains and can be spread via atmospheric transport. Aside from local sources, such as military radar facilities,[64] the predominant pathway for PCB inputs to this region is thought to be atmospheric transport in gas phase and aerosols. Low temperatures (at higher latitudes) may reduce the revolatilization of PCBs and favor the deposition of persistant organic pollutants from the atmosphere into snow, ice, soil and water,[65] giving rise to the "global distillation" or "cold condensation" effect.[66]

Numerous arctic and subarctic freshwater and terrestrial environments have been contaminated with PCBs. Due to the remote location of some sites, conventional cleanup technologies are very expensive, and bioremediation could result in significant cost savings.[64] However, the cold climate poses substantial technical challenges.

Very little is know about PCB degradation by cold-adapted microorganisms. Aerobic biodegradation of di- and trichlorobiphenyls in river sediments at 4°C has been reported.[64] Active PCB degraders were found in antarctic soils and sediments at an average temperature of -2°C.[44]

Mohn et al.[64] examined the degradation of biphenyl and the commercial PCB mixture Aroclor 1221 by indigenous arctic soil microorganisms to assess both the response of the soil microflora to PCB pollution and the potential of the microflora for bioremediation. It was demonstrated that PCB pollution selects for biphenyl-degraders, and that aerobic indigenous soil microorganisms can degrade biphenyl and PCBs with up to three chlorine substituents at low temperature.

Biphenyl degraders are known to metabolize PCBs, and biphenyl as a cosubstrate can stimulate PCB degradation. Both psychrophilic (growing at 7–15°C) and psychrotrophic (growing at 7–30°C) biphenyl-degrading microorganisms were isolated.[64] Two psychrotrophs degraded Aroclor 1221 at 7°C (54–60% removal). In soil slurries, microorganisms of arctic and temperate soils had similar potentials to mineralize biphenyl. The maximum mineralization rates at 7 and 30°C were typically 0.2–1.0 and 1.2–1.4 mg biphenyl g^{-1} soil d^{-1}, respectively. Mineralization began sooner and was more extensive in slurries of PCB-contaminated than in slurries of uncontaminated arctic soils.[64] In slurries of PCB contaminated arctic soils microorganisms degraded Aroclor 1221 more extensively at 30°C (71–76% removal) than at 7°C (14–40% removal). The limited PCB biodegradation at 7°C was explained as follows:

1. cometabolism of PCBs by biphenyl degradation enzymes may be more sensitive to low temperature than it is biphenyl degradation,
2. distinct organisms responsible for biphenyl degradation at 7°C may lack the capability to cometabolize PCBs,
3. the bioavailability of PCBs is affected by temperature.

While low temperatures severely limited Aroclor 1221 removal in soil slurries, results with pure cultures suggest that more effective PCB biodegradation is possible under appropriate conditions. Psychrotrophic microorganisms with high potentials for degradation of mono- and dichlorobiphenyls could be useful in a sequential process in which PCBs are first reductively dechlorinated by anaerobic processes.[64]

Dehalogenation is expected to detoxify PCBs. Wu et al.[63] investigated the influence of temperature (4–66°C) on the microbial dechlorination of 2,3,4,6-tetrachlorobiphenyl (CB) in anaerobic sediments. Temperature influenced the timing and the relative predominance of parallel pathways of dechlorination, i.e. *meta* versus *para* dechlorination of 2,3,4,6-CB and *ortho* versus *para* dechlorination of 2,4,6-CB and 2,4-CB. *Meta* dechlorination of 2,3,4,6-CB to 2,4,6-CB dominated at all temperatures tested except 18 and 34°C, where *para* dechlorination to 2,3,6-CB dominated in some replicates. The dechlorination of 2,4,6-CB was restricted to 15–30°C in both investigated sediments; *para* and *ortho* dechlorination dominated at 20 and 15°C, respectively. The data reported indicate that field temperatures play a significant role in controlling the nature and the extent of the PCB dechlorination that occurs at a given site.[63]

5
Other compounds

5.1
Ethylene glycol

Ethylene glycol is used for aircraft de-icing and therefore of considerable environmental concern. Cold-adapted biodegraders may prove to offer an effective approach to ethylene glycol bioremediation in cold regions.[22]

Colucci and Inniss[22] isolated a psychrotrophic *Pseudomonas fluorescens* which utilized ethylene glycol (1–5%) as a sole carbon source, with removal efficiencies of 98 and 96% in 20 and 55 d at 25 and 5°C, respectively. The response of the psychrotroph to environmental shifts (temperature and ethylene glycol shock) was investigated and the microorganism demonstrated the ability to adapt both to various temperatures and ethylene glycol concentrations. Ethylene glycol shock induced 14 ethylene glycol shock proteins and ten ethylene glycol acclimation proteins. A number of these proteins may be enzymes responsible for the biodegradation of ethylene glycol. During a 25 to 5°C cold shock, ten cold shock proteins were induced, five cold acclimation proteins were synthesized at 5°C. These proteins may be of significane to both shock recovery and constant growth in a new environment.[22]

5.2
Surfactants

Surfactants are especially noted for their wetting qualities and their effectiveness as emulsifiers; the most widely used surfactants in detergent formulations are those containing anionic groups.[67] Surfactants appear in natural ecosystems through outfalls of wastewater or through direct application (e.g., agrochemical sprays). Possible harmful effects of surfactants in the environment are the re-mobilization of organic pollutants, the inhibition of biological activities, and the toxic effect on organisms.

Modern formulations of surfactants rarely resist biodegradation. Alkyl sulphates, such as sodium dodecyl sulfate (SDS), are readily broken down by bacteria and constitute the most readily biodegradable surfactant in common use.[68] Not all the necessary stages in complete breakdown may be present in a single species, in natural environments it is likely that consortia of microorganisms are involved in the ultimate removal.[67,68]

The capacities of epilithic and planktonic river bacterial populations to degrade SDS in samples from clean and polluted sites in a South Wales river were estimated in die-away tests under simulated environmental conditions. There was a slow disappearance of SDS by populations taken from the clean river site (25 mg SDS l^{-1} were degraded after approximately 300 h at 5°C by epilithic samples), as compared with those from the polluted sites (25 mg SDS l^{-1} were degraded after approximately 60–120 h at 5–10°C). The biodegradative capacity of epilithic bacterial populations towards SDS was shown to be more stable than that of planktonic bacteria.[69]

High concentrations of surfactants are applied in agrochemistry where sprays contain >1,000 ppm surfactant. Recent studies showed that high concentrations of SDS (1,000–2,000 mg l^{-1}) are readily degraded at 10°C by cold-adapted mixed microbial cultures[70] and by indigenous soil microorganisms.[71] After 48–72 h at 10°C, the SDS content was below the detection limit in liquid cultures,[70] the presence of diesel oil delayed SDS biodegradation.[71] No abiotic elimination of SDS was observed. The fact that 2,000 mg SDS l^{-1} was fully degraded over a broad temperature range (4–30°C)[70] is of special importance regarding environmental temperature fluctuations.

The extensive use of surfactants is required for the efficient cleaning of fuel-contaminated vehicles, garages, oil tanks etc. Consequently, wastewaters from garages and car washs are contaminated with both fuel oil and surfactants; the biotechnological decontamination could allow the recycling of such waters (e.g., their re-use for car washing).[19] For such purposes, microorganisms which are efficient degraders of both fuel oil and surfactants are necessary. A study on the biological decontamination of the wastewater from a garage and car wash that was polluted with fuel oil (184 mg hydrocarbons l^{-1}) and anionic surfactants (57 mg l^{-1}) was conducted at 10°C.[19] A comparable temperature prevails in ground and surface water, and in non-heated underground water treatment plants. The wastewater investigated contained enough cold-adapted microorganisms that degraded efficiently both contaminants after biostimulation by inorganic nutrient supply. After 7 d at 10°C, approximately 20 and 14% of the initial concentrations of anionic surfactants and hydrocarbons, respectively, remained. After 35 d, the anionic surfactants had been further reduced to to 3 mg l^{-1}, whereas the hydrocarbon content remained unchanged.

The biodegradability and toxicity of surfactants depend on temperature. Microcosms systems were employed to assess the effect of the surfactants, linear alkylbenzene sulfonate (0.5–10 mg l^{-1}) and soap (200 mg l^{-1}), on an aquatic ecosystem at various temperatures. At all test temperatures (10–30°C) stable ecosystems were formed. Biodegradability of the surfactants differed with temperature and decreased when the temperature was lowered, which changes the effects of surfactants on microorganisms. At 10°C, no-observed-effect concentrations for linear alkylbenzene sulfonate and soap were <0.5 and <10 mg l^{-1}, respectively.[72]

6
Catabolic plasmids of degradative microorganisms

The microbial catabolic pathways responsible for the degradation of petroleum hydrocarbons, including the *alk* (C_5 to C_{12} n-alkanes), *nah* (naphthalene) and *xyl* (toluene) pathways, are generally located on large catabolic plasmids.[73] Microorganisms rarely possess both the *alk* and *nah* catabolic pathways. Many of 200 environmental bacterial strains could mineralize aliphatic or aromatic hydrocarbons, but not both,[74] suggesting that alkane and PAH biodegradation may be mutually exclusive properties.[36] Also, a number of 135 psychrotrophic microorganisms isolated from various Canadian ecosystems mineralized toluene, naphthalene, dodecane and hexadecane at both 23 and 5°C, but none of the strains was capable of mineralizing 2-chlorbiphenyl or pentachlorophenol.[21]

Investigations of hydrocarbon-degrading strains from oil-contaminated sediments after the Exxon Valdez oil spill indicated that genes in the *alk* and *xyl* pathways for aliphatic and aromatic hydrocarbon degradation are not incompatible and can coexist in the same organism:[75] A significant proportion (32%) of the naturally occurring hydrocarbon-degrading populations contained both the *alkB* (encoding alkane monooxigenase) and *xylE* (catechol 2,3-dioxygenase) genes and could convert hexadecane and naphthalene to carbon dioxide; a greater proportion of the population had *xylE* than *alkB*, reflecting the composition of the residual oil at the sampling time. The genotypes of hydrocarbon-degrading populations reflected the composition of the hydrocabons to which they were exposed. Thus, the distribution of genes for hydrocarbon degradation changes in response to the composition of the hydrocarbons in the environment to which bacterial populations are exposed.[75]

Two hydrocarbon-degrading psychrotrophic *Pseudomonas* spp. strains isolated from petroleum-contaminated arctic soil degraded C_5–C_{12} n-alkanes, toluene and naphthalene at both 5 and 25°C. Biodegradation studies of a mixture of naphthalene, octane and toluene by the two strains revealed sequential growth on the three substrates. Preferential utilization of naphthalene occurred by one strain at both 5 and 25°C, and by the second strain at 25°C, coinciding with the apparent lag phases for octane and toluene utilization.[36] The strains possessed both the *alk* catabolic pathway for alkane-biodegradation and the *nah* catabolic pathway for PAH-biodegradation. It could further be demonstrated that both catabolic pathways, located on separate plasmids, can naturally coexist in the same bacterium.[36] The alkane catabolic pathways of both strains were similar to the *alk* pathway of mesophilic *Pseudomonas oleovorans*. It was hypothesized that the transfer of biodegradative pathways between mesophiles and psychrotrophs in nature has occurred or is occurring and that the pathways are expressed in both types of organisms. A previously obtained successful conjugational transfer of the toluene degradation plasmid (TOL) from a mesophilic *Pseudomonas putida* to a psychrotrophic *P. putida*, in which the toluene biodegradative genes were expressed at low temperature,[76] supported this hypothesis. The transconjugant degraded and utilized toluate (1,000 mg l^{-1}) as the sole source of carbon at temperatures as low as 0°C. Comparison of growth rates over 0–30°C indicated that the physiological activity of the transconjugant was not reduced and that the plasmid DNA from the mesophile and its encoded enzymes functioned effectively in the psychrotroph at temperatures well below those at which the mesophile could grow.[76] The genetic manipulation of degradative microorganisms to acquire or to enhance biodegradative capabilities was proposed. Such microorganisms should readily function at low temperatures in chemically contaminated environments or in industrial wastewater treatment systems[21] and could be of interest in cases where biodegradation at low temperature is negligible (e.g., petroleum degradation in arctic marine ice[6] and Norwegian soils[46]).

Genetic exchanges between microorganisms in the environment can occur frequently. Bacteria developed efficient mechanisms which allow them to survive in polluted environments. Conjugation is considered the process of highest ecological importance regarding gene transfer in soil. It is part of the bacterial communication, and a means of accelerating bacterial evolution (for a review see ref. 77). A restricted host range for conjugation within closely related bacteria seemed to be the rule for a long time. Natural gene transfer between distantly related organisms[78] (e.g, between gram-negative and gram-positive bacteria[79]) is of high ecological importance, because it can be assumed that it occurs also under natural conditions in the environment. It appears that the potential of genetic transfer between bacteria is much greater that suspected, which should be considered in the problematics of release of genetically engineered microorganisms.

7
Limitations and conclusions

Cold-adapted biodegraders of organic pollutants play a significant role in the *in-situ* decontamination and in bioremediation processes of cold terrestrial and marine environments where ambient temperatures often coincide with the growth temperature range of these microorganisms.

Bioremediation, the process whereby natural biodegradation rates are accelerated through stimulation of the indigenous microrganisms, is an effective, ecologically and ecomically acceptable reclamation alternative. However, all procedures are site-specific, since they must take into account both the physical environment and microbiological issues. They will also be contaminant-specific. Very often, mixtures of a number of compounds are involved, and there are wide variations in the biodegradability of individual components.[1] The biodegradation of petroleum hydrocarbons has been reported in a variety of terrestrial and marine cold ecosystems, but very little is known about low-temperature biodegradation of non-halogenated and chlorinated aromatic compounds. Biodegradation of many compounds of contaminations may be facilitated by inorganic nutrient supply and oxygen. While indigenous microorganisms are effective degraders and adapt rapidly to the contamination, inoculation does not seem to be promising.

There are several unresolved microbiological issues concerning the limits of biodegradation of organic pollutants:[1]

1. the degree to which the compounds are accessible to microorganisms,
2. the extent to which the microbial populations can be maintained and increased,
3. the relative biodegradability of components of complex mixtures
4. due to possible threshold concentrations below which degradation rates are slow or negligible, the complete removal of the organic pollutants may not be possible by using biological decontamination, even after a considerably prolonged treatment.

Contaminations can persist in cold environments for long periods of time; petroleum hydrocarbons still persist in Alaska 20 years after spillage.[27] At low temperatures, a residual amount of approximately 10–30% of the initial contamination remains in microbial batch culture[39] and in soil.[11] Reasons for the low ultimate degradation include the decreasing bioavailability of the contaminants and the accumulation of recalcitrant components in the course of biodegradation.

8
References

1. Allard AS, Neilson AH. Bioremediation of organic waste sites: a critical review of microbiological aspects. Int Biodeterioration Biodegradation 1997; 39:253-285.
2. Atlas RM, Bartha R. Hydrocarbon biodegradation and oil spill bioremediation. Adv Microb Ecol 1992; 12:287-338.
3. Morgan P, Watkinson RJ. Hydrocarbon degradation in soils and methods for soil biotreatment. CRC Crit Rev Biotechnol 1989; 8:305-333.

4. Hamby DM. Site remediation techniques supporting environmental restoration activities – a review. Sci Total Environ 1996; 191:203-224.
5. Cooney JJ, Silver SA, Beck EA. Factors influencing hydrocarbon degradation in three freshwater lakes. Microb Ecol 1985; 11:127-137.
6. Atlas RM. Microbial degradation of petroleum hydrocarbons: an environmental perspective. Microbiol Rev 1981; 45:180-209.
7. ZoBell CE. Action of microorganisms on hydrocarbons. Bacteriol Rev. 1946; 10:1-49.
8. Gibson DT, Saylor GS. Scientific Foundations of Bioremediation: Current Status and Future Needs. Washington DC: American Academy of Microbiology, 1992.
9. Norris RD, ed. Handbook of Bioremediation. Boca Raton Florida: CRC Press, 1994.
10. Hoff RZ. Bioremediation: an overview of its development and use for oil spill cleanup. Marine Pollution Bull 1993; 26:476-481.
11. Margesin R, Schinner F. Cold-adapted microorganisms – efficient oil-degraders in cold soils. Recent Res Devel Microbiol 1998; 2 (in press).
12. Westlake DWS, Jobson AM, Cook FD. In-situ degradation of oil in a soil of the boreal region of the Northwest Territories. Can J Microbiol 1978; 24:254-260.
13. Jobson AM, Cook FD, Westlake DWS. Microbial utilization of crude oil. Appl Microbiol 1972: 23:1082-1089.
14. Margesin R, Schinner F. Efficiency of indigenous and inoculated cold-adapted soil microorganisms for biodegradation of diesel oil in alpine soils. Appl Environ Microbiol 1997; 63:2660-2664.
15. Margesin R, Schinner F. Laboratory bioremediation exeriments with soil from a diesel-oil contaminated site – significant role of cold-adapted microorganisms and fertilizers. J Chem Technol Biotechnol 1997; 70:92-98.
16. Pritchard H, Popovic M, Bajpai R, Mueller JG. Metabolic capability and bioavailability; delivery, sustainability, and monitoring. In: TUHH Technologie GmbH, ed. Extended Abstracts: Innovative Potential of Advanced Biological Systems for Remediation. Hamburg-Harburg: Technical University, 1998:59-62.
17. Margesin R, Schinner F. Properties of cold-adapted microorganisms and their potential role in biotechnology. J Biotechnol 1994; 33:1-14.
18. Morita RY. Psychrophilic bacteria. Bacteriol Rev 1975; 39:144-167.
19. Margesin R, Schinner F. Low-temperature-bioremediation of a waste water contaminated with anionic surfactants and fuel oil. Appl Microbiol Biotechnol 1998; 49:482-486.
20. Kotturi G, Robinson CW, Inniss WE. Phenol degradation by a psychrotrophic strain of *Pseudomonas putida*. Appl Microbiol Biotechnol 1991; 34:539-543.
21. Whyte LG, Greer CW, Inniss WE. Assessment of the biodegradation potential of psychrotrophic microorganisms. Can J Microbiol 1996; 42:99-106.
22. Colucci MS, Inniss WE. Ethylene glycol utilization, cold and ethylene glycol shock and acclimation proteins in a psychrotrophic bacterium. Curr Microbiol 1996; 23:179-182.
23. Chablain PA, Philippe G, Groboillot A, Truffaut N, Guespin-Michel JF. Isolation of a soil psychrotrophic toluene-degrading *Pseudomonas* strain: influence of temperature on the growth characteristics on different substrates. Res Microbiol 1997; 148:153-161.
24. Owens EH, Harper JR, Robson W, Boehm PD. Fate and persistence of crude oil stranded on a sheltered beach. Arctic 1987; 40 Suppl 1:109-123.
25. Pritchard PH, Costa CF. EPA's Alaska oil spill bioremediation project. Environ Sci Technol. 1991; 25:372-379.
26. Braddock JF, McCarthy KA. Hydrologic and microbiological factors affecting persistence and migration of petroleum hydrocarbons spilled in a continuous-permaforst region. Environ Sci Technol 1996; 30:2626-2633.
27. Braddock JF, Ruth ML, Walworth JL, McCarthy KA. Enhancement and inhibition of microbial activity in hydrocarbon-contaminated arctic soils: implications for nutrient-amended bioremediation. Environ Sci Technol 1997; 31:2078-2084.
28. Kerry, E. Bioremediation of experimental petroleum spills on mineral soils in the Vestfold

Hills, Antarctica. Polar Biol 1993; 13:163-170.

29. Wardell LJ. Potential for bioremediation of fuel-contaminated soil in Antarctica. J Soil Contamination 1995; 4:111-121.

30. Delille D, Bassères A, Dessommes A. Seasonal variation of bacteria in sea ice contaminated by diesel fuel and dispersed crude oil. Microb Ecol 1997; 33:97-105.

31. McKenzie P, Hughes DE. Microbial degradation of oil and petrochemicals in the sea. In: Skinner FA, Carr JG, eds. Microbiology in Agriculture, Fisheries and Food. Society for Applied Bacteriology, Symposium Series No. 4. London: Academic Press, 1976:91-108.

32. Fedorak PM, Westlake DWS. Microbial degradation of aromatics and saturates in Prudhoe Bay crude oil as determined by glas capillary gas chromatography. Can J Microbiol 1981; 27:432-443.

33. Westlake DWS, Jobson AM, Philippe R, Cook FD. Biodegradability and crude oil decomposition. Can J Microbiol 1974; 20:915-928.

34. Bailey NJL, Jobson AM, Rogers MA. Bacterial degradation of crude oil: comparison of field and experimental data. Chem Geol 1973; 11:203-221.

35. Colwell RR, Mills AL, Walker JD, Garcia-Tello P, Campos V. Microbial ecological studies of the Metula spill in the Straits of Magellan. J Fish Res Board Can 1978; 35:573-580.

36. Whyte LG, Bourbonnière L, Greer CW. Biodegradation of petroleum hydrocarbons by psychrotrophic *Pseudomonas* strains possessing both alkane *(alk)* and naphthalene *(nah)* catabolic pathways. Appl Environ Microbiol 1997; 63:3719-3723.

37. Margesin R, Schinner F. Effect of temperature on oil-degradation by a psychrotrophic yeast in liquid culture and in soil. FEMS Microbiol Ecol 1997; 24:243-249.

38. Margesin R, Schinner F. Bioremediation of diesel-oil-contaminated alpine soils at low temperatures. Appl Microbiol Biotechnol 1997; 47:462-468.

39. Margesin R, Schinner F. Oil biodegradation potential in alpine habitats. Arctic Alpine Res 1998; 30:262-265.

40. Dibble JT, Bartha R. Effect of environmental parameters on the biodegradation of oil sludge. Appl Environ Microbiol 1979; 37:729-739.

41. Rosenberg E, Legmann R, Kushmaro A, Taube R, Adler E, Ron EZ. Petroleum bioremediation - a multiphase problem. Biodegradation 1992; 3:337-350.

42. Møller J, Gaarn H, Steckel T, Wedebye EB, Westermann P. Inhibitory effects on degradation of diesel oil in soil microcosms by a commercial bioaugmentation product. Bull Environ Contam Toxicol 1995; 54:913-918.

43. Mac Cormack WP, Fraile E. Characterization of a hydrocarbon degrading psychrotrophic Antarctic bacterium. Antarctic Sci 1997; 9:150-155.

44. Tumeo MA, Wolk AE. Assessment of the presence of oil-degrading microbes at McMurdo Station. Antartic J 1994; 29:375-377.

45. Tumeo MA, Gawde P. Land farming of petroleum-contaminated soils in cold climates. In: Stanley SJ, Ward CJW, Smith DW, eds. Proceedings of the 1997 CSCE/ASCE Environmental Engineering Conference, vol. 2. Montreal, Canada: Canadian Society for Civil Engineering, 1997:1251-1262.

46. Sandvik S, Lode A, Pedersen TA. Biodegradation of oily sludge in Norwegian soils. Appl Microbiol Biotechnol 1986; 23:297-301.

47. Walker JD, Colwell RR. Microbial degradation of model petroleum at low temperatures. Microb Ecol 1974; 1:63-95.

48. Karl DM. The grounding of the Bahia Paraiso: microbial ecology of the 1989 Antarctic oil spill. Microb Ecol 1992; 24:77-89.

49. Horowitz A, Atlas RM. Response of microorganisms to an accidental gasoline spillage in an Arctic freshwater ecosystem. Appl Environ Microbiol 1977; 33:1252-1258.

50. Pritchard PH, Müller JG, Rogers JC, Kremer FV, Glaser JA. Oil spill bioremediation: experiences, lessons and results from the Exxon Valdez oil spill in Alaska. Biodegradation 1992; 3: 315-335.

51. Lindstrom JE, Prince RC, Clark JC, Grossmann MJ, Yeager TR, Braddock JF, Brown EJ.

Microbial populations and hydrocarbon biodegradation potentials in fertilized shoreline sediments affected by the T/V Exxon Valdez oil spill. Appl Environ Microbiol 1991; 57:2514-2522.

52. Bragg JR, Prince RC, Harner EJ, Atlas RM. Effectiveness of bioremediation for the Exxon Valdez oil spill. Nature 1994; 368:413-418.

53. Anon. In-situ bioremediation of contaminated land. Biotech Lab Int 1997; November/December:10-12.

54. Keyser P, Pujare G, Eaton RW, Ribbons DW. Biodegradation of the phthalates and their esters by bacteria. Environ Health Perspect 1976; 18:1159-1166.

55. Benckiser G, Ottow JCG. Metabolism of the di-n-butyl phthalate by *Pseudomonas pseudoalcaligenes* under anaerobic conditions with nitrate as the only electron acceptor. Appl Environ Microbiol 1982; 44:576-578.

56. Zhang G, Reardon KF. Parametric study of diethyl phthalate biodegradation. Biotechnol Lett 1990; 12:699-704.

57. Chauret C, Mayfield CI, Inniss WE. Biotransformation of di-n-butyl phthalate by a psychrotrophic *Pseudomonas fluorescens* (BWG) isolated from subsurface environment. Can J Microbiol 1995; 41:54-63.

58. Chauret C, Inniss WE, Mayfield CI. Biotransformation of di-n-butyl phthalate in subsurface microcosms. Groundwater 1996; 34:791-794.

59. Pignatello JJ, Johnson LK, Martinson MM, Carlson RE, Crawford RL. Response of the microflora in outdoor experimental streams to pentachlorophenol. Can J Microbiol 1986; 32:38-46.

60. Trevors JT. Effect of temperature on the degradaion of pentachlorophenol. Chemosphere 1982; 11:471-475.

61. Baker MD, Mayfield CI, Inniss WE. Degradation of chlorophenols in soil sediment and water at low temperatures. Water Res 1980; 14:1765-1771.

62. Järvinen KT, Melin ES, Puhakka JA. High-rate bioremediation of chlorophenol-contaminated groundwater at low temperatures. Environ Sci Technol 1994; 28:2387-2392.

63. Wu Q, Bedard DL, Wiegel J. Effect of incubation temperature on the route of microbial reductive degradation of 2,3,4,6-tetrachlorobiphenyl in polychlorinated biphenyl (PCB)-contaminated and PCB-free freshwater sediments. Appl Environ Microbiol 1997; 63:2836-2843.

64. Mohn WW, Westerbeg K, Cullen WR, Reimer KJ. Aerobic biodegradation of biphenyl and polychlorinated biphenyls by arctic soil microorganisms. Appl Environ Microbiol 1997; 63:3378-3384.

65. Stern GA; Halsall CJ, Barrie LA, Muir DCG, Fellin P, Rosenberg P, Rovinsky FYA, Kononov EYA, Pastuhov B. Polychlorinated biphenyls in arctic air. 1. Temporal and spatial trends: 1992-1994. Environ Sci Technol 1997; 31:3619-3628.

66. Muir DCG, Omelchenko A, Grift NP, Savoie DA, Lockhart WL, Wilkinson P, Brunskill GJ. Spatial trends and historical deposition of polychlorinated biphenyls in Canadian midlatitude and arctic lake sediments. Environ Sci Technol 1996; 30:3609-3617.

67. van Ginkel CG. Complete degradation of xenobiotic surfactants by consortia of aerobic microorganisms. Biodegradation 1996; 7:151-164.

68. White GF, Russell NJ. Biodegradation of anionic surfactants and related molecules. In: Ratledge C, ed. Biochemistry of Microbial Degradation. Dordrecht Boston London: Kluwer Academic Publ, 1994:143-175.

69. Anderson D, Day MJ, Russell NJ, White GF. Die-away kinetic analysis of the capacity of epilithic and planctonic bacteria from clean and polluted river water to biodegrade sodium dodecyl sulfate. Appl Environ Microbiol 1990; 56:758-763.

70. Margesin R, Schinner F. Biodegradation of the anionic surfactant sodium dodecyl sulfate at low temperatures. Int Biodeterioration Biodegradation 1998; 41:139-143.

71. Margesin R, Schinner F. Biodegradation of diesel oil by cold-adapted microorganisms in presence of sodium dodecyl sulfate. Chemosphere 1999 (in press).

72. Takamatsu Y, Nishimura O, Inamori Y, Sudo R, Matsumara M. Effect of temperature on biodegradation of surfactants in aquatic microcosm system. Water Sci Technol 1996; 34:61-68.

73. Sayler GS, Hooper SW, Layton AC, Henry King JM. Catabolic plasmids of environmental and ecological significance. Microb Ecol 1990; 19:1-120.

74. Foght JM, Fedorak PM; Westlake DWS. Mineralization of [^{14}C]hexadecane and [^{14}C]phenanthrene in crude oil: specifity among bacterial isolates. Can J Microbiol 1990; 36:169-175.

75. Sotsky JB, Atlas RM. Frequency of genes in aromatic and aliphatic hydrocarbon biodegradation pathways within bacterial populations from Alaskan sediments. Can J Microbiol 1994; 40:981-985.

76. Kolenc RJ, Inniss WE, Glick BR, Robinson CW, Mayfield CI. Transfer and expression of mesophilic plasmid-mediated degradative capacity in a psychrotrophic bacterium. Appl Environ Microbiol 1988; 54:638-641.

77. Smit E, van Elsas JD. Determination of plasmid transfer frequency in soil: consequences of bacterial mating on selective agar media. Curr Microbiol 1990; 21:151-157.

78. Mazodier P, Davies J. Gene transfer between distantly related bacteria. Annu Rev Genet 1991; 25:147-171.

79. Margesin R, Schinner F. Heavy metal resistant *Arthrobacter* sp. – a tool for studying conjugational plasmid transfer between Gram-negative and Gram-positive bacteria. J Basic Microbiol 1997; 3:217-227.

Biohydrometallurgical processes and temperature

G. Rossi

University of Cagliari, Geoengineering and Environmental Technologies Department,
Piazza d'Armi 19, I-09123 Cagliari, Italy

1
Introduction

Most biohydrometallurgical processes involve the solubilization of certain minerals whose overall kinetics are catalyzed by suitable microflora. These processes are carried out in three-phase systems composed of two main subsystems: the abiotic subsystem, where the chemical reactions occur and whose kinetics are governed by the laws of inorganic chemistry, and a second subsystem whose kinetics are strongly affected by the presence of a microflora. The components of these subsystems are intimately mixed and hence their influence is reciprocal.
The three-phase systems consist of:

1. a solid phase, namely the mineral, containing the metals to be extracted or, more general, the required elements,
2. a liquid phase, i.e. the microflora suspended in an aqueous solution of nutrient salts required by its metabolism,
3. a gaseous phase, usually a mixture of atmospheric oxygen and carbon dioxide.

From the point of view of their commercial application, biohydrometallurgical processes can be divided into two classes, depending on the relative motion of the solid and fluid phases: percolation processes and suspension processes.

Percolation processes are characterized by the movement, usually *per descensum*, of the fluid phase through the solid phase, a mass of broken rock with sizes ranging from some decimetres down to a few micrometres. The microflora catalyzes the leaching action of the percolating solutions on the exposed mineral particles. Typical examples are the *in-situ* operations conducted in underground mining stopes (Fig. 1), and bioleaching carried out in dumps or heaps (Fig. 2). A variant of this process is the so-called "flooding" which consists of simply filling the stopes with water, usually recovered from the mine drainage. A technically important feature of these processes is the alternation of wetting and dry cycles. Because the microorganisms are aerobic, percolation or flooding must be periodically interrupted and the broken ore mass allowed to drain the liquors so that atmospheric oxygen can permeate all the interstices. The length of flooding and rest cycles depends upon a number of factors, including growth rate of the microorganisms.

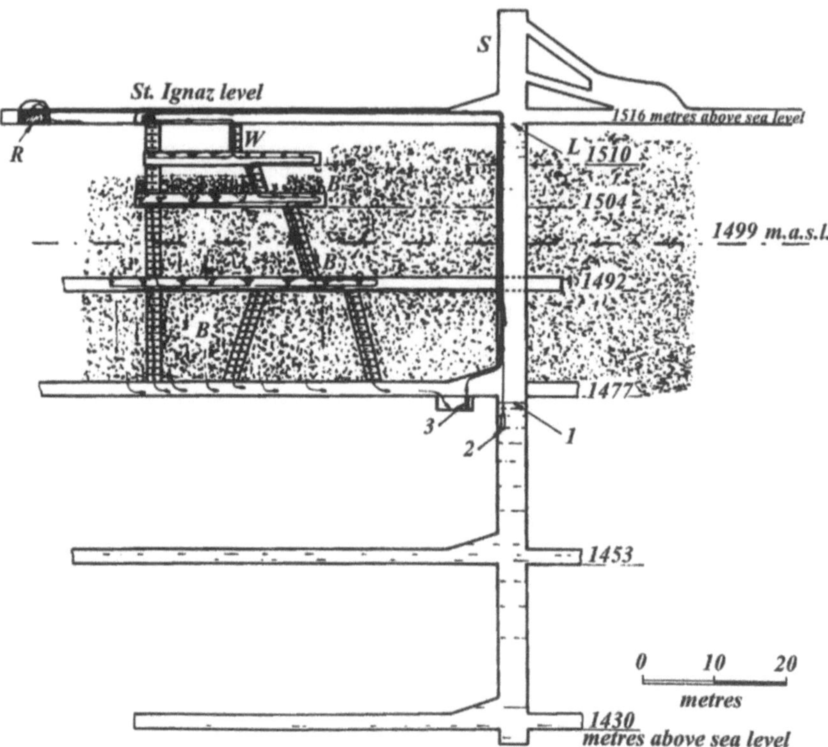

Fig. 1. Schematic of *in-situ* bioleaching of a lens-shaped partly depleted orebody (longitudinal section). The Figure refers to the case history of San Valentino di Predoi mine (reprinted with permission from ref. 22)
1 water level during bioleaching operation; 2 main drainage pump; 3 tank with pump; S Archduke Johann shaft; W raise; R reservoir of pregnant solutions; B old stopes filled with low grade rock; L water table when the mine is flooded

In suspension processes the solid phase is very finely ground to grain sizes smaller than tenths of millimetres and in some cases down to a few micrometres, and suspended in the liquid phase contained in devices called *bioreactors*. Here the suspension is stirred to achieve thorough mixing and solubilization of the gas mixture which is continually injected through spargers. Mixing ensures a high frequency of encounters between the mineral particles and the microbial cells, and the most effective mass transfer of the solubilized ions and of the atmospheric oxygen to the liquid phase.

The bioreactors most commonly used in laboratory and commercial operations are the so-called stirred tank reactors (STR), borrowed from chemical engineering and shown in Figure 3, and airlift reactors (ALR), of which the most widely used is the *Pachuca tank* borrowed from hydrometallurgy (Fig. 4). A new, very effective, type of bioreactor called *Biorotor*, custom built for biohydrometallurgical processes, has been proposed recently.[1]

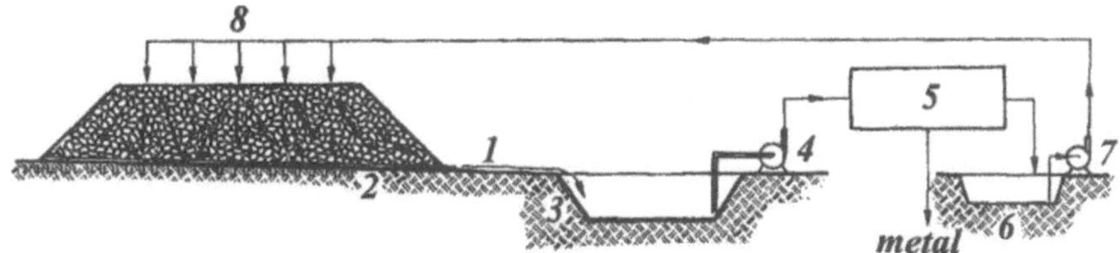

Fig. 2. Schematic of heap leaching operation (reprinted with permission from ref. 22)
1 heap; 2 ground surface; 3 pregnant solution collecting pond; 4 pump; 5 cementation launders;
6 spent solution pond; 7 pump; 8 dump irrigation system

So far, biohydrometallurgical processes have been applied chiefly for recovering base metals, particularly copper, from sulfide minerals, and uranium from uranium ores associated to iron sulfides. They have also been adopted for the pretreatment of gold-bearing complex metal sulfides with the aim to improve exposure and hence accessibility of the gold particles finely interspersed therein ("invisible gold") to the cyanide solutions. A coal biodepyritization test carried out in a semi-commercial pilot plant designed, built and operated in the framework of Project Joule, funded by the Commission of European Communities, has demonstrated the potential of the process for removing sulfur from coal.

Up to now, commercial copper bioleaching operations have resorted mainly to percolation on the so-called "waste" produced during stripping of orebodies and contained in the dumps of large North- and South American open pit mines or to flooding or percolation through the often submarginal ores left in the stopes of worked-out underground mines. Likewise, uranium bioleaching has mainly been achieved by percolation or flooding in the stopes of depleted uranium mines. Only one case history is known of a mine where uranium is extracted by bioleaching alone.

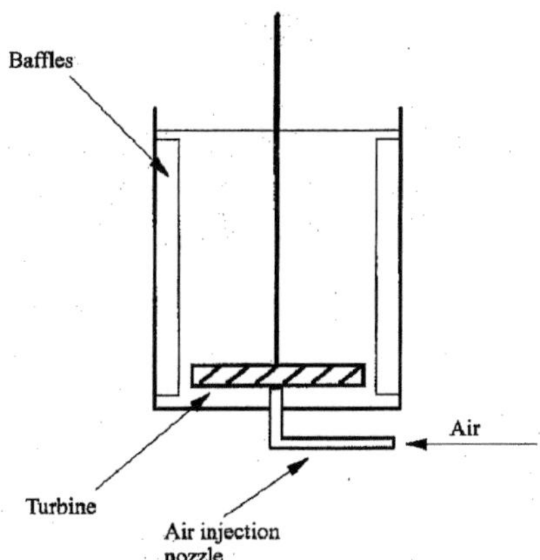

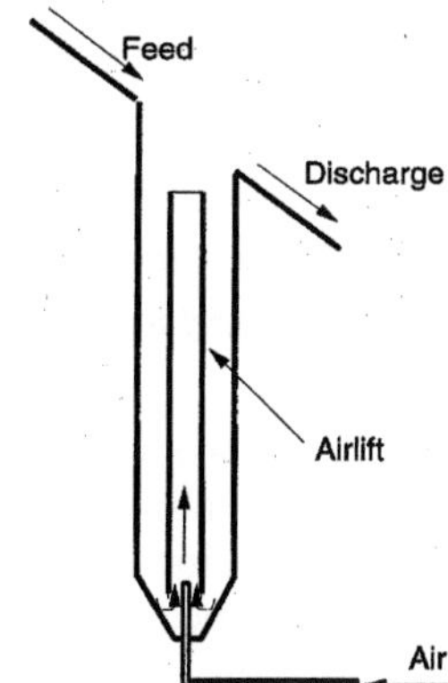

Fig. 3. Schematic of a stirred tank reactor
1 tank or vessel; 2 baffles; 3 impeller

Fig. 4. Schematic of a Pachuca type air lift bioreactor

The application of biohydrometallurgy for enhanced gold recovery from complex sulfides is quite successful, and five plants are currently operating in Africa, Australia and Brasil. The catalytic action of the microflora consists in enhancing the destruction kinetics of the crystal lattices of the sulfide minerals with which the often micron-size gold particles are intergrown. The process requires maximum surface exposure of the sulfides, hence a very fine grind, compatible obviously with the profitability of the operation. Similarly, the success of biodepyritization requires a very fine grind of the solids. Thus, both these processes have to be carried out in bioreactors.

There is general consensus that the oxidation and solubilization of metal sulfides in general and of pyrite (monometric FeS_2), pyrrhotite (Fe_mS_n), marcasite (rhombic FeS_2) and chalcopyrite ($CuFeS_2$) in particular, are the result of direct contact between the bacterial cells and the crystal lattices (the direct attack theory) combined with the purely chemical action of the ferric sulfate that is formed in the liquid phase containing these minerals. The oxidation of iron sulfide minerals produces sulfuric acid and ferrous iron that is converted to ferric iron by the catalytic action of the microorganisms leading to the formation of ferric sulfate. The latter is a strong oxidizing agent of metal sulfides. The close interaction of the two subsystems of the oxidation process is clearly highlighted by the brief description of the process given above.

Laboratory testing has demonstrated that the effectiveness of the microflora used in biohydrometallurgical processes is enhanced when elements or their salts are present in the liquid phase, and this has led to the development of suitable nutrient media. However, there is no information in the literature about additives to the leach solutions circulating in commercial bioleaching operations, probably because the rocks themselves are a source of several salts. Some components or compounds contained in these salts may however form, together with other ingredients in the solutions, compounds harmful to the process. A typical example of such as a process is the formation of jarosites under certain conditions. They tend to precipitate on the mineral surfaces forming a coating that hinders the access of the microorganisms and of ferric ions, slows down, and may even halt the process.

Temperature may affect the kinetics of these reactions, the interactions of the subsystems and eventually the effectiveness of the overall process.

2

Optimum growth temperatures and viability temperature ranges of microorganisms used in biohydrometallurgy

The importance of the marked influence that temperature exerts on the interaction of the biohydrometallurgical subsystems is well documented. A powerful tool for quantitatively assessing how temperature affects the chemical, physico-chemical and biochemical processes is the well known Arrhenius plot. This plot relates the logarithm of the rate constant of a given reaction to the reciprocal of the absolute temperature at which it takes place, providing indications as to its sensitivity to temperature. The empirical Arrhenius equation is:

$$K = A\, e^{-\frac{\Delta E}{RT}} \tag{1}$$

where K is the rate constant, A is the frequency factor, ΔE is the activation energy of the reaction, and R is the gas universal constant.

When this equation is applicable – the case of chemical reactions – taking the logarithm of both members we get the following expression:

$$\ln k = \frac{-\Delta E}{R}\,\frac{1}{T} + \text{constant} \tag{2}$$

The plot of $\ln k$ against 1/T is a straight line of negative slope -(ΔE/R) since ΔE must be positive. The value of ΔE is obtained by multiplying the measured slope by R. It may be readily proven that the temperature sensitivity of a reaction increases with the activation energy, and that the temperature sensitivity of a given reaction decreases as temperature increases.

To find out if a given reaction fits an Arrhenius line, the rate constant – or a parameter proportional to it – must be determined by carrying out the reaction at various temperatures. Then the pairs of values of lnk and the corresponding reciprocals of the absolute temperatures are plotted. In the temperature ranges where external factors alter the regular course of the reaction, the corresponding points fall ouside the Arrhenius line. The conditions corresponding to these points therefore warrant careful interpretation and may suggest the existence of important situations.

Based on the assumption that microbial growth is the result of a complex and highly regulated set of chemical reactions, the influence of temperature on microbial growth is directly related to the influence of temperature on the kinetics of that set of reactions.[2] To support this claim, Ingraham et al.[2] proposed a general Arrhenius plot for microbial growth (Fig. 5), where the cardinal temperatures – the maximum, optimum and minimum for growth – and the temperature ranges where growth rates are high, normal and low, can be clearly identified. The curve is concave toward the x-axis and exhibits only one straight branch corresponding to the normal temperature range. Hence, it is in this range that the system evolves according to the model of a chemical reaction. Above and below the optimum temperature growth rates are lower than those that would be obtained by extrapolating the straight branch of the Arrhenius plot. Above the optimum temperature thermal death of the microorganisms occurs rapidly with activation energies as high as 230 MJ mol^{-1}, that represent the thermal deactivation energies of microbial enzyme systems.[3] In addition, temperature sensitivity is much greater in the high and low ranges than in the normal one and attains its maximum close to 0°C. Growth and microbial activity may decline drastically below the optimum temperature.[4] Whereas protein denaturation provides a likely explanation of thermal death at high temperatures, the causes of growth slowdown and ultimately thermal death at the low temperatures resulting from decreases down to temperatures around 0°C on both sides, are not yet fully understood.

Some *Thiobacillus ferrooxidans* strains isolated from drainage waters of mines, where temperatures range from 15 to 20°C and more, have almost always been found to be highly sensitive to cooling to -40°C (for storage purposes) at the Bio-

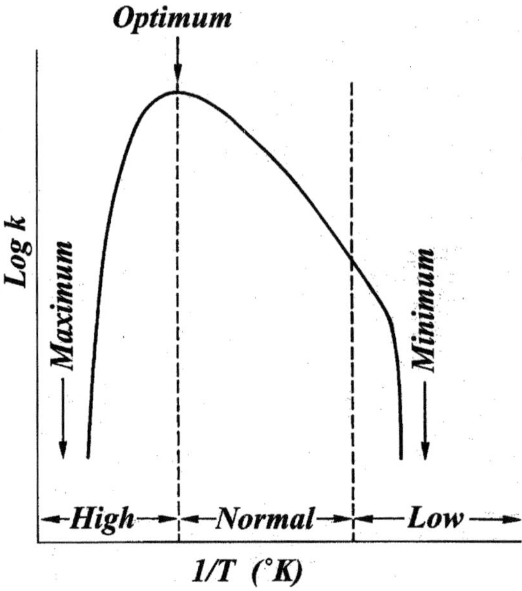

Fig. 5. General form of an Arrhenius plot of bacterial growth. The cardinal temperatures (the maximum, optimum and minimum for growth) and the growth temperatures ranges (high, normal and low) are shown (reprinted with permission from ref. 2)

hydrometallurgy Laboratory of the University of Cagliari. In most cases death occurred and the iron-oxidizing kinetics of a few of these microorganisms was much slower than that of identical strains not cooled. The strains isolated from the San Valentino di Predoi mine, where the temperature of drainage waters is constantly 8°C, behaved differently. After cooling down to -40°C they never lost their iron-oxidizing ability, though in some instances the kinetics appeared to be slightly slower than that of the uncooled strains.

A complex multiplicity of causes determines the sensitivity of bacterial cells to low temperatures. According to Ingraham et al.[2] a decisive role is played by slight changes in conformation of proteins produced by alteration of their primary structure and by the weakening of hydrophobic bonds, caused by changes which solvation water structure undergoes when the temperature is lowered. In consistency with the findings of some researchers, the possibility that strain adaptation alters the type of fatty acids of the cell membranes cannot be ruled out. It is well known that the presence and number of double bonds regulate the physical state of fatty acid molecules at low termperatures.

However, it should be noted that although growth and activity of mesophilic microorganisms of the genus *Thiobacillus* decline significantly as temperature decreases below the optimum, they still appear to be viable at 0°C and below (Fig. 6).[5] According to Figure 6, and in agreement with Figure 5, the microorganisms of the *Thiobacillus* genus maintain pronounced viability, estimated to be approximately 60% of that measured at optimum temperature, even around 0°C.

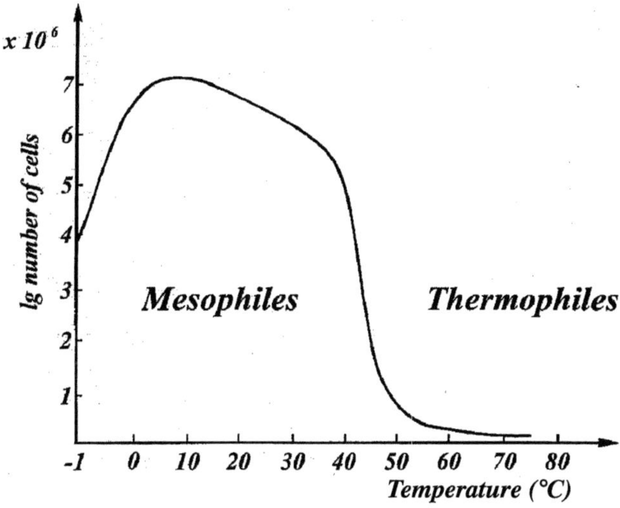

Fig. 6. Effect of temperature on growth of Thiobacilli (after ref. 5)

Figure 7 shows the results of investigations aimed at defining the lower temperature bounds for iron and sulfur oxidizing acidophiles by characterizing the kinetics of bivalent iron[6,7] and elemental sulfur[8] oxidation at temperatures considered as suboptimal. *T. ferrooxidans* strains were isolated from the drainage waters of underground stopes and from tailings ponds suspensions of the Finnish complex sulfides mine of Keretti[7,8] where, during the winters that last up to 6 months, the temperature remains constantly below 0°C, and from waters of the Canadian uranium mine at Denison, Elliott Lake, Ontario.[6] The temperatures of the water samples the strains were isolated from are shown in Table 1.

The two research teams calculated the growth parameters of the microorganisms for very similar temperature ranges (from 4 to 37°C and from 2 to 35°C) and proved that the *Thiobacillus* strains tested (i) may adapt and grow on ferrous sulfate over a wide temperature range, including that prevailing in the underground mines; (ii) exihibit constant growth rates, determined at various temperatures, that fit the Arrhenius equation with the aligned points falling on straight lines having slopes corresponding to activation energies of 95 and 83±3 kJ mol^{-1} respectively; (iii) are viable at temperatures down to 0°C, as opposed to collection strains isolated

Table 1. Temperatures of water samples containing the *Thiobacillus ferrooxidans* strains used for investigating the effects of low temperatures on iron and sulfur oxidizing Thiobacilli

Mining area	Temperature		(°C)
	Drainage waters from underground stopes	Waters tailing ponds	from
Keretti[7]	8	5	
Denison[8]	13÷18	0.5÷5	

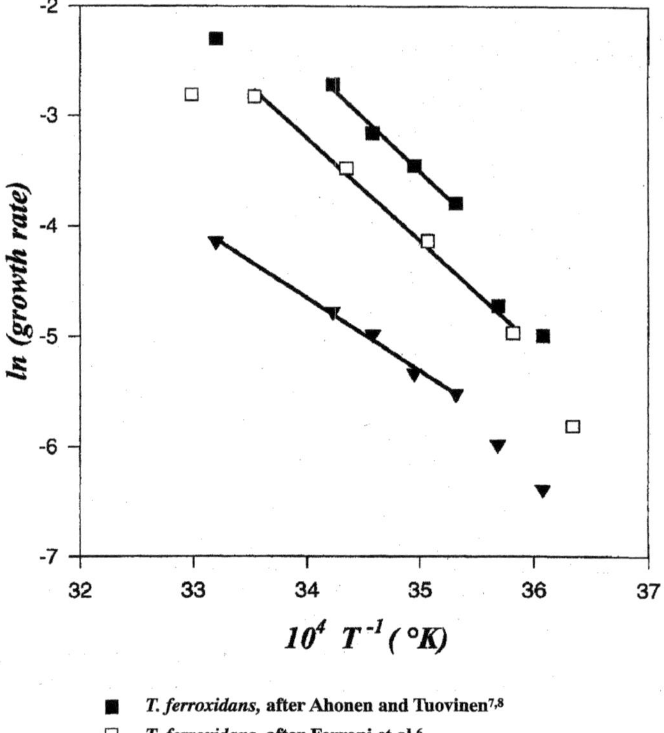

Fig. 7. Arrhenius plot of rate constants determined for elemental sulfur oxidation by acidophilic chemolithotrophic bacteria and for growing cultures of iron-oxidizing thiobacilli by various researchers. The regression lines were drawn limitedly to the aligned points

from environments characterized by higher temperatures.

A similar trend is exhibited in the Arrhenius plot constructed for sulfur oxidation which, for the aligned points, indicates activation energies of 65 kJ mol[-1]. Activation energies in the same order of magnitude (56 kJ mol[-1]) over the temperature range 10–25°C were obtained for bioleaching of the chalcocite (Cu_2S) run of mine ore at Quebrada Blanca mine in Chile.[9]

The activation energies of chemical reactions in solution fall in the range of 40÷105 kJ mol[-1], whereas physical processes, such as plastic flow or diffusion are characterized by activation energies of the order of 20 kJ mol[-1].[10] Zero activation energies appear only to be observed for reactions evolving without breakage of chemical bonds.[11] Hence, the activation energies found for bioleaching processes suggest a biochemical limitation at low temperatures. Further results confirm the above data and concern the outcome of chalcopyrite ($CuFeS_2$), sfalerite (ZnS) and pentlandite ($Fe_2Ni_9S_8$) column bioleaching tests carried out at suboptimal temperatures.[12] Meaningful tests have been carried out on the microflora isolated from the Denison mine with the aim to ascertain the existence of strains that are more psy-

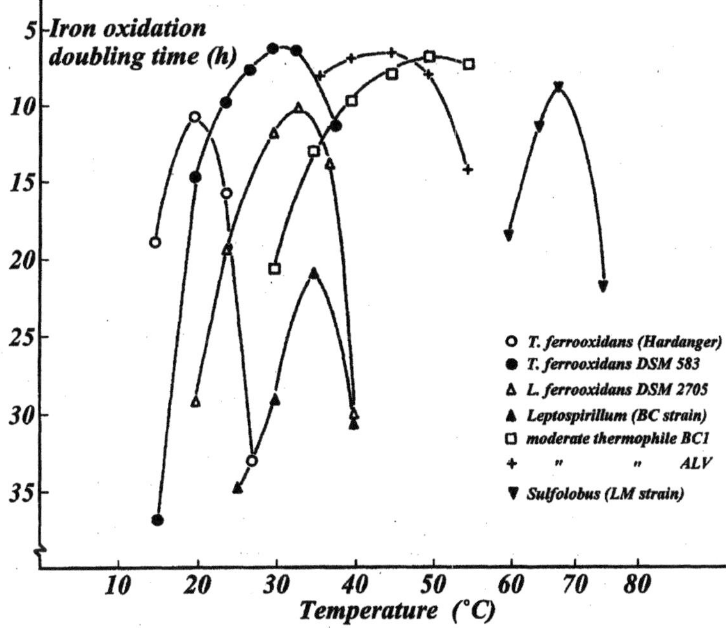

Fig. 8. The effect of temperature on the growth-associated iron oxidation by acidophilic bacteria. All bacteria were grown in shaken flasks on suitable media (reprinted with permission from ref. 14)

chrophilic than others.[13] A particularly active strain, isolated near a ventilation shaft and thus in an environment subjected to strong temperature variations, still grew satisfactorily at 8°C, showing much greater temperature tolerance than other strains which were unable to grow at temperatures below 15°C. Another interesting finding of this research was that the mean generation time of several *T. ferrooxidans* strains roughly halves at every 6°C temperature increase in the interval from 2 to 25°C.

Sensitivity to low temperatures probably depends upon a multiplicity of factors, such as the strain and the substrate on which it grows and possibly also the DNA-DNA homology, as pointed out by Norris[14] (Fig. 8).

The influence of the substrate on the requirements for the microbial growth at the minimum temperature seems to be confirmed by reactor bioleaching tests[15] on a flotation concentrate of a refractory gold-bearing arsenical sulfide which demonstrated that the bioleaching process only initiated when the temperature rose above 23°C.

Extensive measurements carried out in Russian commercial operations[16] provided interesting quantitative data on the effect of temperature on the activity of iron and sulfur oxidizing microorganisms in commercial bioleaching processes. Data collected from several bioleaching operations located in northern and central regions of the former Soviet Union, and confirmed by laboratory tests, demonstrated that a temperature drop from 26 to 15°C causes a 1.5- to 2.5-fold decrease of the mean specific growth rate of *T. ferrooxidans* and a 1.1- to 2.3-fold decrease of the mean oxidation rate of ferrous iron; further drops in temperature from 15 to 5.5°C produce a 4.2- to 16.3-fold decline in microbial growth rate depending on the other conditions.

Therefore, one may think that commercial processes should be designed to be operated at the highest temperature compatible with the physiology of the microflora. However, this option is not so straightforward as it might first seem, because a fundamental distinction must be made between the two classes of possible commercial bioleaching processes described above. The processes taking place within the broken rock mass involve the percolation of leach solutions with flow rates as high as thousands of cubic metres per day through volumes of rock that may amount to as much as several thousands of tons. Under these conditions, heating the whole system is not possible. For a profitable operation, the process should be optimized by regulating all the other parameters that influence its performance, such as the presence of nutrients in the leach liquors, the wetting cycle, and of course the selection of the most suitable microbial strain. The microbial growth rate should correspond to the temperature ranges prevailing in the mine environment.

The possibility of heating the ground mineral suspensions treated in bioreactors may be more realistic, viewed from the process as a whole, but must be carefully evaluated to avoid detrimental interactions in the abiotic subsystems. It is well documented that temperature rises enhance jarosite formation rates to such an extent that the bioleaching process is practically hindered. This drawback, together with the additional power costs, may ultimately outweigh the benefits of heating the system.

As far as the author of this chapter is aware, no commercial biohydrometallurgical operation exists where suspensions processed in bioreactors are heated. At the semi-commercial coal biodepyritization pilot plant built and operated at Porto Torres (Sardinia, Italy) for more than one year, the stirred tank reactors were provided with a suspension heating system, but owing to the mild ambient temperatures (seldom below 10°C) and to the exceptionally high performance of pyrite bioleaching it was never activated. With regard to bioleaching of broken rock masses, only one example (see below) is known of a commercial heap leaching operation where some form of heating or, more precisely, prevention of heat irradiation from the heaps, is being tested.

3

Case histories of biohydrometallurgical operations carried out at low temperatures

The case histories described below have been selected on account of their significance, but are by no means special cases: for more details on the commercial processes the interested reader should consult the references.

3.1
The El Teniente mine

The El Teniente copper sulfides and oxides mine is located 55 km east of the city of Rancagua, and 90 km south of Santiago, the capital of Chile. The orebody, where the copper minerals are associated to iron sulfides, is mined using the so-called "block caving stoping method". This consists of isolating a parallelepiped of rock from the surrounding rock mass by excavating vertical slots into its sides and two sets of tunnels perpendicular to each other at its base. In this way, the block of rock is supported simply by the pillars left after the tunnel network has been excavated. The base of the block may be as large as 25x25 m and more than 100 m high. Simultaneous blasting of the pillars causes the block to collapse and to undergo a sort of autogenous crushing as drawing of the broken ore, from a system of drawing chutes and tunnels previously prepared underneath its base, progresses. The stoping method is obviously non-selective, insofar as the fragmentation and extraction of only the rock containing a sufficiently high metal value is not possible. For this reason rock extraction is interrupted as soon as the copper assay of the run of mine ore becomes so low as to make the mining operation unprofitable (in mining economics this run of mine ore is defined as "waste" or "submarginal ore"). In 1987 a run of mine ore at El Teniente mine was considered submarginal when its copper assay was lower than 0.7%. Usually, the collapse of the rock blocks propagates up through the overburden to the surface, and as mining progresses a crater forms that is also a sinkhole for atmosferic waters. Annual precipitation in the area, where the elevation ranges from 2,500 and 3,700 m a.s.l., amounts to 1,025 mm of which 169 mm is rain and 856 mm equivalent snow.[17] This has resulted in a natural percolation process through the fragmented rock mass and in the development, in an apparently favorable environment, of an acidophilic microflora within the rock mass that triggered off a spontaneous bioleaching process as long ago as 1906. The author was not able to obtain information on the microflora. However, from conversations with local researchers and technicians the author understood that one of its major components is a *T. ferrooxidans* strain. In 1987 the volume of water flowing into the crater amounted to as much as 2,767,500 m³.

The importance of the economic viability of this process after suitable rationalization led the mining company to start controlled irrigation of the crater surface and to design and build a plant for recovering the copper contained in the leach liquors flowing out of the broken rock mass. The plant went into operation in 1985 and yearly copper cathode production ranges from 5,000 to 6,000 tons. During a visit of the crater and the plant just after it had been built in 1985, the water temperature was not above 8°C. However, statistical data provided in 1987[17] suggest that there is not significant seasonal influence on copper production. Two explanations are possible, that do not exclude each other: (i) *T. ferrooxidans* and associated acidophilic microflora already existed prior to mining operations and were well adapted to the local environment and/or (ii) within the fragmented rock mass ecological niches have formed where temperatures are much higher than in the percolating water, as a result of the heat generated by oxidation of the iron sulfides accompanying the copper minerals.

3.2
The Quebrada Blanca mine

The Quebrada Blanca orebody[9,18] is located at an altitude of 4,300 m a.s.l., 250 km south-east of the city of Iquique, in northern Chile. Measured reserves amounted in 1991 to 77 million tons of chalcocite ore with an average total copper assay of 1.41%. The orebody is mined with the open pit method. Rock blasting is designed on the basis of indications provided by preliminary diagnostic drillings. So it is possible to determine in the blasted rock those volumes assaying more than 1% Cu, those assaying from 0.8 to 0.4% Cu and those assaying less than 0.4% Cu. The copper production flowsheet consists of three major stages: (i) comminution of the run of mine ore to achieve adequate mineral exposure for the access of leaching agents, i.e. iron sulfate and microorganisms; (ii) agglomeration of the fine fractions with sulfuric acid and formation of 6 m high heaps with 1 m thick layers of -8 mm crushed rock and agglomerated fines to ensure good permeability followed by heap bioleaching; this consists in sprinkling the flat upper surfaces of the heaps with leach solutions containing microorganisms and nutrient salts and collecting the copper-laden liquors at the toes of the heaps; (iii) concentration of the leach liquors through solvent extraction and production of 98.88% cathode copper in an electrolysis plant.

The daily production schedule is 15,000 tons of run of mine ore and 75,000 tons per year of cathode copper. The extreme weather conditions existing in the area on account of its elevation, are the cause of significant variations in daily temperatures which range from -15 to 20°C. The winds, the intensive solar radiation during the day and the strong night radiation favored by clear skies give rise to thermal flows which had to be taken into account in the plant design. Annual rainfall is variable, with a predominance of dry years (less than 150 mm of water) over years defined as "wet" (more than 300 mm water).

The microflora active within the heaps consists of a variety of autothrophs and heterothrophs. Only scant information was provided about the microflora that was admittedly considered a highly confidential matter of decisive commercial importance. There are, however, good reasons to believe that the main biocatalyst in the process belongs to a species of *T. ferrooxidans*. It was not possible to ascertain whether this microorganism is indigenous, i. e. whether it was already present in the orebody prior to commercial operation, but it is very likely that its activity is largely affected by daily and seasonal temperature variations. This is testified by the fact that transparent polypropilene sheets are used to cover the flat tops of the heaps and that the possibility of blowing warm air into the heaps through pipes running along their bottom is being considered.

The copper content of metal laden liquors can be as much as 3.5 kg m^{-3}. Such a high metal content may be related to the excellent adaptation of the microorganisms to the hostile thermal environment or to thermal compensation within the heaps provided by oxidation processes.

3.3
The San Valentino di Predoi mine

The San Valentino di Predoi mine[22,23] is located in the Alps, a few hundred metres north-east of the town of Predoi (Prettau, in German, in this bilingual region) on the left slope of the Aurino Creek, midway between the towns of Predoi-Prettau and Casere-Kasern. The town of Predoi is located at the entrance of the High Aurino Valley, in the autonomous province of Bolzano-Bozen at an altitude between 1,413 and 1,491 m a.s.l. The orebody consists of arrays of equally oriented, vertical lenses, elongated and irregular in shape, striking N70°E. The most important ore mineral is pyrite, with which chalcopyrite, pyrrhotite, and minor amounts of bornite (Cu_5FeS_4), cubanite ($CuFe_2S_3$), valleriite [$CuFeS_2$ with interlayers of $Mg(OH)_2$, $Ni(OH)_2$ and $Fe(OH)_2$], sfalerite, chalcocite and covellite (CuS) are associated. The copper assay is very variable, ranging from 0.5 to 3%.

Mining activity dates back to the XV century and continued uninterruptedly until 1894.[24,25] Over this period the orebody was worked between the altitude of 1,516 m a.s.l., where the water table is located, and the outcrops, at 2,059 m a.s.l. Access to the stopes was provided by eight drifts of varying length: the St. Nicholas drift, at 1,624 m, was excavated in spring 1761 and is 1,086 m long; the St. Ignaz drift, at 1,516 m, was excavated in 1805 and is 1,165 m long. Both are still in excellent static conditions. The orebody was also developed 88 m below the water table, in a lens (Fig. 9) that was only partially mined. The cavities left by the stopes were filled, for static purposes, with submarginal ore. A shaft, called Archduke John shaft, connected the St. Ignaz level to the drifts excavated for access to the stopes. Being located below the natural water table, the stopes remained flooded from 1972 to 1984.

Climate in the area is characterized by long winters lasting from early October to late April, during which the temperature often falls below -10 to -15°C, and heavy snowfalls. However, the temperature of the air and drainage water underground is constantly 8°C all year around.

Interestingly, it is reported in the chronicles that already in 1530 copper was not only produced by pyrometallurgical processing of the hand sorted high grade run of mine ore, but also by cementation[20] (contact precipitation of copper from solutions dripping from the upper levels) on scrap iron. Technicians of the XVI century were certainly unaware of the causes underlying the process whereby the water percolating from the caved outcrops through the mined out stopes became acidic and leached the copper minerals left, but were clever and skillful enough to exploit the fenomenon. This shows that the microflora responsible for leaching the iron and copper sulfides is characterized by a psychrophilic behavior.

In 1984 the situation above the St. Ignaz level could be summed up as follows:

Where the outcrops once stood at 2,060 m a.s.l., following stoping of the parts of the orebody immediately beneath them, subsidence occurred and the are caved in forming a sort of large sinkhole a few tens of metres wide and a hundred metres long. During rainy periods or when the snow melted, this sinkhole provided a means of access of large and irregular flowrates of water that percolated through the abandoned stopes.

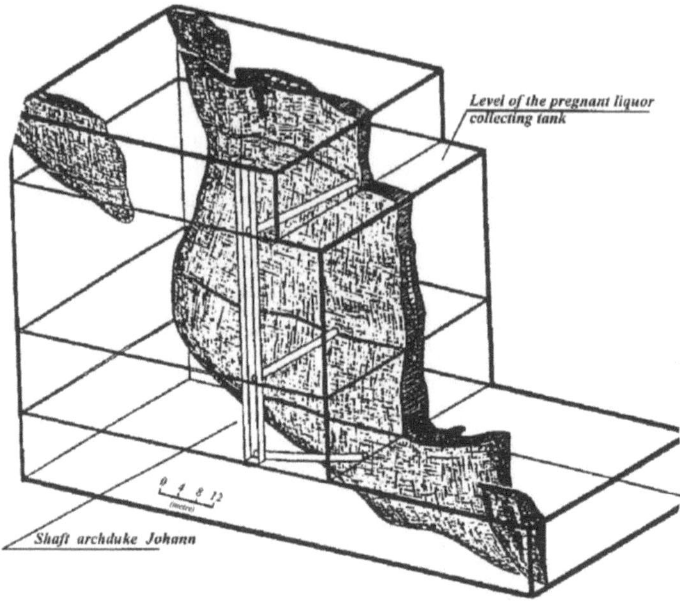

Fig. 9. Isometric view of the lens of the San Valentino di Predoi mine where the bioleaching operation described in the text has been carried out (reprinted with permission from ref. 22)

- There were several cavities created by worked out stopes, partially filled with low grade rock and raises still filled with ore, sills and pillars of ore.
- The St. Ignaz level comprised a network of several hundred metres of wide tunnels (cross section of 15–20 m²) in excellent static conditions.
- From several spots in the tunnel rooves and from the raises connecting the level with the overlying zones water dripped out at a rate often in the order of magnitude of several dm³ min⁻¹. The average flowrate of the water coming through the roof is about 35 m³ d⁻¹. Microflora consisting of acidophilic autotrophs, like *T. ferrooxidans* and *T. thiooxidans*, and several strains of heterotrophs was isolated.

The structural and textural features of both the country rock and the orebody along with the geometry of the lens located below the St. Ignaz level suggested it would be suitable for a semi-commercial integrated bioleaching and copper cementation pilot plant. The water flooding the mine workings was pumped out and the water level was lowered to 40 m below the St. Ignaz level, corresponding to a tunnel where a tank was installed. The acidic pregnant liquors flowing out of the lens after percolation through the low grade backfills and the developed stopes were collected in this tank and pumped to a buffer reservoir with a capacity of more than 1,000 m³, made by closing with a bulkhead one of the tunnels at the St. Ignaz level (Fig. 1). The pregnant liquor was conveyed by gravity flow from this reservoir to the launders installed in the tunnels. In this way as temperature remained constantly at 8°C the cementation plant could also be operated in winter without any risk of freezing. The test continued for about ten years and the plant is still operating.

The leach liquors, whose temperature is 8°C, are not as rich in copper as those of El Teniente mine. The copper content varies over the year, but is never more than 0.5 kg m⁻³. This testifies nonetheless that the microflora is viable and active. An economic analysis carried out over the last 5 years of operation indicated that this copper content is sufficient to achieve a cement copper production that pays off operation costs.[23]

It is highly unlikely that the constant temperature in the mine is the result of heat being generated by pyrite oxidation. However it is a very reliable indication of a psychrophilic response of the microflora that undoubtedly adapted, over periods of time of the geological order, to the extreme environmental conditions, showing the ability to catalyze sulphide oxidation and solubilization when man began exploiting the orebody.

3.4
Experiences in Canadian mines

The results of investigations being carried out in Canadian uranium mines[6,19–21] located in the Elliott Lake region since the sixties are particularly interesting because they provide information on bioleaching operations in environments characterized by temperatures as low as 10°C.

3.4.1
The Denison mine

The Denison orebody is on the Quirke Lake syncline and forms part of the quartz-pebble conglomerates of the Huronian sediments in the region.[19] The uranium-bearing minerals brannerite (UTi_2O_6) and uraninite (UO_3) are concentrated at the pebble/quartz interfaces of these conglomerates and are accompanied by 7% by mass of pyrite and pyrrhotite. The minerals actually bioleached are pyrite and pyrrhotite: uranium-bearing minerals are oxidized and leached by the acidic ferric sulfate produced by oxidation of the iron sulfides catalyzed by *T. ferrooxidans* and the associated microflora.

Stoping consisted of excavating cavities by conventional rock blasting, removing the excess rock and leaving in the stopes as much broken rock as possible though still ensuring access to the miners. Bioleaching was carried out underground by flooding the stopes. The leach solutions, consisting of the acidic mine drainage, with a pH of 2.3 and a constant temperature of 12°C all year round, were pumped into the cavities to submerge the broken rock mass. Complete flooding took 3 d. The bioleaching cycle consisted of 3 d flooding, followed by drainage of the leach liquors and then 3 weeks rest. After the rest period, the cycle was repeated. Bioleaching proper, with production of soluble salts, took place during the rest period. Flooding produced the pregnant liquors and the first metal recovery step consisted of drainage with pumping to the surface. Temperature in the underground stopes ranged from 0 to 15°C during winter to reach 22°C at the end of summer. The microflora consisted of indigenous strains, among which two psychrophilic *T. fer-*

rooxidans strains that were the only ones able to remain viable below 15°C. At Denison Mine too, similarly to what had been observed at Predoi, the drainage water temperature remained constant all year round, in spite of the fact that during the long winters the temperature at the surface dropped to well below 0°C. Like at Predoi, bioleaching performance was satisfactory though temperatures were low and always much lower than the optimal values determined in laboratory tests. However, with milder air temperatures the uranium extraction rate was three times that registered in the colder periods.

3.4.2
The Agnew Lake mine

This mine is the only known example of a "custom built" operation for biohydrometallurgical mining by *in-situ* bioleaching. The orebody is located south-east of the Sudbury basin, 60 km west of the town of Sudbury in Ontario, and consists of a highly fractured oligomictic conglomerate containing $[(U)(Th)SiO_4]$, accompanied by pyrite and pyrrhotite with minor amounts of chalcopyrite, galena and sfalerite in two zones stratigrafically 180 m apart. The ore was blasted in underground workings with the sublevel stoping method. As the volume of the fragmented rock was greater than that of the rock in place, the excess was hoisted to the surface and piled up in heaps. Bioleaching was carried out simultaneously underground and in the heaps. In 1978 at Agnew Lake 181,417 kg of U_3O_8 were produced of which 113,398 kg came from the heaps at the surface.

From the circulating solutions, that had a practically constant temperature of 10°C in the underground stopes, a microflora consisting of iron-oxidizing Thiobacilli and glucose-oxidizing heterotrophs that did not exhibit any seasonal changes was isolated.[21] This commercial operation thus appears to confirm that the indigenous, probably psychrophilic, strains are suitable for an effective oxidation of mineral iron sulfides.

4
Conclusions

It may be concluded that bioleaching of metal sulfides at temperatures below 15°C is mediated by microorganisms belonging to the genus *Thiobacillus* and particularly by psychrophilic strains of the species *T. ferrooxidans*. This statement is borne out by the case histories described in this chapter and by repeated observations made in commercial plants of the former Soviet Union in areas where temperatures fall below 15°C for several months of the year, according to which no leaching of metal sulfides occurs when these strains are not present in the leach liquors.[26]

5
References

1. Loi G, Trois P, Rossi G. Biorotor: a new development for biohydrometallurgical processing. In: Vargas T, Jerez CA, Wiertz JV, Toledo H, eds. Biohydrometallurgical Processing, vol 1. Santiago: Chile, University of Chile, 1995:263-271.

2. Ingraham JL, Maaløe O, Neidhart F. Growth of the Bacterial Cell, 1st ed. Sunderland, MA: Sinauer Associates Inc, 1983:251.

3. Aiba S, Humphrey AE, Mills NF. Biochemical Engineering, 2nd ed. New York: Academic Press, 1973.

4. Atlas RM, Bartha R. Microbial Ecology. Fundamentals and Applications, 2nd ed. Menlo Park, California: Benjamin/Cummings Publishing Co Inc, 1987:236.

5. Karavaiko GI, Lyalikova NN, Pivovarova TA. The microooorganisms in orebodies, their physiology and geochemical activity. In: Research Centre for Biological Phenomena, ed. Ekologiya i Geochimicheskaya Deiatelnost Microorganismov. Pushkino, 1976:25.

6. Ferroni GD, Leduc LG, Todd M. Isolation and temperature characterization of psychrotrophic strains of *Thiobacillus ferrooxidans* from the environment of a uranium mine. J Gen Microbiol 1986; 32:169-175.

7. Ahonen L, Tuovinen OH. Microbiological oxidation of ferrous iron at low temperatures. Appl Environ Microbiol 1989; 55:312-316.

8. Ahonen L, Tuovinen OH. Kinetics of sulfur oxidation at suboptimal temperatures. Appl Environ Microbiol 1990; 56:560-562.

9. Montealegre R, Bustos S, Rojas J, Neuburg H, Araya C, Yañez H, Tapia R, Rauld J. Application of the bacterial thin layer leaching process to Quebrada Blanca ores. In: Torma AE, Wey JE, Lakshmanan VI, eds. Biohydrometallurgical Technologies – Bioleaching Processes, vol 1. Warrendale, Penna, USA: The Minerals, Metals and Materials Society, 1993:1-27.

10. Wadsworth ME. Hydrometallurgical processes. In: Sohn HY, Wadsworth ME, eds. Rate Processes of Extractive Metallurgy. New York London: Plenum Press, 1979:133.

11. Laidler KJ. Chemical Kinetics. London: McGraw-Hill Publ Co Ltd, 1971:566.

12. Ahonen L, Tuovinen OH. Bacterial oxidation of sulfide minerals in column laching experiments at suboptimal temperatures.Appl Environ Microbiol 1992; 58:600-606.

13. McCready RGL. Progress in the bacterial leaching of metals in Canada. In: Norris PR, Kelly DP, eds. Biohydrometallurgy, Proceedings of the International Symposium Warwick 1987. Kew, Surrey: Science and Technology Letters 1987:177-195.

14. Norris PR. Acidophilic bacteria and their activity in mineral sulfide oxidation. In: Ehrlich HL, Brierley CL, eds. Microbial Mineral Recovery. New York: McGraw-Hill Publ Co Ltd, 1990:3-27.

15. Marchant PB. Commercial piloting and the economic feasibility of plant scale continuous biological tank leaching at Equity Silver Mines limited. In: Lawrence RW, Branion RMR, Ebner HG, eds. Fundamental and Applied biohydrometallurgy. Amsterdam: Elsevier, 1986:53-76.

16. Karavaiko GI. Evaluation of microbial distribution and activity in the process of dump and underground leaching of metals. In: Karavaiko GI, Rossi G, Avakyan AA, eds. Biohydrometallurgy. Proceedings of the International Seminar on Dump and Underground Bacterial Leaching Metals from Ores. Moscow: Centre For International Projects, USSR State Committee for Environment Protection and United Nations Environment Programme, 1990:218-234.

17. Ovalle AW. In-place leaching of a block caving mine. In: Cooper WC, Lagos GE, Ugarte G, eds. Copper '87, vol 3: Hydrometallurgy and Electrometallurgy of Copper. Santiago: The University of Chile, 1987:17-37.

18. Trois P. La Miniera di Quebrada Blanca, Cile: il più recente esempio di produzione industriale di rame per biolisciviazione. Resoconti delle Sedute dell'Associazione Mineraria Sarda, Iglesias, 1994.

19. McCready RGL, Gould WD. Bioleaching of uranium at Denison Mines. In: Salley J, McCready RGL, Wichlacz P, eds. Biohydrometallurgy, Proceedings of the International Symposium held at Jackson Hole, Wyoming. Ottawa: CANMET SP89-10, 1989:477-485.
20. Rossi G. Biohydrometallurgy. Hamburg: McGraw-Hill GmbH, 1990.
21. Tuovinen OH, Silver M, Martin PAW, Dugan P. The Agnew Lake uranium mine leach liquors: chemical examinations, bacterial enumeration, and composition of plasmid DNA of iron oxidizing Thiobacilli. In: Proceedings of the International Conference on Use of Microorganisms in Biohydrometallurgy. Hungary: Hungarian Academy of Sciences, Local Committee of Pécs, 1980:59-69.
22. Rossi G, Trois P, Visca P. In-situ pilot semi-commercial bioleaching test at the San Valentino di Predoi mine (Northern Italy). In: Lawrence RW, Branion RMR, Ebner HG, eds. Fundamental and Applied Biohydrometallurgy. Amsterdam: Elsevier, 1986:173-189.
23. Loi G, Rossi G, Trois P, Zuffardi P. Biolixiviation in situ d'un minerai cuprifère. Exemple de l'ancienne mine de San Valentino di Predoi (Province de Bolzano, Italie). Chron Rec Min 511, 1993:33-40.
24. Tasser R. Die Geschichte des Kupferbergwerkes Prettau von den Anfängen bis 1676. In: 25 Jahre Gemeinde Prettau, Township Predoi, 1983:21-42.
25. Tasser R. Geschichte des Kupferbergwerkes Prettau von den Anfängen bis 1676. PhD thesis, University of Innsbruck, 1970.
26. Khalezov BD. Case histories. In: Karavaiko GI, Rossi G, Agate AD, Groudev SN, Avakyan ZA, eds. Biogeotechnology of Metals - Manual. Moscow: Centre For International Projects, USSR State Committee for Environment Protection and United Nations Environment Programme, 1988:277-303.
27. Ismay A, Rosato L, Mckinnon D. Engineering prefeasibility for in-place bacterial leaching of copper. In: Lawrence RW, Branion RMR, Ebner HG, eds. Fundamental and Applied Biohydrometallurgy. Amsterdam: Elsevier, 1986:191-213.

Applications of biological ice nucleators

R. Lundheim[1]* and K. E. Zachariassen[2]

[1] Allforsk Biology, Gryta 2, N-7010 Trondheim, Norway
[2] Norwegian University of Science and Technology, Department of Zoology, N-7034 Trondheim, Norway

1
Sources of biological ice nucleators and their classification

In a sample of pure water at a pressure of 1 bar (10^5 Pa) and a temperature of 0°C, ice and liquid water are in thermodynamic equilibrium. At lower temperatures pure liquid water is metastable, and often referred to as supercooled water. The initiation of the transformation from supercooled water to ice is called nucleation.[1] The term nucleation temperature describes the temperature at which nucleation takes place.

Homogenous nucleation is characterized by the water molecules themselves forming an embryo crystal. As temperature is lowered, water molecules are clustering to form ice-like water aggregates of variable sizes. When an aggregate reaches a critical size, a rapid ice crystal growth occurs. The probability of a water aggregate reaching the critical size is temperature dependent. Because of the strong dependence on temperature, homogenous nucleation will usually take place at around -40°C. According to Vali,[2] the nucleation rate increases by a factor of 50 for each 1°C lowering of temperature. The critical size of a water aggregate is 70 molecules at -40°C and 650 at -20°C.[2]

Heterogenous nucleation is characterized by the water molecules aggregating on the surface of a substance different from water. This is the form of nucleation most relevant in living organisms where water is in close contact with a variety of substances. A site that facilitates ice-like aggregation of water molecules is termed an ice nucleator or ice nucleation site. A nucleation site may consist of one molecule or may be a result of several molecules aggregating to form the nucleation site. Since many substances can facilitate ice nucleation at low temperatures it is convenient to restrict the use of the term ice nucleators to those substances that facilitate nucleation at relatively high temperatures, for example above -12°C.

Ice nucleators of inorganic origin such as silver iodide have long been known.[3] The early works to identify ice nucleators were mainly performed by meteorologists focusing on particles inducing ice nucleation in clouds. Several potent ice nucle-

---

* Corresponding author

ators were identified, most of them mineral particles. In 1970 Russell C. Schnell discovered that the very potent ice nucleators derived from decaying leaves were of microbiological origin,[4] and in 1972 *Pseudomonas syringae* was identified as the organism causing the high nucleation temperatures.[5] Almost simultaneously another research group interested in frost damage to crop plants realized that plants infested by microorganisms where susceptible to frost damage. Arny et al.[6] discovered that the susceptibility to frost of plants infested by *P. syringae* was caused by ice nucleation at high temperatures rather than by a reduced resistance to stress caused by the infection. In the same issue as Arny et al. published their work, Zachariassen and Hammel[7] reported their findings of ice nucleators in the hemolymph of insects tolerant to freezing. The presence of ice nucleators has later been reported in organisms from different taxa such as Bivalvia,[8] Gastropoda,[9] Lichenes,[10] Reptilia,[11] Amphibia,[12] Tardigrada,[13] Fungi,[14] Angiospermae,[15] Chilopoda,[16] Phaeophyta and Rhodophyta.[17]

Ice nucleators have been classified by their nucleation temperatures and their chemical composition. Yankofsky et al.[18] divided bacterial ice nuclei into three classes based mainly on results from *P. syringae*, *Erwinia herbicola* and some unidentified bacteria: group I nucleates at temperatures higher than -5°C, group II nucleates between -5°C and -7°C and group III nucleates at temperatures lower than -7°C.

Turner et al.,[19] studying ice nucleators from *P. syringae*, *E. herbicola* and recombinant *Escherichia coli*, also divided nucleators into three classes based partly on their nucleation temperatures. In addition, they also used the ability to nucleate D_2O and the effects of organic solvents and pH on nucleation temperature as criteria for their classification. They named their classes of nucleators A (nucleating above -4.4°C), B (nucleating at temperatures between -4.8 and -5.7°C), and C (nucleating at temperatures below -7.6°C).

Because relatively few ice nucleators have been characterized chemically, it is difficult to draw conclusions of general differences in ice nucleator composition between taxa. The presence of several ice nucleators in the cranefly *Tipula trivittata*[20] shows that even in one species different types of ice nucleators can be present. Even if data are scarce, some indication of similarities and differences between taxa may be extracted from Table 1. Both the nucleator from the lichenized fungus *Rhizoplaca chrysoleuca* and from the fungus *Fusarium avenaceum* are heat stable at temperatures up to 60–70°C. The highest nucleation temperature of both is about -2°C, they are relatively unaffected by pH, and pass through a 0.22-μm filter. Both these fungi belong to the phylum Ascomycota so the similarities between the ice nucleators can be explained by a possible common origin. The nucleator from *R. chrysoleuca* has been shown to be proteinaceous.[21]

Nucleators from the vascular plant winter rye[22] and sea buckthorn[23] show similarities with bacterial nucleators in their lipid requirements, phospholipid content and their temperature sensitivity, even though the temperature sensitivity of the ice nucleators from winter rye is not investigated in detail. The nucleator from *Prunus* seems to be of a totally different nature than the others since it is probably not proteinaceous.[15] The contention that it is not proteinaceous is, however, not supported by its heat sensitivity and the finding that it is sensitive to guanidine, which

Table 1.   Chemical composition of some biological ice nucleators. If nothing else is indicated, protein molecular weights are given for sodium dodecyl sulphate (SDS) denatured proteins and are therefore not representative for the native size of nucleation sites

Organisms	Protein molecular weight (kDa)	Lipids	Phospholipids	Carbo-hydrates	Heat sensitivity
Bacteria	120–180[25]	Yes[25]	Phosphatidylinositol and others[25]	Yes[25]	Yes, 20–30°C[26]
Insects	74,[27] 265,[28] 81,[28] 77[29]	Yes,[28] no[26,27,35]	Phosphatidylinositol[28]	Yes[28]	<100°C[7,30] 70–90°C[31] 40–70°C[35]
Molluscs	16.6,[32] 17.7[32]	No[32]	No[32]	No[32]	<100°C[8,9]
Tardigrada	Native>200[13]				60–70°C[13]
Vertebrates	Native>300[33]				85–90°C[33]
Vascular plants	Native>300[23] probably not protein[15]	Yes[22,23]	Yes,[22,23] other than phosphatidylinositol[23]	Yes[22]	20–30°C[23] <90°C[22] 50°C[15]
Lichenes		No[21]	No[21]	No[21]	60–70°C[21]
Fungi (*Fusarium*)		No[34]	No[34]	No[34]	60–70°C[14] 40–50°C[34]

denaturates proteins. Both of the *Tipula* apo proteins are essential for nucleation, and both crossreact with antibodies raised against bacterial ice nucleator proteins from the bacterium *P. fluorescens*, indicating similarities in structure.[20] Not all ice nucleators found in insects are proteinaceous; in *Eurosta solidaginis* crystals of tribasic calcium phosphate induce ice nucleation at -7.8°C.[24]

2
Applications

2.1
Food processing

There are several reasons for inducing ice nucleation at high temperatures during processing of foodstuff:

Energy saving
If ice is formed at a high temperature, energy is saved because of the transport of heat from the food to the cooling device takes place over a steep thermal gradient.

Freeze concentration

Almost all solutes will be excluded from the ice lattice when ice crystals are formed in a solution. A liquid fraction of water and solutes is in melting point equilibrium with the ice. Therefore the concentrations of solutes in this liquid fraction will be a function of the temperature of the ice/water mixture. By freezing watery foodstuffs under controlled temperature conditions one can therefore remove pure water as ice crystals, for example by centrifugation. Adding ice nucleators will increase the size of the ice crystals that are formed. Large ice crystals are desired during freezing because they are more easily removed. The effects of adding bacterial ice nucleators on food quality have been tested for different food systems, e. g. egg white,[36] milk,[37] and fruit juice.[36,38] Ice nucleators obtained from fruit juice of sea buckthorn *(Hippophae rhamnoides)* have been used to enhance freeze concentration of coffee.[39] One of the greatest advantages of using freeze concentration to remove water is that the flavor and color of the food items are kept closer to the untreated food item.

Freeze drying

Watanabe and Arai[40] found that adding bacterial ice nucleators to food made the freeze drying of foodstuffs more efficient. It also enhanced a desirable powdering of the freeze-dried product.

Texturing of food items

The anisotropic growth of ice crystals can alter the texture of foodstuffs so that it becomes more appealing to the customer. Examples of this are the texturing of egg white that was obtained by Arai and Watanabe[41] when bacterial ice nucleators were added. Without ice nucleators the frozen product had a sponge-like texture when it was melted. Jann et al.[39] found that a soy bean concentrate frozen with the addition of ice nucleators obtained from sea buckthorn had a fibrous and lammelar texture after freezing, freeze drying and heating. Soy bean concentrate without ice nucleators from sea buckthorn, but otherwise treated in the same way, had a spongy texture.

2.2
Production of artificial snow

The use of bacteria containing ice nucleators (Ice$^+$) to enhance the production of artificial snow is probably the most widespread commercial use of ice nucleators. This application of ice nucleators has created a demand for large quantities of bacteria containing the gene for ice nucleation (Ina$^+$). Large-scale fermentation of Ina$^+$ bacteria has been the subject of several patents.[42–45] Ina$^+$ bacteria are sold under the tradename SNOMAX and are added to the water used in snow-making machines, for example in ski areas. This will raise the critical temperature for artificial snow-making by several degrees Centigrade.[46] The use of ice nucleators in snow-making machines can also be beneficial when ice is produced as construction material for installations in arctic and antarctic areas.[47]

2.3
Cloud seeding

In order to produce a rainfall, water vapor in the atmosphere must condense to form water droplets or ice crystals. This condensation is facilitated by cloud condensation nuclei (nucleating vapor/liquid transformation) or ice nucleators (nucleating liquid/ice transformation). The ice crystals formed will grow in the supersaturated atmosphere and snowflakes will be formed. If the saturation of water vapor in the atmosphere is high enough, the snowflakes will reach the ground as snow or rain depending on temperature.[48] Nucleating agents such as silver iodide[3] have traditionally been used to nucleate supercooled droplets in clouds and thereby increase the probability of rainfall. This method of inducing precipitation is called cloud seeding. The discovery of bacterial ice nucleators has provided promising sources of ice nucleators for cloud seeding. The bacterium *P. syringae* is now available as the commercial product (SNOMAX Weather Manager, Genencor International Inc.) for use in cloud seeding.[49]

2.4
Controlling insect pests

Many insects survive exposure to cold by extensive supercooling. Ice formation in the body fluids is lethal to most insects. Some insects, however, tolerate freezing and ice nucleating microorganisms may be a normal part of the flora in their guts.[50–54]

Strong-Gunderson et al.[54] showed that the supercooling points of the freeze susceptible beetle *Hippodamia convergens* increased from -16 to -2°C when the beetles were fed water suspensions of the ice nucleating bacteria *P. syringae*. Controls fed suspensions of *Escherichia coli* showed no increase in supercooling points. Similar effects have later been demonstrated when different insect species (many of them living in stored products) have been treated with Ina+ bacteria or the fungi *Fusarium acuminatum* and *F. avenaceum*.[55–57] Even external application of microorganisms to insects with sealed mouths increases the insects' supercooling point, making them more susceptible to cold.[58,59] The Colorado potato beetle *(Leptinotarsa decemlineata)* has also been shown to become more vulnerable to low temperatures after the application of Ice+ micro-organisms.[60]

2.5
Prevention of plant frost injury by controlling ice nucleation active bacteria

Many important crop plants are frost sensitive and many survive moderate cold spells by supercooling.[61] Bacteria producing ice nucleators (Ice+ bacteria) are often responsible for ice nucleation in these crop plants under field conditions.[62] If the Ice+ bacteria could be removed from the plants, the plants would survive lower temperatures by supercooling. Several approaches can be made to control Ice+ bacteria. One approach is the use of bactericides. Different copper salts and antibiotics are effective in reducing the populations of Ice+ bacteria and thereby increasing the cold hardiness of the host plant (for a review, see ref. 63).

A problem associated with the use of bactericides is the development of strains resistant against the bactericides. An alternative approach to using bactericides is to use bacteria without the *ina* gene (coding for ice nucleators) to compete with the Ice⁺ bacteria and thereby reduce the amount of Ice⁺ bacteria.[62,64,65] The introduction of Ice⁻ strains (not capable of inducing ice nucleation) of *P. syringae* has been shown to reduce the amount of Ice⁺ *P. syringae* on crop plants.[66] Recombinant Ice⁻ *P. syringae* strains were one of the first genetically engineered organisms to be used in field studies. It has, however, been shown that naturally occurring Ice⁻ strains of *P. syringae* are as effective as the recombinant strains in reducing frost damages by outcompeting Ice⁺ strains.

2.6
The use of genes coding for ice nucleators as reporters

A reporter gene is a gene that is coding for an easily detectable property. If it is expressed together with other genes that code for functions that are not easy to detect or monitor, the reporters will show the activity of both genes. Lindgren et al.[67] used a gene coding for the production of ice nucleators as a reporter for identification of inducible pathogenicity genes in *P. syringae* pv. *phaseolicola*. The use of *ina* genes as reporters has been reviewed by Panopoulos.[68] A special case of using *ina* genes as reporters is the use of species-specific bacteriophages to transduce *ina* genes to pathogenic bacteria.[69,70] By doing this, the presence of a pathogenic organism can be detected by the ice nucleation spectra of a sample that is to be investigated for the presence of a certain bacterium: a species-specific bacteriophage capable of transducing an *ina* gene is added to the food sample. If the target bacterium is present, the expression of the *ina* gene can be detected as a shift in the nucleation spectra of the test samples after a short time.[71]

3
References

1. Knight CA. The freezing of supercooled liquids. New York: D Van Nostrand Company Inc, 1967.
2. Vali G. Principles of ice nucleation. In: Lee REJr, Warren GJ, Gusta LV, eds. Biological Ice Nucleation and its Applications. St. Paul: APS Press, 1995:1-28.
3. Vonnegut B. The nucleation of ice formation by silver iodide. J Appl Phys 1947; 18:593-595.
4. Upper CD, Vali G. The discovery of bacterial ice nucleation and its role in the injury of plants by frost. In: Lee REJr, Warren GJ, Gusta LV, eds. Biological Ice nucleation and its Applications. St. Paul: APS Press, 1995:9-39.
5. Maki LR, Gaylan, EL, Chang-Chien M, Caldwell DR. Ice nucleation induced by *Pseudomonas syringae*. Appl Microbiol 1974; 28:456-459.
6. Arny DC, Lindow SE, Upper CD. Frost sensitivity of corn increased by application of *Pseudomonas syringae*. Nature 1976; 262:282-284.
7. Zachariassen KE, Hammel HT. Nucleating agents in the haemolymph of insects tolerant to freezing. Nature; 1976: 262:285-287.
8. Aunaas T. Nucleating agents in the haemolymph of an intertidal mollusc tolerant to freezing. Experientia 1982; 38:1456-1457.
9. Hayes DR, Loomis SH. Evidence for a proteinaceous ice nucleator in the hemolymph of the

pulmonate gastropod, *Melampus bidentatus*. Cryo Lett 1985; 6:418-421.

10. Kieft TL. Ice nucleation activity in lichens. Appl Environ Microbiol 1988; 54:1678-1681.

11. Costanzo JP, Claussen DL, Lee RE. Natural freeze tolerance in a reptile. Cryo Lett 1988; 9:380-385.

12. Wolanczyk JP, Storey, KB, Baust JG. Ice nucleating activity in the blood of the freeze-tolerant frog, *Rana sylvatica*. Cryobiology 1990; 27:328-335.

13. Westh P, Kristiansen J, Hvidt A. Ice nucleating activity in the freeze tolerant tardigrade *Adorybiotus conifer*. Comp Biochem Physiol 1991; 99A:401-404.

14. Pouleur S, Richard C, Martin JG, Antoun H. Ice nucleation activity in *Fusarium acuminatum* and *Fusarium avenaceum*. Appl Environ Microbiol 1992; 58:2960-2964.

15. Gross DC, Proebsting ELJr, MacCrindle-Zimmerman H. Development, distribution, and characteristics of intrinsic, nonbacterial ice nuclei in *Prunus* wood. Plant Physiol 1988; 88:915-922.

16. Tursman D, Duman J G, Knight CA. Freeze tolerance adaptations in the centipede *Lithobius forficatus*. J Exp Zool 1994; 268:347-353.

17. Lundheim R. Ice nucleation in seaweeds in relation to vertical zonation. J Phycol 1997; 33:739-742.

18. Yankofsky SA, Levin Z, Bertold T, Sanderman N. Some basic characteristics of bacterial freezing nuclei. J Appl Meteorol 1981; 20:1013-1019.

19. Turner MA, Arellano F, Kozloff LM. Three separate classes of bacterial ice nucleation structures. J Bacteriol 1990; 172:2521-2526.

20. Duman JG, Wu DW, Wolber PK, Mueller GM, Neven LG. Further characterization of the lipoprotein ice nucleator from freeze tolerant larvae of the cranefly *Tipula trivittata*. Comp Biochem Physiol 1991; 99B:599-607.

21. Kieft TL, Ruscetti T. Characterization of biological ice nuclei from a lichen. J Bacteriol 1990; 172:3519-3523.

22. Brush RA, Griffith M, Mlynarz, A. Characterization and quantification of intrinsic ice nucleators in winter rye *(Secale cereale)* leaves. Plant Physiol 1994; 104:725-735.

23. Lundheim R. Adaptive and incidental biological ice nucleators. PhD Thesis. Faculty of Chemistry and Biology, Norwegian University of Science and Technology, 1997.

24. Lee RE, Mugnano JA, Taylor RT. Endogenous crystalloid spheres regulate supercooling point of the gall fly *Eurosta solidaginis*. Cryobiology 1992; 29:750-751.

25. Fall R, Wolber PK. Biochemistry of bacterial ice nulcei. In: Lee REJr, Warren GJ, Gusta LV, eds. Biological Ice Nucleation and its Applications. St. Paul: APS Press, 1995:63-83.

26. Obata H, Takinami K, Tanishita J, Hasegawa Y, Kawate S, Tokuyama T, Ueno T. Identification of a new ice-nucleating bacterium and its ice nucleation properties. Agric Biol Chem 1990; 54:725-730.

27. Duman JG, Morris JP, Castellino, FJ. Purification and composition of an ice nucleating protein from queens of the hornet *Vespula maculata*. J Comp Physiol 1984; 154B:79-83.

28. Neven LG, Duman JG, Low MG, Sehl LC, Castellino FJ. Purification and characterization of an insect hemolymph lipoprotein ice nucleator: Evidence for the importance of phosphatidylinositol and apolipoprotein in the ice nucleator activity. J Comp Physiol 1989; 159:71-82.

29. Duman JG, Olsen MT, Yeung KL, Jerva F. The roles of ice nucleators in cold tolerant invertebrates. In: Lee REJr, Warren GJ, Gusta LV, eds. Biological Ice Nucleation and its Applications. St. Paul: APS Press, 1995:201-219.

30. Duman JG, Patterson JL. The role of ice nucleators in the frost tolerance of overwintering queens of the bald faced hornet. Comp Biochem Physiol 1978; 59A:69-72.

31. Duman JG. Factors involved in the overwintering survival of the freeze tolerant beetle, *Dendroides canadensis*. J Comp Physiol 1980; 136:53-59.

32. Madison DL, Scrofano MM, Ireland RC, Loomis SH. Purification and partial characterization of an ice nucleator protein from the intertidal gastropod *Melampus bidentatus*. Cryobiology 1991; 28:483-490.

33. Storey KB, Baust JG, Wolanczyk JP. Biochemical modification of plasma ice nucleation activity in a freeze-tolerant frog. Cryobiology 1992; 29:374-384.

34. Hasegawa Y, Ishihara Y, Tokuyama T. Characteristics of ice-nucleating activity in *Fusarium avenaceum* IFO-7158. Biosci Biotechnol Biochem 1994; 58:2273-2274.

35. Hirai M, Tsumuki H. Characterization and localization of the ice-nucleating active agents in larvae of the rice stem borer *Chilo suppressalis*. Cryobiology 1995; 32:209-214.

36. Watanabe M, Watanabe J, Kumeno K, Nakahama N, Arai, S Freeze concentration of some foodstuffs using ice nucleation-active bacterial cells entrapped in calcium alginate gel. Agric Biol Chem 1989; 53:2731-2735.

37. Watanabe M, Makino T, Kumeno K, Arai S. High-pressure sterilization of ice nucleation-active bacterial cells. Agric Biol Chem 1991; 55:291-292.

38. Watanabe M, Arai E, Kumeno K, Arai, S. A new method for producing a non-heated jam sample: The use of freeze concentration and high-pressure sterilization. Agric Biol Chem 1991; 55:2175-2176.

39. Jann A. Lundheim R. Niederberger P. Richard M. Increasing freezing point of food with sea buckthorn ice nucleating agent. US patent 5665361, 1997.

40. Watanabe S, Arai S. Freezing of water in the presence of the ice nucleation active bacterium *Erwinia ananas* and its application for efficient freeze drying of foods. Agric Biol Chem 1987; 51:557-563.

41. Arai S, Watanabe M. Freeze texturing of food materials by ice-nucleation with the bacterium *Erwinia ananas*. Agric Biol Chem 1986; 50:169-175.

42. Lindsey CB. Recovery of microorganism having ice nucleating activity. US patent 4706463, 1987.

43. Lynn SY, Noto GD. Fermentation of microorganisms having ice nucleation activity using a defined medium. European patent 272669, 1988.

44. Hendricks D, Ward PJ, Orrego SA. Production of microorganisms having ice nucleation activity. US patent 5137815, 1992

45. Lawless RJ, LaDuca RJ. Fermentation of microorganisms having ice nucleation activity using a temperature shift. US patent 5413932, 1995.

46. Woerpel MD. Snow making. US patent 4200228, 1980.

47. Owen LB, Masterson, DM, Green SJ. Rapid construction of ice structures with chemically treated sea water. US patent 4637217, 1987.

48. Dennis AS. Weather Modification by Cloud Seeding. San Diego: Academic Press, 1980.

49. LaDuca RJ, Rice AF, Ward PJ. Applications of biological ice nucleators in spray ice technology In: Lee REJr, Warren GJ, Gusta LV, eds. Biological Ice Nucleation and its Applications. St. Paul: APS Press, 1995:337-350.

50. Strong-Gunderson JM, Lee REJr, Lee MR. New species of ice nucleation active bacteria isolated from insects. Cryobiology 1990; 27:691.

51. Lee REJr, Strong-Gunderson JM, Lee MR, Grove KS, Riga TJ. Isolation of ice nucleating active bacteria from insects. J Exp Zool 1991; 257:124-127.

52. Kaneko j, Kita K, Tanno K. Ice nucleating bacteria isolated from the diamondback moth, *Plutella xylostella* L. pupae (Lepidoptera: Ypononautidae). Jpn J Appl Entomol Zool 1991; 35:7-11.

53. Tsumuki H, Konno H, Maeda T, Okamoto Y. An ice-nucleating active fungus isolated from the gut of the rice stem borer *Chilo suppressalis* Walker (Lepidoptera: Pyralidae). J Insect Physiol 1992; 38:119-125.

54. Strong-Gunderson JM, Lee RE, Lee MR, Riga TJ. Ingestion of ice nucleating active bacteria increases the supercooling point of the lady beetle *Hippodamia convergens*. J Insect Physiol 1990; 36:153-157.

55. Fields PG. The control of stored-product insects and mites with extreme temperatures. J Stored Prod Res 1992; 28:89-118.

56. Fields PG. Reduction of cold tolerance of stored-product insects by ice-nucleating-active bacteria. Environ Entomol 1993; 22:470-476.

57. Lee RE, Lee MR, Strong-Gunderson JM. Review: Insect cold-hardiness and ice nucleating active microorganisms including their potential use for biological control. J Insect Physiol 1993; 39:1-12.

58. Strong-Gunderson JM. Lee RE, Lee MR. Ice regulating bacteria promote transcuticular nucleation in insects. Cryobiology 1989; 26:551.

59. Strong-Gunderson JM, Lee RE, Lee MR. Topical application of ice nucleating active bacteria decreases insect cold tolerance. Appl Environ Microbiol 1992; 58:2711-2716.

60. Lee RE, Costanzo JP, Kaufman PE, Lee Mr, Wyman JA. Ice nucleating active bacteria reduce the cold-hardiness of the freeze-intolerant Colorado potato beetle (Coleoptera: Chrysomelidae) J Econ Entomol 1994; 87:377-381.

61. Chen THH, Burke MJ, Gusta LV. Freezing tolerance in plants: An overview. In: Lee REJr, Warren GJ, Gusta LV, eds. Biological Ice Nucleation and its Applications. St. Paul: APS Press, 1995:115-135.

62. Lindow SE. Population dynamics of epiphytic ice nucleation active bacteria on frost sensitive plants and frost control by means of antagonistic bacteria. In: Li PH, Sakai A., eds. Plant Cold Hardiness. San Diego: Academic Press, 1982:395-416.

63. Lindow SE. Control of epiphytic ice nucleation-active bacteria for management of plant frost injury. In: Lee REJr, Warren GJ, Gusta LV, eds. Biological Ice Nucleation and its Applications. St. Paul: APS Press, 1995:239-256.

64. Lindow SE, Arny, DC, Upper CD. Biological control of frost injury I: An isolate of *Erwinia herbicola* antagonistic to ice nucleation-active bacteria. Phytopathology 1983; 73:1097-1102.

65. Wilson M, Lindow SE. Interactions between the biological control *agent Pseudomonas fluorescens* A506 and *Erwinia amylovora* in pear blossoms. Phytopathology 1993; 83:117-124.

66. Lindow SE. Competitive exclusion of epiphytic bacteria by Ice⁻ mutants of *Pseudomonas syringae*. Appl Environ Microbiol 1987; 53:2520-2527.

67. Lindgren PB, Frederick R, Govindarajan AG, Panopoulus NJ, Staslawocz BJ, Lindow SE. An ice nucleation reporter gene system: identification of inducable pathogenicity genes in *Pseudomonas syringae* pv. *phaseolicola*. EMBO J 1989; 8:2990-3001.

68. Panopoulus NJ. Ice nucleation genes as reporters. In: Lee REJr, Warren GJ, Gusta LV, eds. Biological Ice Nucleation and its Applications. St. Paul: APS Press, 1995:271-281.

69. Wolber PK, Green RL. Detection of bacteria by transduction of ice nucleation genes. Trends Biotechnol 1990; 8:276-279.

70. Gutterson NI, Tucker WT, Wolber PK. Transducing particles and methods for their production. US patent 5187061, 1993.

71. Wolber PK, Green RL. Bacterial detection by phage transduction of ice nucleation and other phenotypes. US patent 5447836, 1995.

Applications of antifreeze proteins

K. E. Zachariassen[1]* and R. Lundheim[2]

[1] Norwegian University of Science and Technology, Department of Zoology, N-7034 Trondheim, Norway
[2] Allforsk Biology, Gryta 2, N-7010 Trondheim, Norway

1
General features of antifreeze proteins

Antifreeze proteins have unique effects on ice in frozen solutions. The most striking property of antifreeze proteins is their ability to prevent growth of ice crystals upon cooling and thus they produce a thermal hysteresis, i.e. a separation of the equilibrium freezing point of a solution and the temperature where an ice crystal can grow.[1]

1.1
What is thermal hysteresis?

When an aqueous solution freezes, only water molecules enter the ice structure, and the solutes remain trapped and concentrated in the fluid fraction that surrounds the ice. The amount of ice formed is determined by the temperature, in that at any subfreezing temperature ice will freeze out of the solution and concentrate the fluid fraction until the vapor pressure of the fluid fraction equals that of ice, i.e. the melting point (see below) of the fluid fraction is equal to the actual temperature of the system.[2] Accordingly, if a frozen aqueous solution is slowly heated, the amount of ice will gradually diminish, and eventually only a small ice crystal is left in the sample, as illustrated in Figure 1. Further heating will cause even the last crystal to disappear, and the temperature where this happens is termed the equilibrium freezing point or melting point (MP) of the solution. Cooling the sample with an ice crystal present will cause the ice to grow and concentrate the liquid fraction as described above. However, if the solution contains antifreeze proteins, the antifreeze proteins will inhibit the growth of the crystal when the sample is cooled, allowing the fluid fraction to become gradually more supercooled as the temperature is lowered (Fig. 1). Sample heating still leads to a diminishing crystal size as in normal solutions.[1] The ice growth inhibition is a non-colligative effect in that the MP of the system is only very moderately affected by the antifreeze proteins. The

---

* Corresponding author

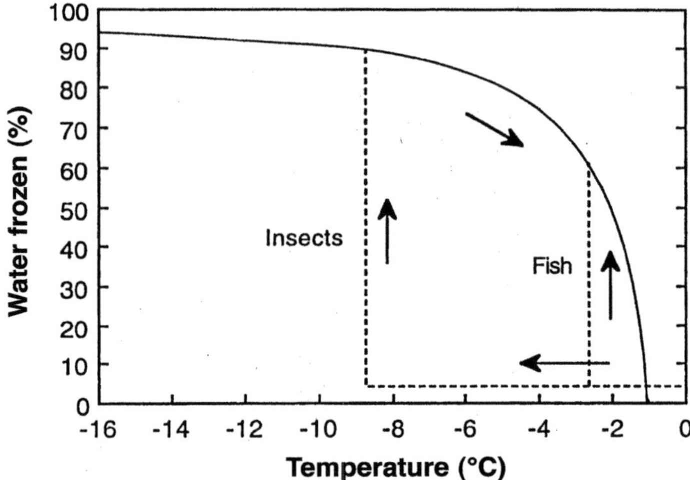

Fig. 1. Curved line describes the relationship between temperature and percentage of ice upon cooling and heating in an ordinary solution with melting point of -1°C, i. e. solution without antifreeze proteins. Arrows indicate the changes when the same solution is cooled and warmed with antifreeze proteins present

reduction in the ice growth can be up to 500 times higher than what could be expected from the colligative MP depression by the proteins in solution.[3] The antifreeze effect is limited in that upon sufficient cooling there is a sudden and rapid ice growth. The temperature where this sudden growth takes place is called the hysteresis freezing point (HFP).[4] The difference between the MP and the HFP is termed the amount of hysteresis or the hysteresis activity.[4]

1.2
Mechanism of thermal hysteresis: the adsorption inhibition model

The ice crystal growth is inhibited, because antifreeze proteins adhere to the ice surface, thus restricting ice growth to the exposed ice surface between adjacent antifreeze molecules[5] as illustrated in Figure 2. The surface of these exposed areas is more convex than the surface of an ice crystal without adsorbed antifreeze proteins. The convex surface has a higher free energy than ordinary ice, and ice growth is thus depressed to lower temperatures.[6]

Growth of ordinary ice crystals takes place along the six so-called a-axes, giving the ice crystal its typical flat hexagonal appearance. This reflects the fact that the free energy of the crystal surface makes a-axes the preferred axes of ice growth, instead of the c-axis, that goes perpendicularly on the crystal plane in the center of the crystal.[7] In systems containing antifreeze proteins or glycoproteins from blood plasma of polar fish the freezing takes place as a single ice spear or as bundles of parallel ice spears that shoot out from the seeding ice crystal. This suggests that the ice growth in this case takes place along the c-axis and that the antifreeze proteins

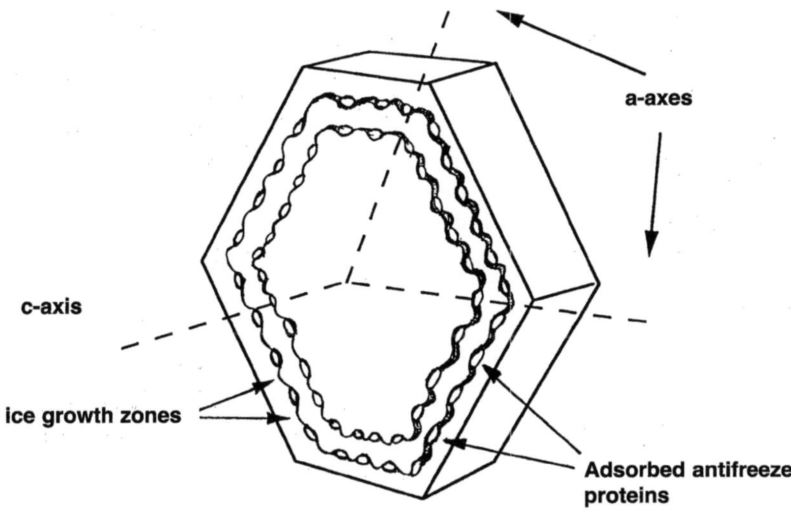

Fig. 2. Illustration of how antifreeze proteins inhibit ice crystal growth according to the adsorption inhibition theory. Antifreeze proteins are adsorbed to the ice surface and restrict ice growth to the exposed zones between adjacent antifreeze molecules. Here the growth zones are more convex than the ordinary ice surface, and ice growth will consequently occur first at lower temperatures

of fish inhibit growth along the a-axes. While antifreeze proteins from fish plasma produce a hysteresis of only up to about 2°C, antifreeze proteins in insect hemolymph may cause a hysteresis of more than 5°C,[8] and the ice growth in insect hemolymph seems to take place along the six a-axes.[9]

Zachariassen and Husby[10] demonstrated that the hysteresis in the hemolymph from hibernating freeze-avoiding beetles of the species *Rhagium inquisitor* depends on the size of the seeding ice crystal, diminishing ice crystal size leading to increasing hysteresis. In fish plasma the hysteresis seems in many cases to be independent of the size of the seeding ice crystal (Zachariassen, unpubl. data).

Yeh et al.[11] measured the effect of antifreeze proteins on the activation energy for ice grain growth. The rate of grain growth was measured as the change in grain diameter in time at different temperatures and glycoprotein concentrations. To be able to measure the rate of grain growth, the experiments were carried out with antifreeze proteins present at only low concentrations, providing only a moderate antifreeze effect. By making an Arrhenius plot of the measured growth rates as a function of $1/T$, the activation energy comes out as the slope of the curve. The most active purified antifreeze glycoproteins (AFGPs; AFGP-4) had activation energies of about 6.5×10^5 J mol^{-1}, whereas the less active AFGPs (AFGP-8) had slightly lower values (5.8×10^5 J mol^{-1}). Ice in solutions without antifreeze glycoproteins had substantially lower activation energy than ice in solutions containing antifreeze proteins. Also the activation entropy differed in the same manner between pure solutions and solutions containing AFGPs. The authors concluded that the antifreeze proteins have a great capacity, both energetically and entropically, to inhibit ice grain growth.

1.3
Inhibition of recrystallization

Since antifreeze proteins have the capacity to prevent growth of ice crystals, one might expect that they are also able to inhibit recrystallization of ice in partly frozen solutions. The first demonstration of a recrystallization inhibitory effect of antifreeze proteins was made by Knight et al.,[12] who tested the effect of an AFGP from winter flounder *(Pseudopleuronectes americanus)* on the ice growth in frozen solutions kept between two glass microscope slides. The investigators found that even at very low concentrations (10^{-9} g g^{-1} solution) the AFGPs were extremely effective in inhibiting recrystallization of ice.

Knight et al.[13] studied the effects of antifreeze proteins from the antarctic fish *Dissostichus mawsoni* on the growth of ice crystals formed by so-called "splat freezing", which implies that a droplet of a solution is dropped on a metal plate previously cooled to a very low temperature by for example dry ice (-78°C). The procedure leads to the formation of a great number of ice grains of variable small size. The results confirmed that the antifreeze proteins have an exceptionally strong inhibitory effect on recrystallization.

Knight et al.[14] obtained evidence that the recrystallization inhibitory capacity of AFPs from winter flounder requires that the system contains a sufficient amount of liquid water, i.e. the system must contain salts or other solutes which reduce ice formation in a colligative manner, and the temperature must be sufficiently high. The authors ascribed the effect to the fact that the antifreeze proteins bind to the ice surface in the ice–solution interfaces, and that the antifreeze proteins need sufficient mobility to establish the required structural fit.

The mechanism of the recrystallization inhibition effect of the antifreeze proteins is not known, but it is likely to be related to the mechanisms behind their general capacity to prevent ice crystal growth.

1.4
Reduced permeability through lipid membranes

Rubinsky et al.[15] found that antifreeze glycoproteins from antarctic fish inhibit Ca^{++} and K^+ currents through cell membranes, probably by blocking Ca^{++} and K^+ channels. In this way the antifreeze proteins may contribute to maintain the membrane potential and the normal ionic distribution between intracellular and extracellular fluid compartments in frozen cells and tissues.

2
Sources of antifreeze proteins and their classification

Thermal hysteresis was observed in fish blood by Scholander and co-workers[16] as early as in 1957, and in the hemolymph from larvae of the beetle *Tenebrio molitor* by Ramsay and co-workers[17] in 1964. In 1968, the proteinaceous nature of the substances responsible for this effect was demonstrated by Grimstone et al.[18] DeVries and Wohlschlag[19] purified antifreeze glycoproteins in 1969, and since then several types of antifreeze proteins have been discovered. Four main types of antifreeze proteins from fish have been identified and classified (Table 1).

Antifreeze proteins from sources other than fish have often been called thermal hysteresis factors (THFs) or thermal hysteresis proteins (THPs). Such proteins were first isolated from insects,[20] later they have been isolated from other phyla such as vascular plants[21,22] and bacteria.[23] Thermal hysteresis (indicating the presence of THFs) has been demonstrated in diverse phyla such as fungi,[24] collembola,[25] spiders[26,27] and mites.[28]

3
Functions of antifreeze proteins

Antifreeze proteins appear to have several functions in cold exposed organisms. To some extent these functions may differ from species to species, but they may also have differentiated functions within the same organism. The fact that more than five different antifreeze proteins are produced in the same organism may, at least to some degree, reflect the diversity of these functions.

3.1
Inhibition of inoculative freezing

Antifreeze proteins in fish plasma and insect hemolymph appear to have important functions in preventing external ice from penetrating the body wall and spread in the body fluids. In fish this function was reflected in experiments made by Scholander et al.,[16] who placed an ice crystal in contact with the skin of a supercooled polar fish with antifreeze proteins in its blood plasma. They found that this did not affect the fish, but fish without antifreeze proteins immediately froze and died following the same treatment.

Hibernating insects may be completely embedded in ice in their hibernacula, and are threatened by lethal inoculative freezing. In this situation hibernating insects benefit from the presence of antifreeze proteins in their body fluid in a similar manner as polar fish. Gehrken[37] demonstrated that when *R. inquisitor* beetles with antifreeze proteins in their body fluid were cooled in contact with external ice, the ice did not penetrate their body wall, and the insects survived cooling to below -20°C. When *R. inquisitor* beetles lacking antifreeze proteins were cooled under the same conditions, the exernal ice inoculated their body and caused a lethal freezing of their body fluids at a few degrees below the body fluid MP.

3.2
Inhibition of spontaneous ice nucleation

Hibernating insects which seek to avoid freezing of their body fluids may remain supercooled at temperatures well below -20°C, in some cases even below -50°C. At these low temperatures there is a high probability of a spontaneous lethal ice nucle-

Table 1.   Properties of different types of antifreeze proteins. The data on fish AFP are mainly based on reviews[29–31]

Properties	Type of protein						
	Fish AFGP	Fish AFP type 1	Fish AFP type 2	Fish AFP type 3	Plant THP	Bacterial THP	Insect THP
Molecular mass (kDa)	2.6–33	3.3–4.5	14–24	6.5–7	67[21] 19–36[22]	164[23]	9[32] 14–20[24]
Chemical composition	Contains a disaccharide o-linked to the threonyl side chains	~65mol% Ala	More than 8 mol% Cys. Ala, Asn, Gln and Thr are also dominating	No particular amino acid dominates	Gly rich[21] Presence of carbohydrate, probably galactose[21] Rich in Asp/Asn, Glu/Gln, Ser, Thr and Ala[22]	Contains both lipids and carbohydrates	Generally a high percentage of amino acid residues with hydrophilic side chains[35]
Primary structure	Repeats of the triplet Ala-Ala-Thr	11 residue repeats	C-type lectin	No repeated structure			
Other features	Tertiary structure probably extended and rodlike	Secondary structure predominantly a-helix (85% at -1°C)	Small a-helix content, random coils and b-sheets dominate	Unusual b-sandwich secondary structure, tertiary structure globular	Low specific activity[21]	Has also ice nucleator activity[23]	High specific activity[32]
Sources	Antarctic notothenioid fish and arctic cods	Sculpins and righteye flounders	Smelt, atlantic herring, sea raven	Zoarcid fish	Several vascular plants[24,33,34]	Several bacteria[23,24]	Several insects[35,36]

AFGP antifreeze glycoprotein, *AFP* anitifreeze protein, *THP* thermal hysteresis protein

ation in the body fluids, and insects would benefit substantially from mechanisms that would inhibit ice nucleation when they are highly supercooled. Antifreeze proteins from insects have been shown to counteract ice crystal growth down to about -10°C, which is far from sufficient to protect insects below -20°C. However, antifreeze proteins may still have the capacity to protect highly supercooled insects. Zachariassen and Husby[10] observed that the hysteresis in the hemolymph of cold-hardy *R. inquisitor* beetles increases when the size of the seeding ice crystal diminishes.[10] There was a linear relationship between the hysteresis and the logarithm of the ice fraction, and extrapolation of the observed curve to the size of water molecule aggregates critical for nucleation at -25°C (the supercooling point of the beetles) gave a hysteresis freezing point corresponding to the supercooling point. The authors concluded that the antifreeze proteins may be able to counteract spontaneous nucleation and thus stabilize the beetles in the supercooled state over their entire supercooling range.

3.3
Masking of biological ice nucleators

There is also evidence indicating that antifreeze proteins can inhibit not only growth of formed ice crystals but also the activity of biological ice nucleators, which are proteins that have the capacity to organize water molecules in an ice-like pattern and hence cause freezing at high subzero temperatures. An effect of this kind might be of great physiological importance, in that hibernating insects would not have to remove ice nucleating proteins to attain a high supercooling capacity, but instead mask their activity with antifreeze proteins.

Parody-Morreale et al.[38] demonstrated that antifreeze glycoproteins from polar fish are able to inhibit the activity of bacterial ice nucleators. Duman et al.[35] demonstrated that antifreeze proteins from larvae of the beetle *Dendroides canadensis* inhibited some, but not all, of the nucleating proteins present in the hemolymph. Baust and Zachariassen[39] tested the effect of antifreeze proteins from *R. inquisitor* hemolymph on the nucleating activity of whole body homogenates of the same species, but did not find any nucleation inhibitory effect. Bremdal and Zachariassen[40] found that *R. inquisitor* beetles about to leave their hibernation state could have a low supercooling capacity and hence a high ice nucleator activity in spite of high thermal hysteresis activity in their body fluid. Consequently, although antifreeze proteins have a capacity to reduce the activity of biological ice nucleators, the physiological importance of this effect remains unclear.

3.4
Maintenance of ionic gradients

The capacity of antifreeze proteins to block ionic channels through biological membranes[15] may be of great importance in helping organisms exposed to extreme cold to maintain ionic gradients across cell membranes in the absence of significant activity of ionic pumps. This would allow the organisms to have intact membrane potentials and transmembrane ionic energy gradients at the moment they wake up from a long period of cold exposure. However, the actual physiological importance of this capacity needs more investigations to be clarified.

3.5
Inhibition of recrystallization in freeze tolerant insects

Although the main function of antifreeze proteins in fish and insects appears to be the prevention of ice formation and ice growth, antifreeze proteins are also found in the hemolymph of certain freeze tolerant insects, in which extracellular ice formation appears to be protective and actively promoted by the presence of extracellular ice nucleating agents. This apparent paradox led Knight and Duman[41] to carry out a study on the possible function of the antifreeze proteins in freeze-tolerant larvae of the beetle *Dendroides canadensis*. The authors assumed that recrystallization of ice in the frozen insects might lead to the formation of fewer and larger ice crystals which eventually might grow so large that they would injure the organisms. The authors demonstrated that solutions of NaCl, glycerol and sorbitol with a composition similar to that of *D. canadensis* hemolymph underwent rapid recrystallization during a 24 h observation period, while hemolymph of *D. canadensis* with antifreeze proteins present did not recrystallize. The recrystallization inhibitory effect was evident even after the hemolymph was diluted by a factor of 10^4, but collapsed following another 10-fold dilution. The authors assumed that the recrystallization inhibitory capacity was based on the same mechanism as the previously known inhibition of ice growth.

4
Applications of antifreeze proteins

Antifreeze proteins may have a variety of practical applications, in fields such as food production, cryopreservation of tissues and organs, and in industry. Examples of actual and possible practical applications are given below.

4.1
Food processing

Freezing is used in several manners in the processing of food stuffs. Food products such as ice cream, which are to be consumed in the frozen state, must be frozen in a manner that secures the best possible quality and texture. This requires that the ice is formed as a large number of tiny ice crystals. For food products which are frozen for storage purposes, the thawing process as well as the freezing process are important. Also in this case freezing must provide a multitude of small ice crystals, but it is also important to prevent the formations of large ice crystals due to recrystallization during the thawing phase. The use of antifreeze proteins may offer great advantages in the processing of both these categories of food products. The use of antifreeze proteins may offer interesting perspectives also for the possible freeze storage of berries and fruits, but in this case the antifreeze proteins must be introduced into the food stuffs by the use of special techniques, such as gene technology. The perspectives of the possible use of antifreeze proteins in food processing have been dealt with in several experimental studies and review articles.

By the use of light microscopy and scanning electron microscopy Payne et al.[42] investigated how antifreeze proteins from polar fish (antarctic cod and winter flounder) influenced the freezing of bovine and ovine meat. For studies with light microscopy the bovine meat samples were soaked in saline with and without antifreeze proteins. The meat was then kept frozen at -20°C or chilled at 2°C for 3 d. Light microscopy studies of freeze-substituted and sectioned samples indicated that meat frozen in the presence of antifreeze proteins contained smaller ice crystal than control samples frozen in the absence of proteinaceous antifreezes. The ovine muscle samples were soaked in saline with different concentrations of antifreeze proteins, thereafter they were kept frozen at -20°C or chilled at 2°C for 5–7 d. Again the authors found that the presence of antifreeze proteins in the frozen samples reduced the size of the ice crystals formed, increasing concentrations of antifreeze proteins leading to smaller ice crystals. In none of the studies did the presence of antifreeze proteins affect the structure of chilled meat samples.

The authors[42] also reported that the use of antifreeze proteins at high concentrations (about 20 mg ml^{-1}) was associated with the formation of very destructive ice spicules, which caused the frozen samples to be structurally more damaged than control samples. Hence, in order to improve the quality of the frozen meat, antifreeze proteins have to be used at moderately high concentrations, i.e. below 1 mg ml^{-1}.

The use of antifreeze proteins in frozen foods has been reviewed by Griffith and Ewarth.[31] These authors pointed out that antifreeze proteins may improve the quality of stored food products in several ways. The proteins may be added to inhibit recrystallization and improve the texture of foods which are eaten while in the frozen state, such as ice cream and popsicles. Their recrystallization inhibitory effect may also prevent the formation of large ice crystals and thus improve the quality of frozen food products which are to be thawed before they are consumed. Similar inhibitory effects of recrystallization have been obtained by the use of gelatin,[43] but antifreeze proteins may have the capacity to give substantially better results.

Griffith and Ewarth[31] hypothesized two manners by which antifreeze proteins may improve the quality of cold preserved fruits and berries, such as strawberries, rasberries and tomatoes, which cannot be frozen without a considerable loss of quality. If antifreeze proteins in low concentrations are somehow added to the extracellular fluid, they may reduce the amount of ice and thus the solute concentrations in the extracellular fluid. This is likely to reduce cellular dehydration and membrane damage during freezing. If they are added at higher concentrations, they may promote supercooling and thus allow the products to be kept cooled and unfrozen at lower temperatures than would otherwise be possible. The authors indicated that the addition of antifreeze proteins to these food products can be achieved by introducing the genes for production of antifreeze proteins to their gene pool. The authors also discussed criteria for selection of antifreeze proteins for use in food processing.

4.2
Cryopreservation

A vast number of studies have been carried out to investigate cryopreservation of cells, tissues and organs, but only a few of these studies are dealing with the use of antifreeze proteins in cryopreservation. So far the conclusions appear to some extent to be contradictory.

Carpenter and Hansen[44] found that the presence of low concentrations of antifreeze proteins in the freezing media improved the post thawing survival of vitrified erythrocytes, probably due to an anti-recrystallization effect. Similar conclusions were drawn by Payne et al.,[45] who found that antifreeze proteins increased the survival of ram spermatozoa frozen in liquid nitrogen.

Arav et al.[46] found that antifreeze proteins from four species of polar fishes (the antarctic notothenioid fish *Dissostichus mawsoni*, winter flounder, sea raven and ocean pout) protect pig oocytes against freezing damage during rapid cooling in vitrifying solutions. All four types of antifreeze proteins displayed the protective effect. Freezing damage appears to be located to the outer oocyte membrane, i.e. the oolemma, and the freeze protection provided by the antifreeze proteins seems to have its site at this membrane. The authors contended that the protective mechanism involves the binding of the antifreeze proteins to membrane structures.

Rubinsky et al.[47] tested the effect of solutions with various concentrations of glycerol and antifreeze proteins on the survival of slowly frozen whole rat livers. The investigators found that the bile production and microscopic structure of the liver were more intact in livers frozen with antifreeze proteins and glycerol than in livers frozen only with glycerol.

In contrast to these results, several authors found that the addition of antifreeze proteins to the freezing media has strong negative effects on post freezing survival of cells and tissues. The presence of fish antifreeze proteins caused substantial hemolysis of erythrocytes, even in the presence of glycerol at concentrations which alone would normally be cryoprotective.[48] Type I antifreeze proteins from winter flounder gave rise to increased intracellular ice nucleation and reduced post-freezing survival of fibroblasts from Chinese hamster and of human lymphocytes and granulocytes.[49] It was contended that the binding of the antifreeze proteins to ice is likely to allow the ice to interact physically with the plasma membrane, and that this might promote intracellular nucleation either by facilitating ice penetration through the cell membranes, by "catalyzing" intracellular ice nucleation, or by destabilizing the membranes through hydrophobic interaction between ice-bound antifreeze proteins and the membrane.[49] The catalyzation of intracellular nucleation may also be due to the fact that the inhibition of extracellular ice growth following extracellular ice nucleation is likely to counteract the increase in the osmolality of the partly frozen extracellular fluid and thus reduce the osmotic outflux of water from the cells. This would allow the intracellular fluid to become so supercooled that it eventually nucleates. Finally, certain antifreeze proteins have been shown to have cytotoxic effects in plants[50] as well as in animals.[42]

Costanzo et al.[51] pointed out that the effect of antifreeze proteins used in cell and tissue cryopreservation depends on the type of cells and tissues used. The effect is likely to depend on a variety of parameters, such as the manner in which the particular antifreeze proteins block ice growth, the ice nucleating potential of intracellular and extracellular fluid, and the properties of the cell membranes. Another and complicating factor is the documented capacity of antifreeze proteins from fish to affect the permeability of inorganic ions through lipid membranes.

The capacity of antifreeze proteins to reduce ionic permeability through lipid membranes may offer great advantages in the cryopreservation of tissues and cells. A reduced passive ionic flux through cell membranes would help to maintain transmembrane ionic gradients, membrane potentials and electrochemical potential differences of ions across membranes in tissues stored at so low temperatures that active transport mechanisms are slowed substantially down.

In order to take full advantage of the cryoprotective potential of antifreeze proteins in practical tissue-cryopreservation it is necessary to learn more about the properties of the various antifreeze proteins and their interaction with living cells.

4.3
Pumping of ice-water mixtures

Ice-water mixtures, or so-called ice slurries, have interesting applications in cold storage and heat transportation. Due to the high heat of fusion of ice, ice slurry has an exceptionally great capacity to store and release heat. This allows transportation systems based on ice slurries to have smaller dimensions and to flow more slowly than conventional fluid systems. However, storage of ice-water mixtures may involve problems related to recrystallization, which may cause the larger crystals to grow and eventually reach dimensions that would block the pipes through which the slurry is to flow. This makes it important to develop methods that inhibit the recrystallization process in the ice slurry. Due to their well documented capacity to inhibit recrystallization of ice, antifreeze proteins or artificial compounds with similar properties may have an interesting potential to keep the ice slurry able to flow through moderately sized pipe systems.

The effect of antifreeze proteins on the recrystallization process in ice slurries has been investigated by Grandum et al.[52] The authors found that the use of a synthetic antifreeze protein, copied from the proteinaceous antifreeze of the North American winter flounder, inhibited recrystallization of the ice slurries. The slurries had an acceptable low viscosity even when they contained 30% of ice, and the authors concluded that the antifreeze proteins provide a feasible solution to the problem of preventing recrystallization in ice slurries.

5
Conclusions

Antifreeze proteins are extended peptides or glycopeptides which have the capacity to prevent growth of ice crystals in supercooled fluid samples. They promote supercooling in freeze-intolerant organisms by counteracting spontaneous ice nucleation in their body fluids and by preventing inoculation of external ice through the body wall. Antifreeze proteins also seem to offer protection in freeze-tolerant organisms by counteracting recrystallization of ice, thus preventing the development of intolerably large ice crystals in the extracellular fluid. They also seem to reduce ionic permeability of cell membranes, thus helping to maintain transmembrane ionic gradients in cold-exposed organisms.

Antifreeze proteins may be applied to improve the quality of frozen food stuffs and to increase the survival of cyropreserved tissues and organs. Their capacity to prevent recrystallization of ice may also improve the quality of ice slurries by preventing the formation of large ice crystals.

6
References

1. DeVries AL. Glycoproteins as biological antifreeze agents in Antarctic fishes. Science 1971; 172:1152-1155.
2. Zachariassen KE. The water relations of overwintering insects. In: Lee RE Jr, Denlinger D, eds. Insects at Low Temperature. New York: Chapman Hall, 1991:47-63.
3. Yeh Y, Feeney RE. Antifreeze proteins: structures and mechanisms of function. Chem Rev 1996; 96:601-617.
4. Zachariassen KE. Physiology of cold-hardiness in insects. Physiol Rev 1985; 65:799-832.
5. Raymond J, DeVries AL. Adsorption inhibition as a mechanism of freezing resistance in polar fishes. Proc Natl Acad Sci USA 1977; 74:2589-2593.
6. Knight CA, Cheng CC, DeVries AL. Adsorption of alpha-helical antifreeze peptides on specific ice crystal surface planes. Biophys J 1991; 59:409-418.
7. Cheng CC, DeVries AL. The role of antifreeze glycopeptides and peptides in the freezing avoidance of cold-water fish. In: di Prisco G, ed. Life Under Extreme Conditions. Berlin Heidelberg: Springer-Verlag, 1991:1-14.
8. Duman JG, Wu DW, Yeung KL, Wolf EE. Hemolymph proteins involved in the cold tolerance of terrestrial arthropods: antifreeze and ice nucleator proteins. In: Somero GN, Osmond CB, Bolis CL, eds. Water and Life. Berlin, Heidelberg: Springer-Verlag, 1992:282-300.
9. Zachariassen KE, DeVries AL. Effect of varying ice crystal size and sample dilution on the antifreeze activity in haemolymph of the cerambycid beetles *Rhagium inquisitor*. J Exp Biol 1999 (in press).
10. Zachariassen KE, Husby JA. Antifreeze effect of thermal hysteresis agents protects highly supercooled insects. Nature London 1982; 298:865-867.
11. Yeh Y, Feeney RE, Mckown RL, Warren GJ. Measurements of grain growth in the recrystallization of rapidly frozen solutions of antifreeze glycoproteins. Biopolymers 1994; 34:1495-1504.
12. Knight CA, DeVries AL, Oolman LD. Fish antifreeze protein and the freezing and recrystallization of ice. Nature London 1984; 308:295-262.
13. Knight CA, Hallett J, DeVries AL. Solute effects on ice recrystallization: An assessment technique. Cryobiology 1988; 25:55-60.
14. Knight CA, Wen D, Laursen RA. Nonequilibrium antifreeze peptides and the recrystallization of ice. Cryobiology 1995; 32:23-34.
15. Rubinsky B, Mattioli M, Arav A, Barboni B, Fletcher GL. Inhibition of Ca^{2+} and K^+ currents by "antifreeze" proteins. Am J Physiol 1992; 262:R542-R545.
16. Scholander PF, van Dam L, Kanwisher JW, Hammel HT, Gordon MS. Supercooling and osmoregulation in arctic fish. J Cell Comp Physiol 1957; 49:5-24.
17. Ramsay, A. J. The rectal complex of the mealworm *Tenebrio molitor* L. Philos Trans Roy Soc

Ser B 1964; 248:279-314.

18. Grimstone AV, Mullinger AM, Ramsay JA. Further studies on the rectal complex of the mealworm *Tenebrio molitor* (Coleoptera, Tenebrionidae). Philos Trans Roy Soc Sci 1968; 253:343-382.

19. DeVries AL, Wohlschlag EE. Freezing resistance in some antarctic fishes. Science 1969; 163:1073-1075.

20. Patterson JL, Duman JG. Composition of a protein antifreeze from larvae of the beetle *Tenebrio molitor* J Exp Zool 1979; 210:361-367.

21. Duman JG. Purification and characterization of a thermal hysteresis protein from a plant, the bittersweet nightshade *Solanum dulcamara*. Biochim Biophys Acta 1994; 1206:129-135.

22. Hon WC, Griffith M, Chong PL, Yang DSC. Extraction and isolation of antifreeze proteins from winter rye (*Secale cereale* L.) leaves. Plant Physiol 1994; 104:971-980.

23. Xu H, Griffith M, Patten CL, Glick BR. Isolation and characterization of an antifreeze protein with ice nucleation activity from the plant growth promoting rhizobacterium *Pseudomonas putida*. Can J Microbiol 1998; 44:64-73.

24. Duman JG, Olsen TM. Thermal hysteresis protein-activity in bacteria, fungi and phylogentetically diverse plants. Cryobiology 1993; 30:322-328.

25. Zettel J. Cold hardiness strategies and thermal hysteresis in collembola. Rev Ecol Biol Sol 1984; 21:189-303.

26. Duman JG. Subzero temperature tolerance in spiders: the role of thermal hysteresis factors. J Comp Physiol 1979; 131:347-352.

27 Husby JA, Zachariassen KE. Antifreeze agents in the body fluid of winter active insects and spiders. Experientia 1980; 36:963-964.

28. Block W, Duman JG. Presence of thermal hysteresis producing antifreeze proteins in the Antarctic mite *Alaskozetes antarcticus*. J Exp Zool 1989; 250:229-231.

29. Hew CL, Yang DSC. Protein interaction with ice. Eur J Biochem 1992; 203:33-42.

30. Davies PL, Ewart KV, Fletcher GL. The diversity and distribution of fish antifreeze proteins: new insights into their origins. In: Hochachka PV, Mommsen TP, eds. Biochemistry and Molecular Biology of Fishes, vol 2. Elsevier Science Publishers BV, 1993:279-291.

31. Griffith M, Ewart KW. Antifreeze proteins and their potential use in frozen foods. Biotechnol Adv 1995; 13:375-402.

32. Tyshenko MG, Doucet D, Davies PL, Walker VK. The antifreeze potential of the spruce budworm thermal hysteresis protein. Nature Biotechnol 1997; 15:887-890.

33. Urrutia M, Duman JG, Knight CA. Plant thermal hysteresis proteins. Biochim Biophys Acta 1992; 1121:199-206.

34. Griffith M, Ala P, Yang DSC, Hon WC, Moffatt BA. Antifreeze protein produced endogenously in winter rye leaves. Plant Physiol 1992; 100:593-596.

35. Duman JG, Xu L, Neven LG, Tursman D, Wu DW. Hemolymph proteins involved in insect subzero-temperature tolerance: Ice nucleators and antifreeze proteins. In: Lee RE, Denlinger DL, eds. Insects at Low Temperature. New York: Chapman and Hall, 1991:94-127.

36. Duman JG, Wu DW, Olsen TM, Urrutia M, Tursman D. Thermal-hysteresis proteins. Adv Low-Temp Biol 1993; 2:131-182.

37. Gehrken U. Inoculative freezing and thermal hysteresis in the adult beetles *Ips acuminatus* and *Rhagium inquisitor*. J Insects Physiol 1992; 38:519-524.

38. Parody-Morreale A, Murphy KP, DiCera E, Fall R, DeVries AL, Gill SJ. Inhibition of bacterial ice nucleators by fish antifreeze glycoproteins. Nature London 1988; 333:782-783.

39. Baust JG, Zachariassen KE. Seasonally active cell matrix associated ice nucleators in an insect. Cryo Lett 1983; 4:65-71.

40. Bremdal S, Zachariassen KE. Thermal hysteresis factors and supercooling of hibernating *Rhagium inquisitor* beetles. In: Sehnal F, Zabza A, Denlinger D, eds. Endocrinological Frontiers in Physiological Insect Ecology. Wrochlaw: Wrochlaw Technical University Press, 1988:187-191.

41. Knight CA, Duman JG. Inhibition of recrystallization of ice by insects thermal hysteresis

proteins: a possible cryoprotective role. Cryobiology 1986; 23:256-262.

42. Payne SR, Sandford D, Harris A,Young OA. Effects of antifreeze proteins on chilled and frozen meat. Meat Science 1994; 37:429-438.

43. Smith CE, Schwartzberg HG. Ice crystal size changes during ripening in freeze concentration. Biotech Prog 1985; 1:111-120.

44. Carpenter JF, Hansen TN. Antifreeze protein modulates cell survival during cryopreservation: mediation through influence on ice crystal growth. Proc Natl Acad Sci 1992; 89:8953-8957.

45. Payne SR, Oliver JE, Upretic GC. Effect of antifreeze proteins on the motility of ram spermatozoa. Cryobiology 1994; 31:180-184.

46. Arav A, Rubinsky B, Fletcher G, Seren E. Cryogenic protection of oocytes with antifreeze proteins. Molecular reproduction and development 1993; 36:488-493.

47. Rubinsky B, Arav A, Hong JS, Lee CY. Freezing of mammalian livers with glycerol and antifreeze proteins. Biochem Biophys Res Comm 1994; 200:732-741.

48. Petzel DH, DeVries AL. Effect of fish antifreeze agents on cryoprotection of red blood cells in the presence of glycerol and PVP. Cryobiology 1979; 16:585-586.

49. Larese A, Acker J, Muldrew K, Yang H, McGann L. Antifreeze proteins induce intracellular nucleation. Cryo Lett 1996; 17:175-182.

50. Hincha DK, DeVries AL, Schmitt JM. Cytotoxicity of antifreeze proteins and glycoproteins to spinach thylakoid membranes - comparison with cryotoxic sugar acids. Biochim Biophys Acta 1993; 1146:258-264.

51. Costanzo JP, Lee RE Jr, DeVries AL, Wang T, Layne JR Jr. Survival mechanisms of vertebrate ectotherms at subfreezing temperatures: applications in cryomedicine. FASEB J 1995; 9:351-352.

52. Grandum S, Yabe A, Tanaka M, Takemura F, Nakagomi K. Characteristics of ice slurry containing antifreeze protein for ice storage applications. J Thermophys Heat Transfer 1997; 11:461-466.

Subject index

Springer
and the
environment

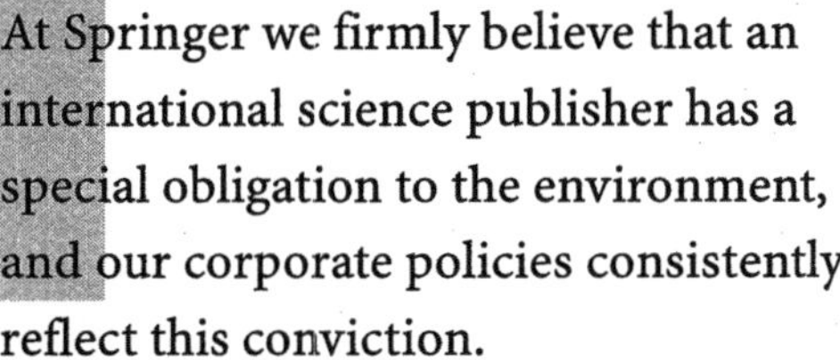

At Springer we firmly believe that an international science publisher has a special obligation to the environment, and our corporate policies consistently reflect this conviction.

We also expect our business partners – paper mills, printers, packaging manufacturers, etc. – to commit themselves to using materials and production processes that do not harm the environment. The paper in this book is made from low- or no-chlorine pulp and is acid free, in conformance with international standards for paper permanency.

Springer

MIX
Papier aus verantwortungsvollen Quellen
Paper from responsible sources
FSC® C105338

If you have any concerns about our products,
you can contact us on
ProductSafety@springernature.com

In case Publisher is established outside the EU,
the EU authorized representative is:
Springer Nature Customer Service Center GmbH
Europaplatz 3, 69115 Heidelberg, Germany

Printed by Libri Plureos GmbH
in Hamburg, Germany